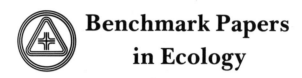

Benchmark Papers
in Ecology

Series Editor: Frank B. Golley
University of Georgia

Published Volumes and Volumes in Preparation

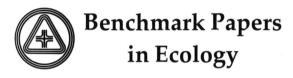

**Benchmark Papers
in Ecology**

——— A *BENCHMARK* ® Books Series ———

CYCLES OF
ESSENTIAL ELEMENTS

Edited by
LAWRENCE R. POMEROY
University of Georgia

**Dowden, Hutchinson
& Ross, Inc.**
Stroudsburg, Pennsylvania

Copyright © 1974 by **Dowden, Hutchinson & Ross, Inc.**
Benchmark Papers in Ecology, Volume 1
Library of Congress Catalog Card Number: 74–4252
ISBN: 0–87933–129–1

Manufactured in the United States of America.

Exclusive distributor outside the United States and
Canada: John Wiley & Sons, Inc.

74 75 76   5 4 3 2 1

Library of Congress Cataloging in Publication Data

Pomeroy, Lawrence R   1925-   comp.
    Cycles of essential elements.

    (Benchmark papers in ecology, v. 1)
    1. Biogeochemical cycles--Addresses, essays, lec-
tures. I. Title.
QH344.P65   574.5'2   74-4252
ISBN 0-87933-129-1

Permissions

The following papers have been reprinted with the permission of the authors and the copyright holders.

ACADEMIC PRESS, INC.—*The Ecosystem Concept in Natural Resource Management*
   A Study of an Ecosystem: The Arctic Tundra

AMERICAN ASSOCIATION FOR THE ADVANCEMENT OF SCIENCE
   International Oceanographic Congress Preprints
      Regeneration of Phosphate by Marine Animals
   Science
      Phosphorus Excretion and Body Size in Marine Animals: Microzooplankton and Nutrient Regeneration
      Residence Time of Dissolved Phosphate in Natural Waters

AMERICAN INSTITUTE OF BIOLOGICAL SCIENCES—*Bioscience*
      The Metabolism of Some Coral Reef Communities: A Team Study of Nutrient and Energy Flux at
      Eniwetok

ANNUAL REVIEWS, INC.
   Annual Review of Ecology and Systematics
      Mineral Cycling: Some Basic Concepts and Their Application in a Tropical Rain Forest
   Annual Review of Microbiology
      Biochemical Ecology of Soil Microorganisms

BLACKWELL SCIENTIFIC PUBLICATIONS LTD. (FOR THE BRITISH ECOLOGICAL SOCIETY)—
      Journal of Ecology
      On the Kinetics of Phosphorus Exchange in Lakes

DUKE UNIVERSITY PRESS (FOR THE ECOLOGICAL SOCIETY OF AMERICA)
   Ecological Monographs
      Effects of Forest Cutting and Herbicide Treatment on Nutrient Budgets in the Hubbard Brook
      Watershed-Ecosystem
   Ecology
      Studies of Cation Budgets in the Southern Appalachians on Four Experimental Watersheds with Con-
      trasting Vegetation
      Systems Analysis of ^{134}Cesium Kinetics in Terrestrial Microcosms

INSTITUTO DE INVESTIGACIONES PESQUERAS—*Investigación Pesquera*
      A Simulation Model of the Nitrogen Flow in the Peruvian Upwelling System

LIVERPOOL UNIVERSITY PRESS—*James Johnstone Memorial Volume*
      On the Proportions of Organic Derivatives in Sea Water and Their Relation to the Composition of
      Plankton

MARINE BIOLOGICAL ASSOCIATION OF THE UNITED KINGDOM—*Journal of the Marine Biological
      Association of the United Kingdom*
      The Phosphate Content of Fresh and Salt Waters in Its Relationship to the Growth of the
      Algal Plankton

NEW YORK ACADEMY OF SCIENCES
  Annals of the New York Academy of Sciences
    Ecological Implications of *in Vitro* Nutritional Requirements of Algal Flagellates
  Transactions of the New York Academy of Sciences
    Essentiality of Constituents of Sea Water for Growth of a Marine Diatom

PEABODY MUSEUM OF NATURAL HISTORY, YALE UNIVERSITY—*Bulletin of the Bingham Oceanographic Collection*
    Mathematical Model of Nutrient Conditions in Coastal Waters

PERGAMON PRESS LTD.—*Deep-Sea Research*
    Nitrogen-limited Growth of Marine Phytoplankton: I. Changes in Population Characteristics with Steady-State Growth Rate

SIGMA XI, THE SCIENTIFIC RESEARCH SOCIETY OF NORTH AMERICA, INC.—*American Scientist*
    Ecological Potential and Analogue Circuits for the Ecosystem

VAN NOSTRAND REINHOLD—*Radioecology*
    Analog Computer Models for Movement of Nuclides Through Ecosystems

WISCONSIN ACADEMY OF SCIENCES, ARTS AND LETTERS—*Transactions of the Wisconsin Academy of Sciences, Arts and Letters*
    Phosphorus Content of Lake Waters of Northeastern Wisconsin

Series Editor's Preface

Ecology—the study of interactions and relationships between living systems and environment—is an extremely active and dynamic field of science. The great variety of possible interactions in even the most simple ecological system makes the study of ecology compelling but difficult to discuss in simple terms. Further, living systems include individual organisms, populations, communities, and ultimately the entire biosphere; there are thus numerous subspecialties in ecology. Some ecologists are interested in wildlife and natural history, others are intrigued by the complexity and apparently intractable problems of ecological systems, and still others apply ecological principles to the problems of man and the environment. This means that a Benchmark Series in Ecology would be subdivided into innumerable subvolumes that represented these diverse interests. However, rather than take this approach, I have tried to focus on general patterns or concepts that are applicable to two particularly important levels of ecological understanding: the population and the community. I have taken the dichotomy between these two as my major organizing concept in the series.

In a field that is rapidly changing and evolving, it is often difficult to chart the transition of single ideas into cohesive theories and principles. In addition, it is not easy to make judgments as to the benchmarks of the subject when the theoretical features of a field are relatively young. These twin problems—the relationship between interweaving ideas and the elucidation of theory, and the youth of the subject itself—make development of a Benchmark series in the field of ecology difficult. Each of the volume editors has recognized this inherent problem, and each has acted to solve it in his or her unique way. Their collective efforts will, we anticipate, provide a survey of the most important concepts in the field. Thus we expect that the Benchmark Series in Ecology will be useful not only to the student who seeks an authoritative selection of original literature but also to the professional who wants to quickly and efficiently expand his or her background in an area of ecology outside his special competence.

This volume, *Cycles of Essential Elements*, has been prepared by Lawrence Pomeroy of the University of Georgia. Dr. Pomeroy has specialized in the study of phosphorus dynamics in marine and estuarine systems. He has published extensively on his studies of this topic, and in 1970 he reviewed the subject in an annual review entitled *The Strategy of Mineral Cycling*. This experience has enabled him to select those papers from the extensive literature which will give the reader a balanced view of this dynamic subject.

Frank B. Golley

Preface

Scientists are not demonstrably more rational than other people. They develop strong emotional attachments to their discoveries and often delude themselves about the way in which the process of scientific discovery works. It should therefore be evident to anyone involved in science that the best we can hope to do is to make successively better approximations of our description of the world and the cosmos. Absolute truth is as elusive as the Holy Grail.

Nonetheless, most scientists grasp at their own discoveries and at those of a select few others as the final truth, and they resist mightily the efforts of others to press on and to find better approximations. The result is that progress in most fields of science is made by discrete jumps from one position of established truth to another. That is, our explanations of the world do not evolve continuously but change suddenly from one explanation to a substantially different one. Kuhn (1962) calls these changes "revolutions." This is an apt name for major changes of scientific paradigms, such as the change from Newton's to Einstein's frame of reference in physics. However, the same process occurs at a more mundane level of scientific activity. The changes we see in the short history of research on the cycles of elements might better be called "quantum jumps," because they represent smaller changes than those of the major scientific revolutions.

As Kuhn has pointed out, changes in paradigms are not recorded in textbooks and treatises. Once a new paradigm is accepted, and not before, it appears in publications to the exclusion of all previous ones. There can be only one version of truth at a time, and scientists neatly bury the past as they progress. Because of these folkways of science, Benchmark Books offer an opportunity and a challenge to look back at former truths and to attempt to recapitulate the process of discovery.

There will be as many versions of both the state of a science and its history as there are active scientists in that field. To some extent each of us carries around his own private version of the truth. The task of presenting a generally acceptable history of one of the specialized fields of ecology is especially difficult in this regard. Ecology is a very young science. In fact, it is really just beginning to be science. Perhaps because ecology is in a state of growth and metamorphosis, its paradigms are not yet well fixed. There is substantial disagreement about what is true and what is important.

Another problem that must be common to all Benchmark volumes is the need to capture the essence of the development of a special field of science in a few hundred pages. There are a number of papers that are undoubtedly benchmarks but that exceed 100 pages each. Therefore, it is necessary to thread together those contributions brief enough to be reproduced, and some portions of books, and only to refer to some of the other longer works. This gives the longer works less than their due. The reader who wishes to get a truly balanced view of the field may have to read some monographs and treatises that are cited but not reprinted here. Hopefully, this volume will serve as an introduction and a guide to studies of the cycles of essential elements.

Lawrence R. Pomeroy

Contents

III. COMPARTMENTAL ANALYSIS

IV. SYSTEMS ANALYSIS

Contents by Author

Introduction

Ecology had its beginnings as a discrete science in the second half of the nineteenth century. Its name was coined by Ernst Haeckel, and its first principles came from the natural historians, who were in full flower in Europe and England at that time. Because ecology evolved out of natural history, it was narrowly biological in its earliest formative years. The chemistry of ecosystems was not of great interest to early ecologists. To trace the development of the first ideas and facts concerning the cycles of elements, we must examine the works of persons who were not ecologists and whose purpose and point of view were not strictly ecological.

The interaction between other sciences and ecology has continued up to the present time. A number of the contributions reprinted here, including some published less than ten years ago, are taken from journals of other disciplines and have as their primary focus something other than ecology in the strict sense. Perhaps this is a phenomenon more prevalent in studies of cycles of elements than in some other areas of ecology simply because it is by its nature an interface area, combining biology with chemistry, as well as lesser amounts of geology, limnology, oceanography, mathematics, agriculture, and forestry. The subject matter is of both academic and applied interest to a diverse group of scientists and technologists.

Each of the fields that have contributed to the development of the study of cycles of elements has had its own focus of interest. Agriculture tends to focus on single species rather than on communities and on fertilizer input rather than on recycling. Geochemistry has sometimes tended to view biological processes as a regrettable complication of chemical equilibria. A truly ecological approach to cycles of elements is just beginning to emerge from the biome studies of the International Biological Program (IBP). These studies of whole landscapes have, at their best, completed the removal of ecology from the old natural history mold and have concentrated on processes of ecosystems, giving appropriate weight to their biological, chemical, and geological aspects. The cycles of elements are embedded in this matrix of ecosystem studies and are perhaps best approached in that context. Since most of the IBP biome study groups are still gestating their data, the full impact of IBP has not been felt and is thus not fully

1

expressed in this volume. What is possible at this time is to show the developments that have made studies of biome scale possible and appropriate.

The cyclic nature of the flow through ecosystems of the chemical elements that make up living organisms has become one of the basic concepts of ecology. Protoplasm is constantly being rebuilt and reworked in all active living organisms. As a result of this and to some extent through death and decay, the elements that make up protoplasm are in continuous flux. In terms of mass, the most important elements are O, C, H, N, Ca, and P. No other element constitutes as much as 1 percent of typical protoplasm, although many elements are essential to the construction and function of all protoplasm. The flux of chemical elements, including minor constituents of protoplasm, may act as a regulating or limiting process, and much of the early research on these elements in nature centered on their influence as factors that limit the growth of organisms and populations. More recently we have come to realize that the flow of elements provides us with a tool for understanding the function of ecosystems. Even in cases where there is no question of regulation or limitation through the limited supply of an element, studies of the cycle of an element yield valuable insights into how an ecosystem functions. Because we can measure the standing stocks and the rates of flow of the elements, we have a quantitative measure of certain kinds of functions of ecosystems. If a species population has a rapid flux of certain elements through it, or if it stores large amounts of an element, that population is clearly important in ecosystem function. Other species that are insignificant as movers or holders of elements may, of course, be important in other kinds of ecosystem functions, such as predator–prey interactions.

The elemental composition of protoplasm, be it bacterial or human, is remarkably uniform. Of course, there are differences among species and among higher taxa. The small protoplasts of bacteria have an even higher phosphorus content per unit mass than do vertebrates with phosphatic bones. Yet the major elements are always the same and occur in nearly the same order of abundance in all instances. What is *not* uniform and is therefore significant is the size of the pool of an element in a population and the rate at which it turns over, moving in from some source in the environment and moving out to other populations or environmental pools. Much of the research of the last twenty years in the field has been concerned with storage and the flux of elements at the population level. The pools of elements in populations can be large or small, and their turnover rate can be large or small. Moreover, these two factors can be combined in various ways. For example, the chemical elements bound up in the wood of a redwood forest have under natural conditions a very long turnover time. Those same chemical elements in a population of bacteria might have a turnover time of a day or less. The consequences of these differences of storage and flux have been made apparent by the application of mathematical modeling, which has begun to open up a new view of the complex interactions of populations as open-ended pools of chemical elements.

In this volume, modeling has been treated as a new and important tool for ecologists. Some critics of modeling claim that it is an electronic chess game that mimics certain aspects of the real world but that tells us nothing believable about it. At a certain level, modeling is just that. When properly used, however, modeling is a tool that can save time and money in research, and at its best it can produce nonintuitive insights into ecosystems that can then be tested in real ecosystems. In my opinion, modeling is here to stay and will become a standard tool of ecologists. It will not solve all our problems, and it certainly does not give us a crystal ball, but it will take its place beside chemistry, radiochemistry, statistics, and systematics as a means to the understanding of ecosystems.

Much of the potential utility of modeling remains to be exploited in the future, after the techniques have been prefected. Predicting the future of science is even more open to controversy than unearthing its past. One obvious limitation to modeling, and one that has drawn criticism from investigators, is that it cannot yet consider many elements at once. Modern chemical methods make it feasible to measure the amounts of many elements in all the components of any ecosystem. To measure all the fluxes between components is more challenging. To model mathematically the fluxes and interactions of many elements simultaneously at the ecosystem level is a taxing problem even for the largest computers now available. Improvements in both chemistry and computing are moving rapidly, however, and soon we may be able to cope with such complexity.

Readers who are familiar with this field of ecology may wonder why the term "mineral cycling" was not used in either the title or my comments. Probably because of the early association with agriculture, those elements which are constituents of protoplasm are often called "nutrients" or "mineral nutrients" by ecologists. Unfortunately, there is ambiguity in this ecological jargon. The term "nutrient" is more commonly understood to mean organic food, an energy source of some sort, rather than a chemical element. The term "mineral nutrient" is perhaps clear with reference to plants but a bit out of place in reference to animals. However, among ecologists the term "mineral cycling" is commonly used to describe what I have chosen to call "the cycles of essential elements." Both these terms encompass the full cycle of an element, wherever it may go, in organisms, water, soil, or the atmosphere. The term "mineral cycling" will appear frequently in the articles reprinted in this volume.

In organizing the material for this volume I have emphasized the development of new ways of studying ecosystems which transcend differences among the ecosystems themselves. Obvious differences exist between terrestrial and aquatic ecosystems, but some of the differences may be more apparent than real. All ecosystems involve water to some degree. Are rain forests and tundra terrestrial when water plays such a major role in them? Is an intertidal salt marsh aquatic when so much metabolic gas exchange with the atmosphere occurs? Perhaps this collection of papers will help to show, among other things, the need for the study of comparative ecology at the biome or ecosystem level. Chemical elements are common threads running through all ecosystems, and they make a useful point of reference.

I
Foundations

Editor's Comments on Papers 1 Through 5

Interactions of species, food webs, and the effects of abiotic factors were among the first considerations of ecologists, but the underlying importance of certain chemical elements in the shaping and limiting of assemblages of organisms was not appreciated so much by the pioneer ecologists as it was by two of their contemporaries. The interaction between the essential elements and the communities of organisms shaped from them was first recognized and explored by two European scientists, who were applying the rapidly rising understanding of chemistry to the solution of problems in other areas of science. They were Justus Liebig and Vladimir Ivanovich Vernadskii.

The dynamic and opinionated Liebig was one of the most influential organic chemists of his time, both as a result of his own discoveries and through his outstanding success as a teacher. His success seems to have resulted from the combination of high intelligence, enormous egotism, and a practical approach to science. In a time when students were never permitted to use laboratory apparatus but only to stand and watch demonstrations, Liebig put each student to work on an original research project (Browne, 1942). At the height of his career as a chemist, the British Association for the Advancement of Science asked Liebig to prepare a treatise on the existing state of knowledge of organic chemistry. The result was his book *Organic Chemistry in Its Applications to Agriculture and Physiology,* which was published simultaneously in German and English. Liebig skillfully reviewed the experimental work in agricultural chemistry of the previous several decades, selected what seemed to him the correct results, and then took credit for discovering it all. Because of his overbearing personality and existing fame, he succeeded in winning both acceptance for his views and credit for their originality.

How does one establish where the benchmark is in cases like this? In fact, this case is unusual only because Liebig was such a controversial figure and because the views he promoted so successfully were of great importance to the further development of agriculture and later to ecology. There are other instances that will be mentioned in this volume where the germ of an idea is published but never followed up by the sort of detailed studies that bring general acceptance. Later the idea is used by someone else who verifies it, extends it, and sells it to a skeptical audience.Probably we should recognize more than one benchmark at such major turns in the course of science. Too much emphasis on the contribution of one individual may be both unfair to others and inaccurate as history, but that is what usually happens. I have chosen for reproduction Liebig's ultimate summation of his position in 1855 rather than his perhaps-more-pivotal opening shot in 1840.

Most agriculturalists and plant physiologists of Liebig's time believed that the principal source of nutrients for plant growth was soil humus, that even carbon came from this source. Liebig showed that other sources were much more significant. Moreover, he showed differences in chemical composition of different plant species. He recommended the design of specific inorganic fertilizers for specific plant populations, a development that revolutionized agriculture. He also formulated his famous "law of the minimum," which stated that the growth of a population would be limited by the supply of that essential element that was present in least abundance relative to the requirement for it. This was a reverse twist to what is generally accepted as the order of events in scientific discoveries. A wholly practical application of chemistry to agriculture resulted in the formulation of a law that later became the central dogma of another basic science, ecology. Ecologists perceived that what Liebig's law said about cultivated crops and managed forests was equally true of all naturally occurring populations: it was as true of animals as of plants. Ecologists took Liebig's law and applied it far beyond the area intended by Liebig. In some instances the applications have been overextended. Probably Liebig's law should not be applied to communities or ecosystems but only to single-species populations. The first response to limiting elements by a community is succession, not cessation. As Liebig himself pointed out, the requirements of species vary, and where one will be limited, another can flourish and take its place in the community. In fact, Liebig's emphasis was on the input of elements and on monoculture, rather than on recycling. In spite of the overapplication of Liebig's law in the first half of the twentieth century, it has been an important influence on ecological theory.

A different kind of foundation was provided for ecology by Vernadskii, an early geochemist who founded biogeochemistry. His interests ranged widely, but a dominant one in his middle years was the influence of the biosphere on the geochemistry of the planet. Vernadskii saw the cycles of chemical elements throughout communities of organisms as processes that redistributed and sometimes concentrated those elements in the lithosphere: "Living matter basically affects the entire chemistry of the earth's crust, imparting direction to the geochemical history of almost all the elements in it" (quoted by Rodin and Bazilevich, 1956).

Vernadskii was one of the first to recognize the large-scale importance of bacteria and other microorganisms in biogeochemical processes. He also recognized the importance of the ocean and its microbial processes on the entire chemistry of the planet. It is startling to read passages written by him fifty years ago in which he describes the earth quite accurately from the viewpoint of a cosmonaut out in space. He describes the surface of the moon as if he had been there, and his description has been verified by recent lunar exploration. Not all of Vernadskii's sanguine assertions of the importance of the biosphere to the chemistry of the earth would be accepted by modern biogeochemists. Purely physical and chemical forces do move and shape much of the earth's chemistry. However, there is recognition that geochemistry is more than equilibrium reactions, and we are seeing a return to the emphasis on biological processes, which originated with Vernadskii.

Although the most direct impact of Vernadskii's thinking was on the new field of geochemistry, his thinking also had important meaning for ecology. It pointed up the importance of cycling between the biosphere and the lithosphere. The ecologist could no longer consider living populations in isolation from their nonliving environment. Liebig's approach looked outward from a cultivated field toward the forests and ultimately to all natural ecosystems. Vernadskii's approach looked inward from cosmic and planetary processes operating on a geological time scale. Modern studies of the cycles of essential elements have brought these two views together by considering a broad spectrum of scales of space and time.

It has been said that progress in science is limited by the development of ideas and that once the proper question is formulated ways will be found to answer it. Although this has much truth in it, the study of essential elements sometimes lagged for many years awaiting development of the necessary analytical tools. Others before Liebig had suggested the ideas that he championed, but he had a rapidly expanding range of analytical expertise to help verify his claims. The additional application of those ideas to natural assemblages of organisms was also delayed by a lack of analytical methods. It was known before Liebig that there were three elements, nitrogen, phosphorus, and potassium, which were likely to be present in limiting amounts. The one that limited growth would depend on the nature of the population. In field crops in Europe, potassium was often limiting, and it is often shown as the limiting element in illustrations of Liebig's law. In natural waters, phosphorus was found to be in very low concentration and possibly limiting. However, analyses were then carried out by gravimetric methods that were not sensitive enough to measure accurately the microgram quantities of phosphate present in natural waters. The development of a colorimetric method for measuring microgram quantities of phosphate by Denigès (1921) started an avalanche of analysis that has not yet wholly ceased. Much of the emphasis, or overemphasis, on phosphorus as a limiting factor resulted from the success of microanalysis.

It should be noted that in the 1920s electronic colorimeters and spectrophotometers were not available. Colorimetry was done by visual comparison using an optical device known as a color comparator, which permitted the investigator to view side by side a standard solution and an unknown solution. By varying the path length through which the standard was viewed until it matched the unknown, a reasonably accurate visual estimate of the unknown could be made (Barnes, 1959). Modern spectrophotometers increase both the sensitivity and the accuracy by at least two orders of magnitude. Improvements have also been made in the reagents used in colorimetry to yield intense, stable colors.

W. R. G. Atkins of the Plymouth Laboratory of the Marine Biological Association of the United Kingdom was one of the first to exploit the Denigès method to examine the phosphorus content of natural waters. Up to this point there was debate over the order of magnitude of the concentration of phosphate in seawater. With the Denigès method not only was there no question about the magnitude, but seasonal cycles and regional differences could be demonstrated. The Plymouth Laboratory became an important center for the study of the cycles of phosphorus, nitrogen, and iron in the sea by means of the new colorimetric methods. In addition to Atkins, important contributors were L. H. N. Cooper and H. W. Harvey. Harvey (1926) made an important observation: "It is a remarkable fact that plant growth should be able to strip seawater of both nitrate and phosphate, and that in the English Channel the store of these nutrient salts formed during Autumn and Winter should be used up at about the same time."

A. C. Redfield of the Woods Hole Oceanographic Institution showed that what was true of the English Channel was also true of the World Ocean. He suggested that the amounts of two essential elements, phosphorus and nitrogen, in surface seawater were influenced by the composition of the plankton and that the repeated cycling of the elements through the plankton in very nearly constant proportions ultimately regulated their proportions in solution in the water. Perhaps this was when the approaches of Liebig and Vernadskii were first brought together. Redfield's observation paved the way for detailed studies of the mechanisms that control the cycles of essential elements. Redfield was by background a physiologist, but his work since the 1934 paper has been increasingly biogeochemical, with full utilization of the techniques of physical oceanography. Be this ecology or oceanography, it has been influential in setting the course of thinking in marine ecology and limnology. Redfield refined and elaborated upon his ideas in several later papers, modifying slightly his estimate of the proportions of the principal essential elements in the plankton and suggesting other possible interactions between atmosphere, biosphere, and lithosphere (Redfield, 1958; Redfield et al., 1963).

The Denigès method for measuring phosphate and total phosphorus in water was quickly taken up by limnologists. Reproduced in this volume is the first of a series of papers on phosphorus and other essential elements in lakes by E. A. Birge, Chauncy Juday, and their collaborators at the University of Wisconsin. They found that phosphorus in temperate, stratified lakes goes through a yearly cycle of changes in abundance similar to that found in the temperate parts of the World Ocean. In their second paper, Juday and Birge (1931) demonstrate the presence of dissolved organic compounds that contain phosphorus, as well as inorganic phosphate, in lake waters. Matthews (1916) discovered dissolved organic phosphorus in the ocean, but the pre-Denigès analytical methods available to him did not permit detailed examination of their cycles of abundance.

These discoveries set the stage for a quarter century of analytical exploration of the abundance and distribution of phosphorus, nitrogen, iron, and some other elements in lakes and in the ocean. A considerable body of data on the amounts of these elements was built, seasonal cycles of the elements were found, and these were interpreted in terms of Liebig's law. Scientists believed that available phosphorus and nitrogen were regenerated slowly in surface water during the winter by bacterial recycling of organic matter and by turbulent mixing of water. In spring these available supplies were utilized quickly by a bloom of phytoplankton that would develop as soon as light intensity increased and the water began to warm and to stratify. When the supply of dissolved phosphate and nitrate was exhausted, the bloom subsided and summer populations of both phytoplankton and zooplankton were limited by the scarcity of those essential elements. This view of the cycles of nitrogen and phosphorus in natural waters was generally held until the development of tracer techniques that permitted us to see the cycles of those elements in greater detail.

The period from the statement of Liebig's law, through the development of biogeochemistry, and up to the beginning of tracer methods was also a period of gathering of extensive analytical data on the concentration of essential elements in soils and terrestrial plants. Much of this material was related to agriculture, forestry, and geochemistry at a time when the basic discipline of ecology was still in a formative stage. However, the body of information that was being gathered would be used by ecologists in years to come. Much of this has been summarized, especially in the works of Rodin and Bazilevich (1956) for terrestrial ecosystems and Clarke (1924) for geochemistry.

1

Reprinted from *Principles of Agricultural Chemistry with Special Reference to the Late Researches Made in England* by J. Liebig, 1885, pp. 17–34

Principles of Agricultural Chemistry with Special Reference to the Late Researches Made in England

JUSTUS LIEBIG

I now proceed to lay down the following propositions, which contain the views I hold and have taught on this subject.

1. Plants in general receive their *carbon* and *nitrogen* from the air, the carbon in the form of *carbonic acid*, the nitrogen in that of *ammonia*. The *water* (and ammonia) yield to plants their *hydrogen ;* the *sulphur* of those parts of plants which contain that element, such as the sanguigenous bodies, is derived from *sulphuric acid.*

2. On the most diversified soils, in the most varied climates, whether cultivated in plains or on high mountains, *plants invariably contain a certain number of mineral substances, and, in fact, always the same substances ; the nature and quality, or the varying proportions of which are ascertained by finding the composition of the ashes of the plants.* The mineral substances found in the ashes were originally ingredients of the soil; all fertile soils contain a certain amount of them; they are never wanting in any soil in which plants thrive.

3. In the shape of the agricultural produce of a

11

field, or in the crop, the entire amount of these ingredients of the soil which have become ingredients of the plants, are removed from the soil. The soil is richer in these matters before seed-time than after harvest; or, in other words, *the composition of the soil after harvest is found to be changed.*

4. After a series of years, and a corresponding number of harvests, the fertility of the soil or field diminishes. While all the other conditions have remained the same, the soil alone has not done so; it is no longer what it was at first. *The change which is found to have taken place in its composition, is the probable cause of its diminished or lost fertility.*

5. *By means of solid and liquid manure, or the excreta of men and of animals, the lost or diminished fertility of the soil is restored.*

6. Solid or farm-yard manure consists of decaying vegetable and animal matters, which contain a certain proportion of the constituents of the soil. The excrements of men and animals represent the ashes of the food consumed, that is, oxidised or burned in the bodies of men and animals; food derived from plants which have been reaped on the supposed soil. The urine contains the soluble, the solid excreta the insoluble, constituents of the

soil derived from the crops used as food, and reaped from the soil. It is clear, that by adding manure, or liquid and solid excreta to the soil, that soil recovers those constituents which have been removed from it in the crops. Thus, the restoration of its original composition is accompanied by the restoration of its fertility. It is therefore certain, that *one of the conditions of fertility in a soil is the presence in it of certain mineral constituents.* A rich, fertile soil contains more of these than a poor, barren one does.

7. The roots of plants, in regard to the absorption of their atmospherical food, behave like the leaves; that is, they possess, like these, the power of absorbing carbonic acid and ammonia, and of employing these, in their organism, in the same way as if the absorption had taken place through the leaves.

8. The *ammonia* which is contained in, or brought by means of rain, &c., into, the soil, *plays the part of a constituent of the soil.* This is true, likewise, of the *carbonic acid* in the soil.

9. Vegetable and animal matters, and animal excreta, when in the soil, undergo putrefaction and decay, or slow oxidation. The nitrogen of their nitrogenised constituents is changed, in the

processes of putrefaction and decay, into *ammonia;* and a small part of this ammonia is converted into *nitric acid*, which is the product of the oxidation or decay of ammonia.

10. There is every reason to believe, that in the process of nutrition of plants, nitric acid can replace ammonia as a source of nitrogen; that is, its nitrogen can be applied, in the vegetable organism, to the same purposes as that of ammonia.

11. In animal manure therefore, not only are plants supplied with the mineral substances which the soil must yield, but they are also supplied with those parts of their food which the plant obtains from the atmosphere. This latter supply is a clear addition to that which the air at all times affords.

12. The solid and liquid parts of the food of plants contained in the soil, enter the organism of the plant through the roots; their introduction is effected by means of *water*, which gives to them solubility and mobility. Many dissolve in pure water, others only in water which contains *carbonic acid*, or some *salt of ammonia.*

13. All those substances which render soluble those constituents of the soil which are by themselves insoluble in water, have this effect when present in the soil, that they cause the same volume

of rain water to take up and introduce into the plant a greater quantity of these constituents.

14. By the progressive decay of animal manure, the animal and vegetable remains of which it chiefly consists are converted into carbonic acid and ammoniacal salts, and thus constitute an active source of carbonic acid, which renders the air and the water which pass through the soil richer in carbonic acid than they would be without the presence of these remains.

15. Hence, animal manure not only supplies the plants with a certain amount *of their mineral and atmospheric food*, but also provides them, in *carbonic acid* and *ammoniacal salts*—those substances which are the most indispensable for the *introduction into the vegetable organism of the mineral constituents which by themselves are insoluble in water;* and this to a larger amount in the same time than could be effected without the cooperation of decaying organic matter.

16. In warm, dry seasons, plants receive from the soil less water than they do, in the same circumstances, in wet seasons. The harvest in these different seasons is in proportion to this variable supply. A field of the same quality yields in dry years a smaller crop, which increases in more moist

seasons; and, if the average temperature be the same, it increases, up to a certain limit, with the amount of rain.

17. Of two fields, of which the one contains more food for plants of all kinds, taken together, than the other, the richer yields, even in dry seasons, a higher produce than the poorer, other circumstances being the same.

18. Of two fields, of equal quality, and containing equal amounts of mineral constituents adapted to vegetable growth, but one of which contains *a source of carbonic acid*, in the form of decaying organic matter or manure, that one, even in dry years, yields more produce than the other.

The cause of this difference or inequality in the crop, in such cases, is to be found in the unequal supply of mineral constituents, both as to quantity and quality, which the plant obtains from the soil in equal times.

19. All things which oppose or impede the solubility, and consequently the absorbability, of those parts of the food of plants which occur in the soil, diminish in the same proportion the power of these substances to nourish the plant, or, in other words, render the nourishment inefficacious. A certain physical or mechanical quality or state of the soil

is a necessary condition to the efficacy of the food which is present. The soil must admit the free passage of air and water, and allow the roots to spread on all sides in search of food. The term *telluric conditions* comprises all such conditions as depend on the mechanical quality of the soil and on its chemical composition, and are necessary to the development of plants.*

* According to an excellent article in the supplement to the *Augsburg Allgemeine Zeitung* of 28th October, 1854, it appears that, to many persons, the question whether manure only exalts the physical powers of the soil, or serves also, by its constituents, for the nutrition of plants, still requires to be solved. The expression "physical powers" makes the answer difficult, because we do not know what is meant by it. The constituents of a fertile soil have many properties, among which some are physical, by which we understand such as are cognisable by our senses, as colour, density, porosity, stiffness, or lightness. To the other class of properties of the soil, not cognisable by the senses, belong the chemical characters, by which we mean, those properties which accompany chemical combination or decomposition. The absence or the presence of the physical properties impedes or promotes the manifestation of the chemical ones, that is, the processes of chemical combination and decomposition; but considered by themselves, they produce no effect. By the term "nutrition of a plant," we understand the increase of its mass in all its parts. Increase of mass is increase of weight, which can only be effected by the assimilation of ponderable particles. A substance contributes to the nutrition of a plant, or, in other words, contributes, while becoming a constituent of an organ or organs of the plant, by its own mass to this result, that the weight of the plant is increased. It is easy to see that the physical properties of matter by themselves have no direct share in this nutrition. A soil may possess the very best physical qualities and yet be barren: in order to be fertile it must contain substances of certain chemical properties, and its physical

20. All plants, without exception, require, for nutrition, *phosphoric acid, sulphuric acid, the alkalies, lime, magnesia, and iron.* Some important genera require *silica.* Those which grow on the sea-shore and in the sea, require *common salt, soda,* and *iodides of metals.* In some genera, the alkalies may be, in part, replaced by lime and magnesia, or these latter by the alkalies. All these substances are included in the term *mineral food of plants.* Carbonic acid and ammonia are the *atmospheric food* of vegetables. Water serves both as a nutritive substance, and, as a solvent, is indispensable to the whole process of nutrition.

21. The different substances necessary to the growth of a plant, or the different articles of their food, are *all of equal value ;* that is to say, if one out of the whole number be absent, the plant will not thrive.

character must be such as to allow these chemical properties to be manifested. If the soil, from being very stiff, does not allow the roots to spread, the roots cannot reach the substances which they require as food. If it do not permit water to percolate through it, the nutritious substances cannot reach the roots. A piece of meat, as everybody knows, possesses nutritive properties, but it does not nourish by means of its physical properties, its colour, the strength of its fibres, or its cohesion, but because its parts are capable of becoming constituents of the living body. If we lay a piece of meat on the stomach externally, it has no effect, but must be first introduced into that organ, where it is dissolved, and so enters the circulation.

22. The soils which are proper for the cultivation of all sorts of plants, contain all the mineral constituents necessary for these plants.   The words *fertile* or *rich*, *barren* or *poor*, express the relative quantities or qualities of these mineral substances present in the soil.

By difference *in quality*, we understand the unequal state of solubility, or capacity of entering the vegetable organism, in the mineral constituents, which entrance is effected by means of the solvent power of water.

Of two soils which contain *equal* quantities of mineral constituents, one may be considered rich or *fertile*, the other poor or *barren;* if in the latter, these constituents are not free or available, but in a form of combination which renders them insoluble. A substance, chemically combined with another, in consequence of the attraction between its elements, opposes a resistance to any other substance tending to combine with it; and this resistance must be overcome, if the new compound is to be formed.

23. All soils adapted for culture contain the mineral food of plants in these two states.   The whole, added together, constitute the capital of the soil; and the available or soluble portions form the floating or moveable capital.

24. To improve, enrich, or fertilise a soil by proper means, but without adding to it any mineral constituents, is to render moveable, soluble, available for the plant a part of the dead or immoveable capital of the soil.

25. The mechanical preparation of the land has for its object to overcome the chemical resistance in the soil, or to render soluble and available those mineral constituents which are in chemical combination, and thence insoluble. This is effected by the aid of the air, of carbonic acid, of oxygen, and of water; and the effect is called the weathering, or action of the weather on the soil. Stagnant water in the soil, which excludes the air from access to the insoluble compounds, is an obstacle or resistance to the weathering.

26. *Fallow* is the time during which this weathering takes place. During fallow, carbonic acid and ammonia are conveyed to the soil by the rain and the air; the ammonia remains in the soil, if substances be present in due proportion which deprive it of its volatility by combining with it.

27. A soil is fertile for a *given kind* of plant when it contains the mineral food proper to that plant in due quantity, in just proportion, and in a form adapted to assimilation, or available for the plant.

28. When such a soil, by a series of crops grown on it without any replacement of the mineral substances removed in those crops, has become barren for that kind of crop, it becomes, after one or more years of fallow, again fertile for the same plant, provided it contained originally, besides the available mineral food removed, a certain amount of the same substances in an insoluble form, which have been rendered available during the fallow time, by ploughing and weathering. *Manuring with green crops* enables us to attain the same object in a shorter time.

29. Land, on which these necessary mineral constituents are not present in any form, cannot be rendered fertile by fallow or by ploughing.

30. The increase of fertility in a soil by fallowing and mechanical preparation, *if the mineral matters removed in the crops be not restored to the soil*, produces, sooner or later, a permanent barrenness.

31. If the soil is to retain *permanently* its fertility, the mineral constituents removed in the crops must be restored to it from time to time, at shorter or longer intervals, or, in other words, the original composition of the soil must be restored.

32. Different kinds of plants require for their

c 2

development, in some cases, the same mineral substances, but in unequal quantities, or in unequal times. Some cultivated plants must find soluble silica in the soil.

33. When a given piece of land contains a certain amount of all the mineral constituents *in equal quantity* and in an available form, it becomes barren for any one kind of plant when, by a series of crops, one only of these constituents—as, for example, soluble silica—has been so far removed, that the remaining quantity is no longer sufficient for a crop.

34. A *second kind* of plant, which does not require this constituent, for example, silica, may yield, on the same soil, after the former has ceased to thrive, one or a series of crops, because the other mineral substances necessary for it are present,—no longer, indeed, in the same proportion as at first, no longer in equal quantities, but in quantities sufficient for its perfect development. A *third sort* of plant may thrive on the same soil after the second, if the remaining mineral constituents suffice for a crop of it; and if, during the cultivation of these crops, a new quantity of the substance wanting for the first—for example, of soluble silica—has been rendered available by weathering, then, if the

other necessary conditions be fulfilled, the first crop may again be grown on the same land.

35. On the unequal quantity and quality (solubility, &c.) of the mineral constituents, and on the unequal proportions in which they are required for the development of the different cultivated crops, depends the *rotation of crops*, and the varieties of rotation employed in different localities.

36. The growth of a plant, its increase in mass, and its complete development in a given time, all other conditions being equal, are in proportion to the surface of the organs destined to absorb the food of the plant. The amount of the food obtainable from the air depends on the number and surface of the leaves; that of the food obtainable from the soil depends on the number and surface of the root fibres.

37. If, during the formation of the leaves and roots, two plants of the same kind are supplied with *unequal* amounts of food in the same time, their increase in mass is unequal. It is greater in that plant which, in that time, received more food : its development is accelerated. The same inequality of increase in mass is observed, when the same food is supplied to both plants in equal quantity, but *in different conditions of solubility*.

By supplying any plant with the due amount of all the atmospheric and telluric constituents necessary to its nutrition, in the required time and in the proper forms, its development in a given time is accelerated. The conditions which *shorten the time* required for its growth are the same as those which determine its *increase in mass*.

38. Two plants, whose root fibres have an *equal* length and extent, do not thrive so well beside each other, or in succession, as two whose roots, being of *unequal* length, receive their food from different strata or depths of the soil.

39. The nutritive substances, necessary to the life of a plant, must act together within a given time if the plant is to attain its full development in that time. The more rapidly a plant is developed in a certain time, the more food it requires in that time. Thus, annual or summer crops require, in the same time, more food than perennial plants.

40. If one of the co-operating constituents of the soil or of the air be absent or deficient, or do not possess the proper form or state, the plant is either not developed, or only imperfectly developed in its parts.

The *absence or deficiency*, or the want of available form, in that one constituent, renders the others which are present *ineffectual*, or diminishes their efficacy.

41. If the absent or deficient substance be added to the soil, or, if present, but insoluble, be rendered soluble, the other constituents are thereby rendered *efficient*.

By the deficiency or absence of *one* necessary constituent, all the others being present, the soil is rendered barren for all those crops to the life of which *that one* constituent is indispensable. The soil yields rich crops, if that substance be added in due quantity and in an available form. In the case of soils of unknown composition, experiments with individual mineral manures enable us to acquire a knowledge of the quality of the land and the presence of the different mineral constituents. If, for example, phosphate of lime, given alone, is found efficacious, that is, if it increases the produce of the land, this is a sign that that substance was absent, or present in too small proportion, whereas there was no want of the others. Had any of these other necessary substances been also wanting, the phosphate of lime would have had no effect.

42. The efficacy of all the *mineral constituents of the soil* taken together, in a given time, depends on

the co-operation of the *atmospheric constituents* in the same time.

43. The efficacy of the *atmospheric constituents* in a given time, depends on the co-operation of the *mineral constituents* in the same time; if the latter be present in due proportion and in available forms, the development of the plants is in proportion to the supply and assimilation of their atmospheric food. The quantity and quality (available form) of the mineral constituents in the soil, and the absence or presence of the obstacles to their efficacy (physical qualities of the soil), increase or diminish the number and bulk of the plants which may be grown on a given surface. The *fertile* soil takes up from the air, in the plants grown on it, *more* carbonic acid and ammonia than the barren one; this absorption is in proportion to its fertility, and is only limited by the limited amount of carbonic acid and ammonia in the atmosphere.

44. With *equal supplies of the atmospheric conditions* of the growth of plants, the crops are in direct proportion to the amount of *mineral constituents* supplied in the manure.

45. With *equal telluric conditions,* the crops are in proportion to the amount of *atmospheric constituents* supplied by the *air* and the *soil* (including

manure). If, to the available mineral constituents in the soil, ammonia and carbonic acid be added in the manure, the fertility of the soil is exalted.

The union of the *telluric* and *atmospheric* conditions, and their co-operation in due quantity, time, and quality, determine the *maximum* of produce.

46. The supply of more atmospheric food (carbonic acid and ammonia, by means of ammoniacal salts and humus) than the air can furnish, increases the efficacy of the mineral constituents present in the soil, in a given time. From the same surface there is thus obtained, in that time, a heavier produce—perhaps in one year as much as in two without this excess of atmospheric food.

47. *In a soil rich in the mineral food of plants* the produce cannot be increased by adding more of the same substances.

48. *In a soil rich in the atmospheric food of plants*, (rendered so by manuring), the produce cannot be increased by adding more of the same substances.

49. From land rich in the mineral constituents, we may obtain in one year or for a series of years, by the addition of ammonia alone (in its salts) or of humus and ammonia, rich crops, without in any

c 3

way restoring the mineral substances removed in these crops. The duration of this fertility then depends on the supply, that is, the quantity and quality of the mineral constituents existing in the soil. The continued use of these manures produces, sooner or later, an exhaustion of the soil.

50. If, after a time, the soil is to recover its original fertility, the mineral substances extracted from it in a series of years must be again restored to it. If the land, in the course of ten years, has yielded ten crops, without restoration of the mineral substances removed in those crops, then we must restore these, in the eleventh year, in a quantity tenfold that of the annually removed amount, if the land is again to acquire the power of yielding a second time, a similar series of crops.

The preceding fifty propositions are all contained in one proposition; namely, that the nutrition, the growth, and the development of a plant depend on the assimilation of certain bodies, which act by virtue of their mass or substance. *This action is within certain limits directly proportional to the mass or quantity of these substances, and inversely proportional to the obstacles or to the resistance which impede their action.*

2

The Biosphere

V. I. VERNADSKII

This article was translated expressly for this Benchmark volume by Lawrence R. Pomeroy, University of Georgia, from Biosfera, *Leningrad, 1926, pp. 225–226, 227, 241–242, 278–280†*

1. The surface of the earth, seen from the infinite distance of outer space, appears unique, specific, and distinct in appearance from all other neighboring celestial bodies. On the surface of the earth we see its biosphere—its external layer that delimits it from cosmic space. The terrestrial surface can be seen thanks to the light of the stars, especially that of the sun. It receives an infinite number of diverse radiations from all points in space, of which visible light is an insignificant part.

We recognize only a small number of invisible radiations, and are scarcely beginning to enumerate their variety or to understand the plethora of radiations that surround us and penetrate us in the biosphere. Not only the biosphere but all conceivable space is enveloped by radiation of wavelengths from 10^{-5} millimeters to the order of kilometers. They flow all around us and within us, everywhere and at all times; they collide and coincide with one another.

Indeed, they are present throughout all space. It is difficult and perhaps impossible to conceive that environment completely—the cosmic environment of the universe, in which we live and where we are devising and perfecting methods to distinguish and measure the endless radiation.

* * * * * * *

3. Cosmic radiation continuously striking the surface of the earth produces a flow of power that gives new and specific characteristics to the parts of the planet exposed to cosmic space. As a consequence, the biosphere takes on new properties that terrestrial matter does not otherwise have. The surface of the earth is transformed by cosmic forces.

---

†I have translated only a part of the first section, which is somewhat repetitious. I have then selected several sections that serve to indicate the range and depth of Vernadskii's thinking, and they are translated in their entirety.

Editor's Note: A row of asterisks indicates that material has been omitted from the original article.

The matter of the biosphere, penetrated by transmitted energy, is active. It collects and distributes energy in the biosphere received as radiation, and it is able to do work in the terrestrial environment by the transformation of free energy.

That exterior terrestrial layer should not be considered to be the domain of matter alone. It is a region of energy — a means of transformation of the planet by external cosmic forces. These forces transform the surface of the earth; to a great extent they shape it. That surface is not only the result of planetary events, a manifestation of the earth's matter and energy, but it is also a creation of the external forces of the cosmos. Because of this situation, the history of the biosphere is different from that of the other parts of the planet, and its role in the mechanism of the planet is distinctive.

The biosphere is quite as much, but not moreso, the creation of the sun as it is the manifestation of terrestrial processes. The old religions that considered the creatures of the earth, especially mankind, as children of the sun were as near the truth as were those which believed that terrestrial beings were only an ephemeral creation — the blind and accidental modification of matter and terrestrial forces.

Terrestrial organisms are the fruit of a long and complicated cosmic process and form an essential part of a harmonious cosmic mechanism, in which they exist only by chance.

*   *   *   *   *   *   *

20. Should life disappear, it is evident that the continuous chemical processes associated with it will disappear as well, at least in the biosphere, if not in the entire crust of the earth. The free aluminosilicates (clays), the carbonates (limestone and dolomite), the hydrated ferric and aluminum oxides (limonite and bauxite), and certain other minerals are perpetually produced under the influence of life. If life were to disappear, the elements in these minerals would form new chemical combinations in response to the new chemical conditions, whereas the usual minerals would disappear. After the extinction of life, there would be no force on the earth's crust to continue producing new chemical compounds, and stable chemical equilibrium, a chemical calm, would irrevocably result—disturbed from time to time in only a few places by material brought from the depths of the earth, such as gaseous emanations, thermal springs, or volcanic eruptions. These newly arrived materials would be altered more or less quickly into stable molecular forms appropriate to a lifeless crust and henceforth would undergo no further alterations. For, although there are thousands of places where materials emanate from the depths of the earth and are then dispersed over the surface of the planet, they are lost in its immensity. Such processes as occasional volcanic eruptions would be unnoticed in the infinity of terrestrial time.

Following the disappearance of life, only slow changes would take place on the earth's surface. These changes would not be noticed in the course of years or centuries but only over geologic time. Like the radioactive mutations of atomic systems, they would be perceptible only over cosmic time.

The continuously active forces of the biosphere, the heat of the sun, and chemical activity in water alter the range of activities. With the extinction of life, free oxygen would disappear [*sic*] and the amount of carbonic acid would decrease. The agents of superficial lateration which are associated with the biosphere would also disappear. Under the thermodynamic conditions of the biosphere, water is a powerful chemical agent, for "natural" water is rich in centers of chemical activity, thanks to the existence of life and, above all, of microorganisms. Water is modified by oxygen and carbonic acid in solution. However, water stripped of life, free of oxygen and carbonic acid, still having the temperature and pressure of the earth's surface and an environment of inert gas [nitrogen] is a compound of weak chemical activity.

The surface of the earth would become as immobile and chemically inert as that of the moon, that is, covered by fragments of celestial bodies attracted to it by gravity, by meteorites rich in metals, and by cosmic dust from outer space.

* * * * * * *

66. It is remarkable that most of the atoms in living organisms always return immediately to living matter after the destruction of the organism in which they reside. An insignificant fraction of them migrates invariably and constantly through global processes.

This small percentage of matter is not accidental, and is probably constant for each element. Such atoms return to living substance by another route after thousands and millions of years. During that time the materials separated from living matter play a dominant role in the history of the biosphere and even in that of the earth's crust in general, for a large part of those atoms leave the biosphere for a long time.

Here we are concerned with a new kind of process — the slow penetration of the planet by radiant energy that falls on it from the sun. By that process, living matter transforms the biosphere and the crust of the earth. It continually abandons part of the elements that have passed through it, producing great masses of minerals that were not present before and penetrating the inert matter of the biosphere with its detritus. Moreover, these processes by their cosmic energy alter the chemical structure of compounds that were formed independently of the immediate influence of life.

The crust of the earth, to the depth that we can see it, has been changed in this way. Radiant energy penetrates deeply over geological time as a result of the action of living matter. Minerals are altered and also become implements of transport.

The inert matter of the biosphere is largely the creation of life. We have revived here, in a new form, the ideas of the natural philosophers of the nineteenth century — the ideas of L. Ocken, J. Steffens, and J. Lamarck. These scholars were impressed with the idea of the primordial importance of life in geological processes, an idea based on the empirical facts of the history of the crust developed by preceding generations of observers.

It is curious that such influence on all matter in the biosphere, especially on the formation of minerals in the presence of water, should be principally the result of the action of aquatic organisms. The continual shifting of water basins through geological time has scattered over the planet accumulations of free chemical energy of cosmic origin which originated in this way.

31

All these processes seem to have the form of a stable dynamic equilibrium, and the masses of matter that enter into it are as unchangeable as the sun's energy that falls on the earth and affects them.

67. In the final analysis, a considerable amount of material is incorporated into organisms in the biosphere and is transformed by means of solar energy. The weight of the biosphere amounts to some 10^{24} grams. In that surface layer of the planet, living matter receives less than 1 percent of the cosmic energy, probably a few hundredths of 1 percent. Here and there in thin layers such as soils it amounts to as much as 25 percent.

The formation of living matter on our planet is clearly a cosmic phenomenon that is not connected with abiogenesis but with a continuity of living organisms. All organisms are genetically related, and no organism absorbs and converts solar energy independently of previous living organisms.

How did the specific mechanism of the biosphere, the living substance of the terrestrial crust, which has functioned for hundreds of millions of years of geological time, originate? That is a mystery, just as life itself is a mystery in our scheme of knowledge.

3

Copyright © 1925 by the Marine Biological Association of the United Kingdom

Reprinted from *J. Marine Biol. Assoc. U.K.*, **13**, 119–150 (1925)

The Phosphate Content of Fresh and Salt Waters in its Relationship to the growth of the Algal Plankton.

By

W. R. G. Atkins, O.B.E., Sc.D., F.I.C.

Head of the Department of General Physiology at the Plymouth Laboratory.

With Figures 1–8 in the Text.

CONTENTS.

INTRODUCTION AND PHOSPHATE CONTENT OF FRESH WATER SUPPLIES.

ON account of the minute quantities in which they are present and of the fact that they are considered of secondary importance as indicating sewage contamination, phosphates are not usually estimated in analyses of natural waters. The tediousness of the determination also militated against it in the past. As a result, of the numerous analyses recorded by Clarke (1920), but few mention phosphates. C. H. Stone's analysis of the Mississippi in 1905, carried out upon a sample above Carrolton, Louisiana, shows 0·27 per cent of phosphate (PO_4) with a total salinity of 146 parts per million, or 0·39 mgrm. PO_4 per litre, corresponding to 0·29 mgrm. P_2O_5.

The presence of as little as 0·5 part P_2O_5 per million, viz. 0·5 mgrm. per litre, is considered as indicative of sewage contamination (Kenwood, 1911, quoting Hehner), though owing to the rapid removal of phosphates by plants a smaller amount need not necessarily prove the purity of the water. The American Public Health Association's Standard Methods for water analysis do not include one for phosphate (1920).

Recently McHargue and Peter (1921) have carried out a large number of phosphate determinations in small and large streams and some of the great rivers of the United States. Spring water in an Ordovician area was found to contain 0·5–0·8 parts per million of phosphate as pentoxide ; springs in other areas were considerably poorer, containing only 0·1–0·2 p.p.m. Figures for the rivers Ohio, Tennessee, Green River, Cumberland, Missouri, and Mississippi averaged 0·2 p.p.m. Calculating from the mean annual volume of the Mississippi near its mouth these authors conclude that the amount of the element phosphorus carried to the sea in solution amounts each year to 62,188 tons ; to this must be added the phosphorus (0·15 per cent) in 7469 million cubic feet of suspended matter. The concentration of phosphate in the sea is, as will be shown later, far less than 0·2 p.p.m., so, while diluting the general salinity of the ocean, the river raises its concentration as regards phosphates.

In view of the scanty data available as to the quantity of phosphate in natural waters and reservoirs in this country, the following miscellaneous determinations carried out by the writer may be placed on record. The analyses were made by the colorimetric method of Denigès, as described later.

In order the better to characterize the water the pH value and electrical conductivity, which gives an idea of the proportion of total solids, are also tabulated (see Table I).

It may be seen that the phosphate content of uncontaminated streams and fresh water supplies is extremely small in the districts examined, being under 0·05 parts of P_2O_5 per million. These values are considerably below those of McHargue and Peter, obtained in the U.S.A. How small these quantities are may be appreciated from the fact that Matthews (1916–18), when making up artificial sea water from the purest chemicals of Merck and Kahlbaum, found that the mixture contained 0·0286 mgrm. of P_2O_5 per litre, and the writer has found hydrogen peroxide sold as free from phosphoric acid to contain the equivalent of 0·20 mgrm. of P_2O_5 per litre.

The earlier analyses of the phosphate content of sea water are reviewed by Matthews (1916), Raben (1920), and Brandt (1920).

With samples taken just outside Plymouth Breakwater Matthews found a maximum of 0·06 mgrm. per litre at the end of December, 1915, with an irregular fall to a minimum of less than 0·01 in April and May.

He attributes the seasonal variation to the removal of phosphates from solution by the fixed algæ, the diatoms, and Phæocystis.

Raben's analyses extend from 1904–14, and include numerous determinations upon the water of the North Sea, Baltic, Barentz Sea, and North Atlantic Gulf Stream. These, as plotted by Brandt, show minimal values in May and June. After a rise to a peak in September low values are again shown early in October.

Brandt's graph, like that given by Matthews, refers to surface water, though Raben also analysed water from various depths down to 800

TABLE I.

Source of water.	Phosphate as mgrms. P_2O_5 per litre.	Electrical conductivity at 0° C. $\times 10^6$.	pH.
Plymouth tap, May	0·003	26	6·6
Maryfield (Cornwall) tap, June . .	0·023	270	7·2
Basingstoke tap, June	0·032	270	7·2
Peverell (Plymouth) old reservoir, June .	0·278	222	—
Pool in waterlogged pasture, Anglesey .	0·167	290	6·9
Stream, Bodorgan, Anglesey, February .	0·019	192	6·8
Stream, basalt district, S. Scotland, March	0·007	59	6·4
Ditch, calcareous sandstone district, S. Scotland, March	0·016	186	6·9
Stream, S. Scotland, March . . .	0·021	72	6·8
Stream, Yorkshire, March . . .	0·036	227	7·1
Stagnant ditch, meadow, near Plymouth	0·019	213	7·7
Ditch in lane, near Plymouth . . .	0·047	294	7·6
Yard well, Antony, Cornwall . . .	1·25	227	6·4
Sea water, winter	0·049	28,200	8·1
Aquarium tanks, Plymouth . . .	4·81	30,300	7·6

metres in the North Atlantic. There is usually a considerable increase from the surface downwards. None of the values, however, indicate exhaustion of the water as regards phosphate, the minimum recorded figure being 51 mgrm. of P_2O_5 per cubic metre (viz. 0·051 mgrm. per litre) and the maximum 221 mgrm., both values being from North Sea Station N7. These figures are about four times as great as those given by Matthews, whose results it may be added agree well with those obtained by the Government chemist, London, using the same method as Matthews upon samples sent from Plymouth in 1922, and with analyses carried out by the writer, according to an entirely different method.

In view of the importance of phosphates for plant growth it seemed of interest to make a further study of these seasonal changes, both in the sea and in fresh water, and to study the diminution of phosphate in laboratory cultures.

THE UPTAKE OF PHOSPHATE IN A DIATOM CULTURE.

A culture of *Nitzschia closterium* W. Sm., pure save as regards the presence of bacteria, was kindly supplied by Dr. E. J. Allen. This was growing in sea water enriched with Miquel's solution, as described by Allen and Nelson (1910). It was exposed in a north window for periods as given in Table 2, the temperature being about 12°–15° C. The results are shown in Fig. I, and it may be seen that a great increase in diatoms results in the almost complete utilization of the phosphate, which appears to be the factor limiting further multiplication.

TABLE II.

Changes in phosphate in culture flask of *Nitzschia closterium*.

Date.	Days.	Nitzschia, thousands per c.c.	P_2O_5 as milligrams per litre.		
17/3	0	0	—	—	—
27/3	10	510	2·38	—	—
13/4	26	2140	0·55	—	—
26/4	40	3065	0·006	—	—

From the count of 13/4 and the previous one 1630×10^6 diatoms use up 1·83 mgrm. P_2O_5, namely, 1×10^9 require 1·12 mgrm. From the final count 925×10^6 diatoms have appeared at the expense of 0·544 mgrm., which is equivalent to 0·59 mgrm. per 1×10^9 diatoms. This being considerably less, about half, the former value indicates either a reduction in size of the diatoms, which may result from their mode of division, or else a regeneration of phosphate from the protoplasm of dead diatoms ; the hæmacytometer count includes all diatoms, but the number given may not all be alive.

An attempt was made to settle this point by estimating the phosphate content of a known number of diatoms. Accordingly 105 c.c. of Nitzschia culture was filtered through close-grained paper, and evaporated to dryness with hydrochloric acid, in order to decompose organic compounds containing phosphoric acid. The residue was then taken up with water, and since the culture contained $2·9 \times 10^6$ diatoms per c.c., as read from the graph for the date of the analysis, it was ascertained that 0·307 mgrm. of P_2O_5 was yielded by 1×10^9 diatoms. Another portion of the culture was taken later on, and submitted to the more drastic treatment of evaporation to dryness with nitric acid. The residue was then evaporated to dryness after having been taken up with water, and, finally, after the addition of sulphuric acid. The culture at this stage contained 3.06×10^6 diatoms per c.c. and 0.303 mgrm. P_2O_5 per 1×10^9 diatoms was obtained, which agrees closely with the first analysis. Since the amount is, however, only about one-fourth of that taken up by the production of this number of diatoms it appears that the treatment

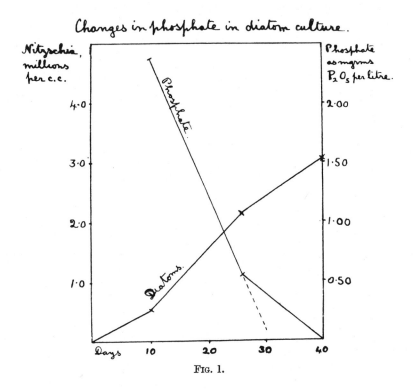

FIG. 1.

has not sufficed to convert all the organically combined phosphorus into the acid, but has split off a more easily hydrolysed fraction of it. The estimation of the total phosphorus has been deferred till a later date.

As 1×10^9 of the diatom require 1.12 mgrm. P_2O_5, one gram of this should suffice for 9×10^{11} Nitzschias. It now becomes of interest to study the seasonal change in phosphate which occurs in sea water, and to estimate the Nitzschia crop that could be produced were the whole amount available for this organism, neglecting any processes that may enrich the sea with phosphate during its period of diminution.

The decrease in Phosphate occurring in stored Sea Water when Insolated.

Open sea water stored in the dark in bottles used for chloride samples, or in Winchester quart bottles, appears to undergo but little change for a couple of weeks in spring. There is, however, always the possibility that owing to the growth of moulds water kept for considerable periods may give low results, or even possibly high results, if bacterial decomposition has been active, though on the latter point there is as yet no direct evidence.

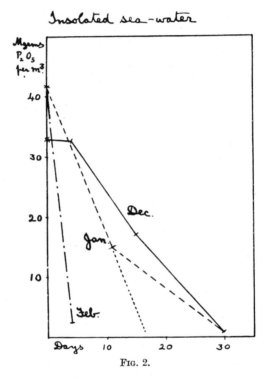

FIG. 2.

In order to test the hypothesis that the vernal decrease observed to occur in the sea was due to the uptake of phosphate by algæ, five Winchester quart bottles of water were exposed in a south window. Of these A and B were taken from Station E1 on December 18th and January 16th respectively, C and D at Station L5, the Eddystone, in quite similar sea water on February 12th and March 8th, whilst E was from L1 in Plymouth Sound. On insolation A and B decreased slowly at first, but only 0·001 mgrm. P₂O₅ per litre was left after thirty days. The others, however, contained only 0·003 mgrm. per litre after four days. The results are shown in Table III and Fig. 2.

TABLE III.

Changes in phosphate content of stored sea water when insolated. Phosphate in milligrams of P_2O_5 per cubic metre. Insolated from 24/3/23.

Sample Source of water: Days.	A E1, 18/12/22.	B E1, 16/1/23.	C L5, 12/2/23.	D L5, 8/3/23.	E L1, 7/3/23.
0	33	42	42	37	43
4	32·5	—	2·5	3	3·5
11	—	15	—	—	—
15	17	—	—	5	—
30	1	1	—	—	—

On account of their similarity to the line given by C, those for D and E have been omitted from the figure. It is evident that the diminution in phosphate becomes increasingly rapid as the spring samples are approached; the explanation appears to be that these contain a larger number of plankton algæ per unit volume, and their multiplication under the favourable light conditions speedily results in the consumption of the small amount of phosphate occurring in sea water. As many as 464 plankton organisms per cubic centimetre have been found by Allen (1919) in sea water in summer. In winter, however, the much smaller numbers present can apparently be doubled or quadrupled with but little effect upon the amount of phosphate as ascertained by analysis.

The figures obtained make it clear that, just as in the Nitzschia culture, which was artificially enriched with nitrates, in sea water also algal growth results in the uptake of phosphate till none remains, for a quantity such as 0·001 mgrm. per litre (viz. 1 in 10^9) is about the limit which can be detected by the extremely delicate method used. Recent work by Pentanelli (1923), of which an abstract only has been seen, claims to show that the development of marine algæ in unchanged sea water is stopped by deficiency in carbon dioxide, nitrogen, and phosphorus, and by an alteration of the water which is independent of the consumption of food.

In this connection it may be remarked that Allen and Nelson (1910) found that the tank water was more favourable, when sterilized, for the cultivation of diatoms than was open sea water. This is, no doubt, due in part at least to its higher phosphate content. It may also be added that the Laboratory supply of open sea water filtered through a Berkefeld candle, as explained by Allen and Nelson, was found, after standing in a covered beaker for a fortnight, to contain less than 0·01 mgrm. P_2O_5 per litre, whereas water freshly drawn contained 0·12 mgrm. Sea water at the time had about 0·049 mgrm. When filtered through a Doulton filter candle, which had been well washed with tap water containing

39

under 0·02 mgrm. P_2O_5 per litre, sea water was deprived of phosphate. After rejecting the first portion likely to be diluted by fresh water, the next 80 c.c. was found to have 0·020 mgrm. per litre. A further 300 c.c. gave 0·026 mgrm. None of the sea water analyses recorded by the writer were made upon filtered water unless expressly stated to the contrary.

As mentioned in the analytical section of this paper, Matthews used ferric chloride solution to precipitate the phosphate of sea water for estimation. It was found by the writer that on adding a few drops of Laboratory reagent ferric chloride all phosphate was removed with the ferric hydroxide precipitate and the filtered solution contained not more than 0·001 mgrm. P_2O_5 per litre.

With water from the Aquarium tanks containing 4·75 mgrm. P_2O_5 per litre the addition of ten drops of ferric chloride to a beaker containing about a litre reduced the phosphate to 0·62 mgrm. and the pH value from 7·6 to 6·7. A further ten drops brought the reaction to pH6·6 and the phosphate down to 0·01. On bringing the total number of drops up to thirty, a great increase in acidity, pH3·4, was found, together with an increase in the phosphate in solution. One drop of 0·880 ammonia, however, made the solution alkaline, about pH10, and reduced the phosphate to 0·005 mgrm. per litre. This action of iron in precipitating phosphate is of much biological importance, and should be considered when culture media are being prepared.

SEASONAL CHANGES IN PHOSPHATE IN SEA WATER, 1922 RESULTS.

Table IV shows seasonal variations of phosphate, expressed in milligrams of P_2O_5 per litre ; the analyses were carried out on surface samples stored for some weeks at the Government Chemists' Laboratory, London, by Pouget and Chouchak's colorimetric method, as used by Matthews.

TABLE IV.

Date.	L2 and L3.	E1.	E2.	E3.	N2.
12/2/22	0·051	—	0·070	—	0·016
15/3	—	0·046	—	—	—
30/3	0·034	0·039	—	—	0·041
25/5	—	—	—	0·022	0·031
6/6	0·012	0·015	—	—	—
12/7	—	0·019	—	0·020	0·019

Aquarium of the Marine Biological Association, east reservoir, about 5·0 mgrm. per litre.

As already mentioned, these results agree well with those obtained by Matthews in 1916, his site, the Knap Buoy, being in between stations L2 and L3. They further show that these changes occur simultaneously in the sea water over a wide area. It should be explained that the L series of stations extend from below the Laboratory, in Plymouth Sound to L6, which is half-way between the Eddystone (L5) and E1. The remainder are the International Hydrographic Stations, E1, E2, and E3, lying on the course from the Eddystone to Ushant, N1, N2, N4, and N5, on the course from Ushant to Cork Harbour. N3 is between the Scilly Islands and Cornwall, E6 being 20 miles to the north in the Bristol Channel. Their positions are shown in the map given by the writer (1922).

The relatively high value 5·0 mgrm. per litre given by the water of the Aquarium is noteworthy, as it indicates the mode, or one mode, whereby the phosphate taken up by the algal plankton is returned again to the sea—namely, through the excretion of phosphate by fish and marine invertebrates. The tanks are well stocked with both, but there is little algal life, so the normal balance of the sea is disturbed.

It may be added that similar values for the tank water have been obtained by the writer, viz. 4·75 mgrm. per litre for both east and west reservoirs on April 10th, 1923, and 4·81 mgrm. on June 29th. The reservoirs had been drained and refilled between these analyses and that of the Government chemist.

SEASONAL CHANGES IN PHOSPHATE AT L STATIONS, 1923.

The work was continued in 1923, all the determinations being made by the writer according to the method of Denigès, upon samples taken the same or the preceding day. The samples were kept in the dark during the interval.

Table V gives the results for the L series from March to August. Certain values for water taken at the east slip, directly below the Laboratory, are also included. Owing to sewage contamination these do not exhibit regular seasonal changes. The effect of sewage upon the L1 values is surprisingly small, judging by the uniformity of the figures with those of other stations. Low values were obtained from the end of April onwards, and Fig. 3 represents the seasonal change at L4, half-way between Rame Head and the Eddystone, about five miles outside the Breakwater.

Within the limits of experimental error the surface values are equal to or less than the bottom, due to the fact that photosynthesis and consequently algal growth and reproduction is more active near the surface. Occasionally, however, one meets with an abnormal surface value, such as that for L4 on May 31st and L6 on August 15th. One can only attribute these results to a local contamination of the water from a ship, as the bucket had been rinsed repeatedly, as were also the bottles.

TABLE V.

Seasonal variations of phosphate, expressed in milligrams of P_2O_5 per litre, surface samples mainly.

Date.	East slip.	L1.	L2.	L3.	L4.	L5.	L6.
7/3/23	0·0485	0·0485	0·049	0·049	—	—	—
12/3/23	—	0·049	0·045	—	0·041	0·033	—
21/3/23	—	0·042	0·040	0·041	—	0·038	—
22/3/23	0·0395	—	—	—	—	—	—
27/3/23	—	—	—	0·020*	—	—	—
28/3/23	—	0·032	0·033	0·039	0·037	—	—
9/4/23	—	0·033	0·033	—	—	—	—
9/4/23	—	0·036B	0·033B	—	—	—	—
11/4/23	—	—	—	—	—	0·031	—
11/4/23	—	—	—	—	—	0·041B	—
16/4/23	—	0·021	0·020	0·013	0·016	0·023	—
16/4/23	—	0·024B	0·024B	0·018B	0·028B	0·021B	—
18/4/23	—	—	—	0·024	—	—	—
18/4/23	—	—	—	0·024B	—	—	—
20/4/23	—	0·024	—	—	0·014	0·024	—
20/4/23	—	0·028B	—	—	0·027B	0·023B	—
24/4/23	—	0·016	0·010	0·015	0·021	0·023	—
3/5/23	0·042	—	—	—	—	—	—
7/5/23	—	0·027	0·025	—	0·023	—	—
7/5/23	—	—	—	—	0·023B	—	—
22/5/23	—	0·0235	0·0155	0·023	0·0105	0·012	0·004
31/5/23	—	—	—	0·0065	0·050†	—	—
31/5/23	—	—	—	0·0065B	0·046†	—	—
31/5/23	—	—	—	—	0·008B	—	—
19/6/23	—	0·0045	—	0·0055	—	0·004	—
19/6/23	—	0·009B	—	0·0115B	—	0·012B	—
23/6/23	0·008	—	—	—	0·009	—	—
2/7/23	—	—	—	—	0·013	—	—
2/7/23	—	—	—	—	0·013B	—	—
10 & 12/7/23	—	—	—	0·013	0·014	0·0135	0·007
10 & 12/7/23	—	0·017B	0·019B	0·017B	0·016B	0·014B	—
15/8/23	—	0·017	0·019	0·017	0·010	0·020	0·032
15/8/23	—	—	—	0·017B	0·021B	0·019B	0·020B

The general trend of the seasonal changes in the L series is illustrated in Fig. 3, in which are plotted the results for L4. The abnormal result for May 30th has been omitted, and the bottom value taken for surface also since L3 had identical values for both on that date; these differed

* Mean of two samples. B indicates bottom sample.
† Abnormal result verified by analysis on two bottles.

only by 0·0015 from the L4 bottom value. The curve is similar to that obtained by Matthews, save that the seasonal changes are about a month later all through. Comparison with the bottom values shows how a low surface value in April may so quickly be followed by one over twice as great; clearly the deeper water acts as a reservoir of phosphate, as is more fully shown in subsequent figures. The higher bottom value found in August indicates that the regeneration of phosphate takes place in the deeper water, or rather that its effect is more evident there since it is being rapidly removed at the surface in summer.

It seemed possible that these changes could be detected in rock pools, exposed for several hours each tide.

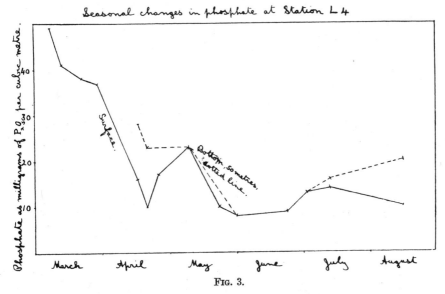

Fig. 3.

It was at first thought permissible to compare the pool water with that taken about a hundred yards to the east at the slip, but results given in Table VI (page 133) show that sewage contamination renders this unreliable. As far as the analyses go they indicate an increase in phosphate in the pools during their separation from the sea on two days, but an appreciable decrease one very sunny day. The pools have an abundance of animal life as well as algæ, so excretion may account for the small increases noted.

SEASONAL CHANGES IN PHOSPHATE AT THE INTERNATIONAL HYDROGRAPHIC STATIONS E1–E3 AND N1–N3.

Table VII (page 133) contains the results of the analyses of sea water taken at E1 from March to August at various depths. From the end of

NEW SERIES.—VOL. XIII. NO. 1. DECEMBER, 1923.

I

May onwards the surface water may be seen to be almost totally devoid
of phosphates. Fig. 4 makes this clear, and an increase in the phosphate
content of the bottom water in August is also noticeable. Fig. 5 illustrates the variations in phosphate with depth; the seasonal change is
here shown by the shifting of the curve to left for diminution or to
right or increase. Bad weather precluded the taking of a February
series, but the sea water was apparently richer in phosphate then than
in March, judging from Matthews' results.

The differences which exist, in the calmer summer weather, between

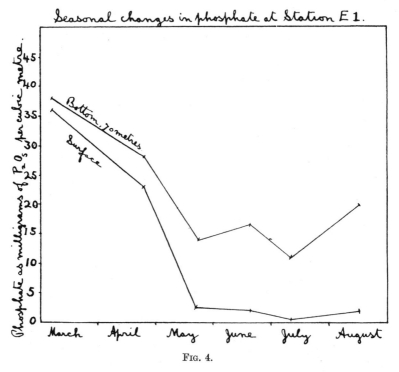

FIG. 4.

surface and deeper water samples show that mixing of the water is not
rapid at E1 at this season. On account of the diminution in the intensity
of the light the phosphate in the deeper water is not used up till it is
brought nearer the surface, or at least it is used up at a greatly reduced
rate.

In Table VIII (page 134) the corresponding data are given for Stations
E2 and E3. The depth series results are plotted in Fig. 6 (page 132).
Samples taken on the cruises to Ushant, etc., have perforce to await
analysis for two to three days, but no appreciable error appears to be
introduced by this as the samples are stored in the dark.

The almost total depletion of the phosphate down to 10 metres is noticeable at E2, and here, as at E1, the minimum value is found in July. At E3, however, the May value is the lowest for the bottom, and the mixing of the water diminishes the surface to bottom gradient.

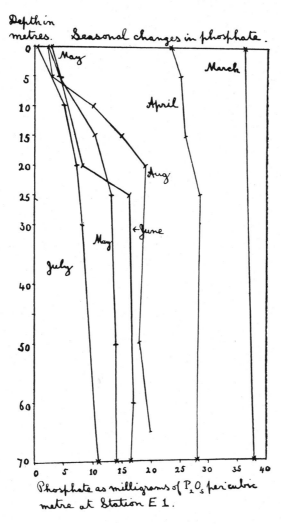

FIG. 5.

This has also been observed as regards temperature and pH gradients at this station, as pointed out by the writer in an accompanying paper in this Journal.

For Stations N1, N2, and N3 no April records are available owing to the renewal of stormy weather during the cruise, and a thick fog

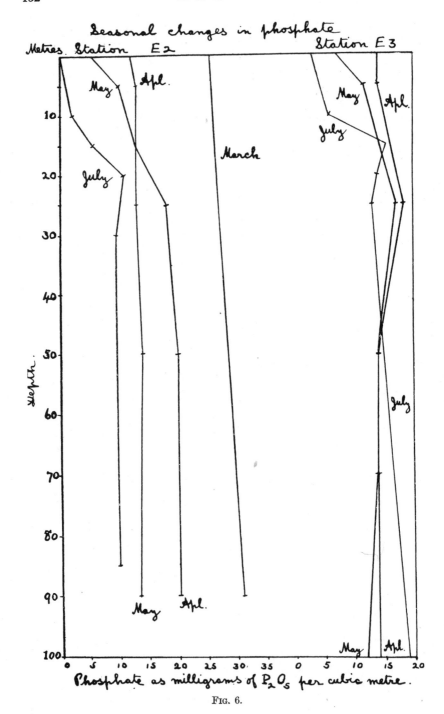

FIG. 6.

prevented the obtaining of samples at N3 in July. The analyses for May and July are given in Table IX (page 134) and plotted in Fig. 7 (page 135). At both N1 and N2 the July values are greater than those for May in the samples from the deeper water. The gradient is also very remarkable, especially the difference between the 15- and 20-metre samples at N1. The settled calm weather appears to account for this. There is a suggestion of regeneration of phosphate at both N1 and N2, or the higher values may be due to a transgression of deeper water moving eastwards over the edge of the west European submarine shelf.

TABLE VI.

Source of water.	Date.	Hour.	Phosphate as mgrms. P_2O_5 per litre.	Notes
East slip	3.4.23	11.40 a.m.	0·031	
Rock pool west of slip	,,	11.50 a.m.	0·037	
Do.	,,	2.50 p.m.	0·040	
East slip	4.4	11 a.m.	0·033	
Do.	,,	5 p.m.	0·0325	
Rock pool	,,	12.50 p.m.	0·036	
Sound, by pool	,,	,,	0·036	
Another pool, close to first one	,,	4 p.m.	0·040	First pool submerged.
East slip	5.4	10.15 a.m.	0·039	
Rock pool	,,	12.30 p.m.	0·039	Pool covered at
Sea by pool	,,	,,	0·039	10.30 a.m.
Pool	,,	4.15 p.m.	0·030	Very sunny day.
East slip	,,	4.15 p.m.	0·055	Sewage effect.

TABLE VII.

Seasonal variations of phosphate, expressed in milligrams of P_2O_5 per litre, Station E1.

Depth in metres.	March 7th	April 24th.	May 22nd.	June 19th.	July 10th.	August 15th.
0	0·036	0·023	0·0025	0·002	0·0005	0·002
5	—	0·025	0·004	0·004	—	0·003
10	—	—	—	—	0·005	0·010
15	—	0·026	0·010	—	—	0·015
20	—	—	—	0·008	0·007	0·019
25	—	0·0285	0·013	0·016	—	—
30	—	—	—	—	0·008	—
50	—	—	0·014	—	—	0·018
60	—	—	—	0·017	—	—
70	0·038	0·028	0·014	0·0165	0·011	0·020

TABLE VIII.

Seasonal variations of phosphate, expressed in milligrams of P_2O_5 per litre, Stations E2 and E3.

Depth in metres.	March 14th.	April 24th.	May 22nd.	July 10th.	April 25th.	May 22nd.	July 10th.
0	—	0·012	0·0055	0·000	0·014	0·007	0·003
5	0·0255	0·013	0·010	—	0·014	0·0115	—
10	—	—	—	0·002	—	—	0·006
15	—	0·013	0·013	0·0055	—	—	0·0155
20	—	—	—	0·011	—	—	0·014
25	—	0·018	0·013	—	0·0185	0·017	0·013
30	—	—	—	0·0095	—	—	—
40	—	—	—	—	—	—	—
50	—	0·020	0·014	—	0·014	0·014	—
60	—	—	—	—	—	—	—
70	—	—	—	—	—	0·014	—
80	—	—	—	0·010	—	—	—
90	0·031	0·020	0·0135	—	—	—	—
100	—	—	—	—	0·014	0·012	0·019

TABLE IX.

Seasonal variations of phosphate, expressed in milligrams of P_2O_5 per litre, Stations

Depth in metres.	N1		N2.		N3.
	May 22nd.	July 11th.	May 22nd.	July 11th.	May 22nd.
0	0·015	0·0045	0·017	0·014	0·016
5	0·016	—	0·015	—	—
10	—	—	—	—	—
15	0·013	0·005	—	0·014	—
20	—	0·022	—	0·015	—
25	0·016	0·021	0·015	0·0235	0·016
30	—	—	—	—	—
40	—	—	—	—	—
50	0·017	—	0·016	—	—
60	—	—	—	0·0235	0·0205
70	0·017	—	—	—	—
80	—	—	—	—	—
90	—	0·023	0·022	—	—
105	0·019	—	—	—	—

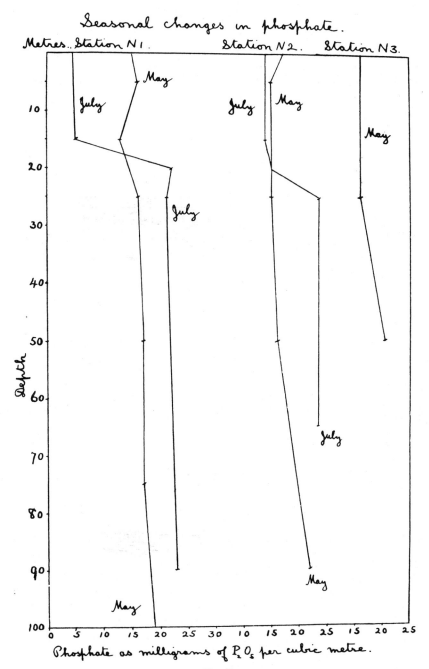

Seasonal changes in phosphate.

FIG 7.

PHOSPHATE CONTENT OF THE NORTH SEA.

Since many of the determinations made by Raben were carried out upon water from the North Sea, it seemed to be of interest to examine samples from that region also in order to see whether values in better agreement with his results would be obtained. This was rendered possible through the courtesy of Dr. E. S. Russell, Director of the Fisheries Laboratory, Lowestoft, and of Mr. J. R. Lumby, who kindly collected the water samples. The analytical results are shown in Tables X and XI.

TABLE X.

Phosphate in North Sea, surface samples, April to July.

Date.	Position.	Lat. N.	Long.	Phosphate as mgrms. P_2O_5 per cubic metre.
13/4	Cross Sand Lightship .	—	—	35
,,	Inner end of Stanford Channel, Lowestoft .	—	—	40
,,	10' E. × N. from Tyne .	55° 4'	1° 8' W.	36
,,	60' E. × N. from Hartlepool	55° 8'	0° 22' E.	36
3/5	Do.	55° 23'	1° 22' W.	15
11/7	Off Newbiggin Point .	55° 15'	1° 20' W.	8
25/7	18' N.E. × E. Tyne . .	—	—	11

It may be seen from Table X that the values found are quite similar to those for the English Channel for the same months.

Table XI also gives figures quite in accord with those found off Plymouth, but far lower than Raben's values for North Sea water. It should be noted that the figures in Table XI are not as uniform as might be expected, as in several instances the surface values are slightly higher than the bottom. This may be connected with the circumstance that there was a delay of one month between the collection and analysis of these samples.

Of especial interest are the results for Stations 24 and 25 in the deeper water off the coast of Norway. The 280-metre sample is two and a half times as rich in phosphate as is the surface water; again, there is over twice as great a concentration of phosphate at the bottom at Station 25 as at Station 23, with a depth of 70 metres. Near a coast there is usually more vertical mixing of the water than there is at stations well out, such as E2 and N1, accordingly one may expect an abundant plankton where deep water approaches the land or a submerged bank which causes upwelling. The phosphate values found support the views put forward by Natterer in this connection.

TABLE XI.

Phosphate in North Sea, England to Norway, May 3rd to 6th, 1923.

Station.	Lat. N.	Long.	Depth in metres.	Phosphate as mgrms. P_2O_5 per cubic metre.	Notes.
1	54° 32′	0° 2′ W.	0	21	Near Tyne.
1	,,	,,	60	19	
2	54° 39′	0° 11′ E.	0	19	
2	,,	,,	65	15	
8	54° 54′	0° 34′	0	17	
8	,,	,,	70	23	
10	55° 23′	1° 22′	0	15	
10	,,	,,	55	17	
13	56° 8′	2° 35′	0	14	
13	,,	,,	70	16	
14	56° 26′	3° 0′	0	17	
14	,,	,,	65	17	
15	56° 38′	3° 24′	0	25	
15	,,	,,	60	19	
16′	56° 31′	3° 37′	0	11	South of usual course.
16	,,	,,	60	16	
18	56° 45′	3° 36′	0	20	
18	,,	,,	50	17	
22	57° 0′	4° 5′	0	18	Course more northerly, heading to Udsire.
22	,,	,,	60	18	
23	57° 30′	4° 15′	0	14	
23	,,	,,	70	17	
24	57° 59′	4° 25′	0	14	
24	,,	,,	100	24	
25	58° 28′	4° 34′	0	14	
25	,,	,,	280	36	

SEASONAL CHANGES IN THE PHOSPHATE CONTENT OF FRESH WATER.

The changes occurring in the sea are naturally not without a parallel in fresh water, the study of which shows how minute is the amount of phosphate left unabsorbed by the plankton during the summer. The fresh waters available for study were as follows :—

Staddon reservoir.—This is a cement-walled tank 22×8 metres and about 2 metres in depth. It receives surface drainage water in wet weather, and at all times it receives through an inlet pipe the overflow from a small spring, which may very well be contaminated as it issues out. This is situated at about 200 feet elevation on the east of Plymouth Sound, upon the Staddon Grits, a formation of the Lower Devonian. There are no trees surrounding it.

Maryfield quarry pond.—This has precipitous slaty sides and seems to depend upon rainfall for its water, though it may be replenished by a small spring below water level, and in very wet weather some surface water may find its way in. It is situated upon Middle Devonian Slates in the Antony district of Cornwall, about five miles east of Staddon and at an elevation of about 150 feet. It is surrounded by trees, which shade it to some extent. The dimensions are roughly 80×80 metres, with a depth of 2 to 3 metres in the middle.

Plymouth tap is supplied from Burrator Reservoir on Dartmoor.

TABLE XII.

Seasonal changes in phosphate, expressed in milligrams of P_2O_5 per cubic metre.

Date.	Staddon Reservoir.	Inlet.	Date.	Maryfield, quarry pond.	Date.	Plymouth, town tap.
19/8/22	32*	—	21/10/22	59*	—	—
2/10	91*	—	21/1/23	57*	27/1/23	19*
4/11	75*	—	18/2	42*	—	—
23/2/23	128*	119	30/3	14	31/3	3
3/4	3·5	—	15.4	5	—	—
16/4	2	82	24/4	0	—	—
1/5	6	112	6/5	0·5	1/5	3
30/5	0	81	2/6	3	5/6	0·5
15/6	11	78	24/6	3	—	—
23/6	19	60	—	—	—	—
29/6	15	86	30/6	3	5/7	5·5
26/7	9	116	26/7	3	30/7	1·5
24/8†	0	100	24/8	0	24/8	0

* Stored till analysed early in April.

† A slight turbidity in all three samples rendered the tint impossible to match with exactness. They were taken after rain.

The phosphate analyses are shown in Table XII, and are plotted in Fig. 8. It might be thought that the phosphate values were largely influenced by dilution with rain water, but electrical conductivity measurements show that this is not the case. That for the Staddon inlet is quite usual for a calcareous water, the reservoir values are some-

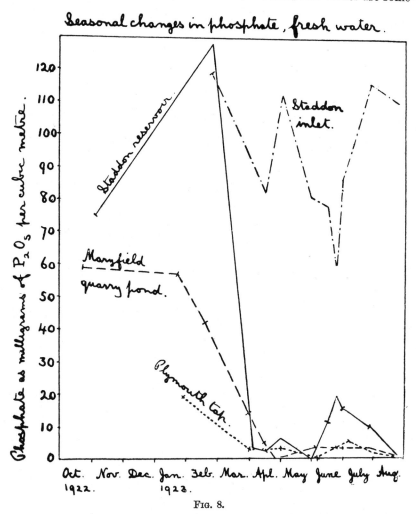

FIG. 8.

what lower. Maryfield quarry pond has, as a rule, a lower conductivity than has Staddon reservoir, and both are alkaline, around pH8 according to the season. Burrator water is at pH6·4–6·8, and its conductivity is only about one-tenth that of Staddon inlet. Though these waters are so different, two being quite " hard " and one very " soft," yet their changes in phosphate are closely similar, as may be seen in Fig. 8. It must be pointed out, however, that the winter values given may be somewhat too high, since it is possible that phosphate was regenerated from plankton —more plentiful near the surface—during the period of storage. In general the Staddon reservoir has a much more abundant plankton during summer than has the Maryfield pond. In it the vernal outburst is followed by a period when but little algal life is found. The former becomes rich owing to its minute plankton and floating masses of Spirogyra, etc., absorbing the phosphate supplied by the inlet pipe.

APPROXIMATE ESTIMATION OF THE ALGAL PLANKTON CROP AND OF FISH PRODUCTION BASED ON PHOSPHATE CONTENT.

In the section dealing with the Nitzschia culture it was shown that the production of 1×10^9 diatoms of the species grown consumed 1·12 mgrm. of phosphate reckoned as the pentoxide.

By measuring the areas between the ordinate and the March and July curves for phosphate at Station E1, as given in Fig. 5, it was ascertained that 2070 mgrm. of P_2O_5 was consumed in the water column from 70 metres to the surface, having one square metre as its base. In other words, an average content of 37 mgrm. per cubic metre on March 7th had by July 10th fallen to 7·4 mgrm. In round numbers there was a consumption of 30 mgrm. per cubic metre or 2·1 grm. in the whole column. If the winter value be taken at 49 mgrm., the value found at L3, the consumption may be taken as 40 mgrm. in the same period.

Taking, however, the lower figure which was directly determined and the phosphate factor for diatom production, namely, 1·12 mgrm. per 1×10^9 diatoms, it may be seen that each litre of water could produce $26,800 \times 10^3$ diatoms of this species, provided nothing else grew in the water. Up to 30,000 diatoms per c.c. were found by the writer in a fresh water pond. These may be compared with the figures 462×10^3 and 464×10^3 given by Allen (1919) as the minimum values for plankton organisms per litre found early in August and September in sea water between Stations L2 and L3. The value found in the sea is only 1·7 per cent of that calculated from the phosphate consumption, because the algal plankton is eaten up by the smaller animal organisms, and serves indirectly as the food of all animal life in the sea.

Converting the above estimate per litre into per cubic metre it is seen that $26·8 \times 10^9$ diatoms could be produced, or in the 70-metre column the enormous number 188×10^9. To be able to convert the numbers into weight it is necessary to either weigh diatoms directly or to know their phosphate content. An approximation to this may be obtained as follows : According to figures quoted by Czapek (1921) and Strasburger (1921) leaves may be taken as containing phosphate as pentoxide equivalent to 15 per cent of their ash. Leaves were chosen as being assimilating organs, and so nearer to algæ than other parts such as wood or roots. Taking the ash as 10 per cent of the dry weight and the latter as 10 per cent of the moist weight, the phosphate content of the fresh plant is 0·15 per cent. Making the approximation 0·2 per cent P_2O_5 as the phosphtae content of diatoms, and using the value 2·1 grm. P_2O_5 as the amount consumed in the whole column, it follows that 1·05 kilograms of diatoms could be produced; as a matter of fact if diatoms are not produced other unicellular algæ are, and their phosphate content must be very similar.

After this estimate was made, data relating to the phosphate content of algæ were found in the Fertilizer Resources of the U.S.A., pp. 225–9. Analyses by Barlow for three species of Fucus give 0·43 per cent phosphate on the dry weight. Determinations quoted from Tom show that Fucus has 24·2 per cent dry weight, which leads to the value 0·11 per cent phosphate on the wet weight. Tom's figure, 17·7 per cent for the dry weight of Laminaria, may be rounded off as 20 per cent, since there is a considerable variation; combining this with the value given by Russell, 0·66 per cent as a maximum for phosphate calculated on the dry weight the value for the wet weight works out at 0·13 per cent. Analyses made at the Connecticut State Experimental Station give as a mean for five algæ 0·14 per cent of phosphate as pentoxide, calculated on the wet weight. These figures, 0·11, 0·13, 0·14, show that the original estimate of 0·15 per cent P_2O_5 as the phosphate content of unicellular algæ was probably fairly correct. Using it, instead of 0·2 per cent, the calculation of the algal plankton in the 70-metre column gives the result 1·4 kilogram, or $1·4 \times 10^6$ per square kilometre.

When this result is compared with the value given as a minimum by a less exact method, the change in alkalinity of the water (Atkins, 1922), the agreement is extraordinarily close when a certain assumption is made, namely, that the carbohydrates of the algal cell, including protein carbon, calculated as a hexose sugar, amount to 15 per cent of the wet weight. This assumption was made as a consequence of Tom's figures for the total dry weight, and before the agreement was found by calculation. The alkalinity results gave an estimate of 1 kilogram per 4 square metres down to a depth of 83·3 metres. Converting this into the wet weight of algæ in a 70-metre column the value reached is 1·4 kilogram. The exact agreement is, of course, fortuitous in view of the assumptions; but it shows that the methods must have a certain degree of reliability, or rather it confirms the alkalinity result, for the phosphate method involves only one assumption, that the percentage of phosphate in the algal plankton is close to that of the larger brown algæ.

Turning now to the question of the phosphate content of marine invertebrates, according to Clarke and Wheeler (1922), only trifling quantities are, as a rule, found ; certain analyses for calcareous algæ quoted by these authors are also very low, usually a trace to 0·00 per cent in the calcareous portions. The highest record is 0·18 per cent. The shells of crustacea are, however, fairly rich in phosphate, 4·07 to 6·70 per cent being the value for a medium-sized lobster, expressed as P_2O_5. Tricalcic phosphate is, indeed, the main constituent of the ash of the minute crustacea, as shown by Clarke and Salkover (1918), from which doubtless young fish obtain much of their phosphate. The shells, too, of certain brachiopods contain 75–90 per cent of tricalcic phosphate, and some worm tubes are notably phosphatic. The various amounts in the hard portions as well as in the softer tissues make it impossible to give even an approximation to the weight of invertebrates that could be produced each year.

With respect to fishes a greater uniformity is found. Atwater (1888) gives 0·514 per cent as the average value for the pentoxide of the flesh of fifty-five species. He quotes Sempolowski as giving the following figures for the phosphoric acid in the whole fish, wet weight :—

Pleuronectes limanda	1·25
Gadus aeglifinus	1·22
Trigla gurnardus	1·78
Raia radiata	0·91
Acanthias vulgaris	0·98
Mean	1·27 per cent.
As P_2O_5	0·95 per cent.

It may be seen that the bony fishes are considerably richer in phosphate than are the cartilaginous. Seeing that they constitute by far the larger amount of fish in the sea one may take as an approximation 1 per cent of P_2O_5 for fishes in general. Now if all the phosphate used up in the 70-metre column were converted into fish it could yield each year 210 grams of fish, or roughly 1 kilogram per 5 square metres. Since there are also vast numbers of plankton and bottom-dwelling animals this is, of course, a very large overestimate ; the figure yields the value 2×10^5 kilograms per square kilometre. In the absence of precise data one may perhaps assume that the fish represent between 1 per cent and 1 per thousand of this possible total quantity, which gives an estimate of between 200 and 2000 kilograms per square kilometre in water 70 metres in depth.

METHODS FOR THE ANALYSIS OF PHOSPHATES.

Details of the usual methods where moderate quantities are involved may be found in the text-books ; but their use for quantities of the order of one milligram or less per litre, reckoned as P_2O_5, involves the use of inconveniently large volumes of liquid. In precipitating with magnesia mixture the resulting ammonium magnesium phosphate is usually weighed after converting into pyrophosphate. Recently, however, Jones and Perkins (1923) have given details of a method in which the double salt may be weighed directly. Using the ammonium molybdate method of precipitation, Kleinmann (1919) has found that it is permissible to weigh as ammonium phosphomolybdate. The work of Posternak (1920) on the variability of this precipitate should, however, be remembered.

A very delicate reaction was developed by Pouget and Chouchak (1909, 1911) into a colorimetric or nephelometric means of estimating phosphates, using strychnine sulphate and sodium molybdate. The reagents produce a yellow opalescence. This has since been used by several workers, notably Kleinmann (1919), Embden (1921), who converted it into a gravimetric method, and by Matthews (1916–18). Embden found it convenient to use the resulting strychnine phosphomolybdate precipitate for work with solutions containing 1·0–4·0 mgrm. P_2O_5, since the precipitate is about thirty-nine times as heavy as the corresponding amount of pentoxide. The precipitation being performed in the cold renders this method specially suitable for the estimation of phosphate in the presence of organic phosphates, which are easily hydrolysed.

Matthews (1916–18) used the Pouget and Chouchak colorimetric method for estimating the phosphate in 500 c.c. of sea water after precipitation as ferric phosphate. The method was adopted after a very careful comparison with others available.

Raben (1916–20), working with Brandt (1916–20) at Kiel, precipitated the phosphate in 10 litres of filtered sea water by means of ferric chloride. After an elaborate purification the phosphate was determined gravimetrically as phosphomolybdate.

The results for sea water from various sources are from 51 mgrm. P_2O_5 per cubic metre in May to 221 in November. It may be said that these values are greater than those obtained by Matthews, 0·06–0·01 or less, expressed in milligrams of P_2O_5 per litre. Matthews also obtained evidence for the existence of a soluble compound of phosphorus, which can be converted into phosphoric acid by oxidising agents. The results obtained by the writer for phosphate in sea water are in complete agreement with those of Matthews, though obtained by an entirely different method. No explanation can as yet be offered as to why these differ so much from the very careful determinations of Raben and his co-workers.

A new method of great delicacy was developed by Denigès (1920, 1921), and was found by him to agree with the gravimetric method of Posternak (1920). The latter showed that the composition of the ammonium phosphomolybdate precipitate varies largely according to the proportions of the various salts present and to the temperature of precipitation ; he accordingly worked out a process in which a barium phosphomolybdate of constant composition may be obtained.

METHOD OF DENIGÈS FOR PHOSPHATES.

Two reagents are required for the "cœruleomolybdic" method of Denigès : (a) 10 per cent ammonium molybdate and pure sulphuric acid in equal parts by volume, and (b) stannous chloride, freshly prepared from 0·1 grm. of tin dissolved in 2 c.c. of hydrochloric acid with one drop of 3–4 per cent copper sulphate and made up to 10 c.c. On mixing a few drops of (a) with 10 c.c. of the liquid to be tested and adding one or two drops of (b), an intense blue appears in the presence of phosphate. Denigès employed this reaction for the analysis of biological products, but it was used in a slightly different form by Florentin (1921) for the determination of the phosphate content of fresh waters. Denigès considers that the maximum delicacy of the method is for solutions containing 0·5–10 mgrm. of phosphorus as phosphoric acid.

Florentin has employed it for the estimation of phosphate equivalent to 0·01–5·0 mgrm. of P_2O_5. He makes up solution (a) with 100 c.c. of 10 per cent ammonium molybdate plus 300 c.c. of 50 per cent (by volume) sulphuric acid. For analysis 10 c.c. of water is taken, to which are added three or four drops of (a) and one drop of (b), or three drops of (b) if more than 2 mgrm. of P_2O_5 is present. The blue colour developed reaches its maximum in less than ten minutes. Comparison is then made with standards containing known amounts of phosphate, or indigo carmine for greater permanency. The acidity prevents the production of blue with molybdate alone. According to Florentin more than 0·1 grm. per litre of Na_2SiO_3 gives a colour. As shown in an accompanying paper by the author no such amount of silicate has been found in any of the natural waters examined, for which 0·006 grm. per litre SiO_2 (or 0·012 grm. approximately of silicate) is a high value. H_3AsO_4 gives a blue colour similar to that given by phosphate, so any traces present are included in the phosphate estimation.

The writer has made use of the reagents according to Florentin's formula for (a), and has found it advisable to use 100 c.c. of the water to be tested owing to the minute traces of phosphate present. To this quantity of fresh or sea water 2 c.c. of (a) are added and five drops of (b), and the blue tint is examined in a graduated 100 c.c. cylinder with a tap near the base. The tint is compared with that given by a convenient strength of phosphate solution, usually one containing the equivalent of 0·05 mgrm. of P_2O_5 per litre. The standard solution falls off somewhat on keeping; a 6 per cent decrease was observed in $2\frac{1}{2}$ hours, by comparison with a fresh solution. This amounts to 1 per cent per half-hour approximately, so when examining a series a fresh standard is mixed after about half an hour. Taken over a twenty-five hour period, however, the decrease was only 1 per cent per hour. Sometimes the solutions quickly develop a turbidity. This trouble has been traced to the stannous solution, which is apt to give the precipitate if added to the sample before the acid molybdate, or if added in too great amount, or if heated for an undue length of time when being prepared. It was, moreover, noticed that the precipitate came more readily in distilled or naturally occurring fresh water than in salt water, in which the sodium chloride apparently lessens hydrolysis by diminishing the percentage ionised.

When adjusting the height of the stronger solution to match that of the lighter at the 100 c.c. level the columns are viewed standing on a thin glass shelf below which is opal glass. The sides and back of the stand are black. Accuracy is assisted by having on the opal glass a white card on which are ruled black lines. This is adjusted so that half of the field of each column is occupied by the card, and half by the opal glass. The tubes are screened in front by cardboard.

Before trying the cylinders, which are now used invariably, Nessler tubes containing 50 c.c. were used; a series from 0·05–0·01 mgrm. P_2O_5 was made up, and it was found that the members could readily be arranged in the correct order. The use of the cylinders increases the accuracy, as it is usually possible to get duplicate readings to within 2 c.c. on the column. Good agreement may also be obtained against a standard of a different strength. Thus sea water tested against a 0·05 standard gave :—

1st reading 66, viz. $\dfrac{0·05 \times 66}{100} = 0·0330$·mgrm. P_2O_5 per litre.

2nd reading 67·5 $= 0·0337$,,   ,,   ,,

Against a 0·04 standard the reading was 82, corresponding to 0·0328 mgrm. P_2O_5 per litre. The colour is not sufficiently intense with such dilutions to permit of the use of the Duboscq colorimeter, on account of the shorter length of liquid column available.

There is, however, one source of error which remains as yet quite unexplained. On standing with the reagents sea water and certain fresh water samples from ponds develop a slight yellowish tint. This is not noticeable as a rule till after five minutes, so the comparison should be made before it has time to develop, and as soon as the blue has reached its maximum intensity. The colour is not given by the acid molybdate alone. An exact match may nevertheless be obtained even in the presence of the yellow tint by adding drops of very dilute Bismarck brown to the standard. The result got by trying to match the tints without the addition of the brown is usually about 0·004 mgrm. per litre too low.

It must be added that blank estimations are made from time to time by adding the reagents to distilled water. With freshly made up molybdate mixture no more than 0·0005 mgrm. per litre need be subtracted for the tint given by the reagents, 0·002 mgrm. is a very usual value for molybdate mixture stored in the dark, and after some time in the light as much as 0·004 mgrm. may have to be deducted.

It should be stated that the standard phosphate solutions were made up by diluting a solution of sodium ammonium hydrogen phosphate equivalent to 5 mgrm. P_2O_5 per c.c. The stock solution was diluted to give 50 mgrm. per litre, and for general use this was further diluted to 0·5 mgrm. per litre. By taking 10 c.c. of this and making up to 100 c.c. the usual standard 0·05 mgrm. P_2O_5 per litre was obtained. Solutions not conveniently matched against this strength were either diluted suitably or else a more concentrated standard was used. Such solutions are very liable to grow moulds or minute green algæ, which, of course, alter their phosphate content. The addition of a little toluene was, however, found to prevent this for some months at any rate.

It is also noteworthy that Florentin pointed out that the presence of the acid prevents the molybdate alone from giving a blue with stannous chloride. On one occasion through an error the acid molybdate solution was made up to contain only 25 per cent of sulphuric acid ; as usual 2 c.c. of this was added to 100 c.c. sea water, followed by five drops of stannous chloride. The intense blue which developed appeared to denote an absurdly large phosphate content, and on repeating the estimation with fresh reagents the mistake was discovered and Florentin's observation was recalled to mind.

As previously mentioned it is possible to get readings in duplicate, when comparing the blue tints in the 100 c.c. cylinders, which agree to 2 c.c. This limit, using a 0·05 mgrm. P_2O_5 per litre standard, corresponds to 0·001 mgrm. per litre. Even taking it that the reading may be 2 c.c. too high or too low, the error only becomes ±0·001 mgrm. per litre. This should not be surpassed in clear solutions in which no yellow tint develops. With slightly turbid solutions or those which are tinted the error may, of course, be greater, though use of dilute Bismarck brown materially reduces it. Matthews, using Pouget and Chouchak's method on the phosphate from 500 c.c., considers that the estimation is accurate to about 0·003 mgrm. per litre. The method of Denigès, as used by the writer, gives results which are in most cases accurate to ±0·001 mgrm.

per litre, and may certainly be considered at least to equal those obtained by the Pouget and Chouchak method in accuracy. Furthermore, since the method of Denigès requires only 100 c.c. the phosphate actually estimated is only one-tenth of the concentration in milligrams per litre.

Matthews found that, using filtered sea water, duplicate determinations required five hours. The filtration, moreover, took upwards of sixteen hours, and was necessary on account of the risk of contamination of the precipitate. Using the method of Denigès an estimation occupies ten minutes, and unless particles of phosphate are suspended in the liquid no error results from the presence of the ordinary amount of algal plankton. It must be concluded that this mode of estimation has many advantages.

It may be added that to convert the conventional P_2O_5 values into the more rational values for the PO_4 ion the factor 1·338 may be used to multiply the former. The factor is very approximately $\frac{4}{3}$. For the converse the factor 0·7474 should be used, which may be taken as $\frac{3}{4}$.

SUMMARY

1. The phosphate content of uncontaminated streams and fresh water supplies examined was under 0·05 parts per million reckoned as P_2O_5. To convert to PO_4 the factor 1·338, very approximately $\frac{4}{3}$, may be used.

2. A pure culture of *Nitzschia closterium* W. Sm., in sea water enriched with Miquel's solution, multiplied in numbers up to over three million per cubic centimetre, when the phosphate was all used up. It was ascertained that 1·12 mgrm., expressed as P_2O_5, is required for the production of 1×10^9 diatoms during the early stage of the culture. One gram of the pentoxide suffices for 9×10^{11} diatoms.

3. Sea water insolated in the Laboratory decreases rapidly in phosphate till none is left. Samples taken in winter show a less rapid decrease than those taken in spring. This is due to their smaller content of algal plankton. Ferric chloride removes phosphate from sea water or culture solutions very completely.

4. The phosphate content of sea water falls from a value of 0·036 mgrm. per litre at the surface at Station E1 in March to zero in July. The bottom value also falls to 0·011 mgrm. in July, so that there is a consumption throughout the column of water to 70 metres of 0·030 mgrm. per litre. Similar changes take place in Plymouth Sound and at the Hydrographic Stations E2, E3, and N1–N3. The surface water is almost free of phosphates from May to August.

5. A few determinations made indicate the same seasonal change in the North Sea. The deep water off the Norwegian coast acts as a reservoir of phosphate, which presumably gets depleted during summer; 0·036 mgrm. per litre was found there on May 6th at 280 metres. The North Sea values for phosphate are much lower than those found by Raben, and the phosphate analyses in general agree well with the results obtained by Matthews. As regards the seasonal change the results are in agreement with both workers.

6. The phosphate of fresh water ponds was found to fall almost to zero early in April, and to continue low throughout summer.

7. An estimate may be made of the total algal plankton crop each year, using the figures recorded in §2 and §4 of this Summary. Since 1·12 mgrm. of P_2O_5 suffices for 1×10^9 diatoms, each litre of sea water could produce 26·8 million diatoms for a consumption of 0·030 mgrm. As many as 30 million diatoms per litre were found by the writer in a fresh water pond, so these large figures, as calculated, need not seem impossible.

Taking it that each cubic metre to a depth of 70 metres loses 30 milligrams of phosphate as P_2O_5 and that the phosphate content of the algal plankton is 0·15 per cent, calculated on the wet weight, it results that the column of water produces 1·4 kilograms algal plankton per square metre of sea. If one assumes that the carbon content of the algæ, reckoned as a hexose sugar, amounts to 15 per cent of the wet weight the calculation made by the writer (1922) from the seasonal change in alkalinity gives an identical value 1·4 kilograms. The exact agreement is fortuitous, but it lends support to the validity of the alkalimetry method.

8. The colorimetric method of Denigès was found very convenient for the analysis of waters containing 0·050 to 0·001 mgrm. of P_2O_5 per litre. An accuracy of $\pm0·001$ mgrm. can be obtained in clear solutions free from tint, and results to within $\pm0·002$ may readily be obtained. For samples which develop a yellowish tint with the reagents it is convenient to add a little Bismarck brown to the standard.

BIBLIOGRAPHY.

ALLEN, E. J. 1919. A contribution to the quantitative study of plankton. Journ. Mar. Biol. Assocn., **12**, 1–8.

ALLEN, E. J., and NELSON, E. W. 1907. On the artificial culture of marine plankton organisms. Journ. Mar. Biol. Assocn., **8**, 421–474, and Q. J. Microscop. Sci., 1910, **55**, 361–431.

AMERICAN PUBLIC HEALTH ASSOCIATION. 1920. Standard methods for the examination of water and sewage. Boston.

ATKINS, W. R. G. 1922. The hydrogen ion concentration of sea water in its biological relations. Journ. Mar. Biol. Assocn., **12**, 717–771.

ATWATER, W. O. 1888. The chemical composition and nutritive values of food fishes and aquatic invertebrates. Rep. of Commissioners of Fish and Fisheries, U.S.A.

BRANDT, K. 1920. Über den Stoffwechsel im Meere. 3 Abhandlung. Wiss. Meeresuntersuch. Abt. Kiel, **18**, 185–430.

CLARKE, F. W. 1920. The data of geochemistry. U.S. Geol. Survey, Bull. 695. Washington.

CLARKE, F. W., and SALKOVER, B. 1918. Inorganic constituents of two small crustaceans. Proc. Washington Acad., **8**, 185.

CLARKE, F. W., and WHEELER, W. C. 1922. The inorganic constituents of marine invertebrates. U.S. Geol. Survey, Professional Paper 124. Washington.

CZAPEK, F. 1913–1921. Biochemie der Pflanzen. Jena.

DENIGÈS, G. 1921. Détermination quantitative des plus faibles quantités de phosphates dans les produits biologiques par la méthode céruléo-molybdique. Compt. Rend. Soc. Biol. Paris, **84**, No. 17, 875–877. Also C.R. Acad. des Sc. 1920, **171**, 802.

EMBDEN, G. 1921. Eine gravimetrische Bestimmungsmethode für kleine Phosphorsäuremengen. Z. physiol. Chem., **113**, 138–145.

Fertilizer resources of the U.S.A. Washington, 1912, 225–229.

FLORENTIN, D. 1921. The determination of phosphates in water. Ann. chim. anal. chim. appl., **3**, 295–6. Cited from Chem. Abstracts.

JONES, W., and PERKINS, M. E. 1923. The gravimetric determination of organic phosphorus. J. Biol. Chem., **55**, 343–51.

KENWOOD, H. R. 1911. Public health laboratory work. London.

KLEINMANN, H. 1919. The determination of phosphoric acid as strychnine phosphomolybdic compound. Biochem. Zeitsch., **99**, 150–89. Cited from Chem. Abstr.

MATTHEWS, D. J. 1916. On the amount of phosphoric acid in the sea water off Plymouth Sound. Journ. Mar. Biol. Assocn., **11**, 122–130. Also Pt. II, *loc. cit.* 1917, 251–257.

McHargue, J. S., and Peter, A. M. 1921. The removal of mineral plant-food by natural drainage waters. Kentucky Agric. Expt. Sta., Bull. No. 237.

Pentanelli, E. 1923. Influenza delle condizioni di vita sullo sviluppo di alcune alghe marine. Arch. di Sci. Biol., **4,** 21–87. Cited from Physiol. Abstr.

Posternak, S. 1920. Variations in the composition of ammonium phosphomolybdate. Comp. rend. Acad. des Sc., 170, 930–3.

Do. The determination of small quantities of phosphoric acid as barium phosphomolybdate in the presence and in the absence of organic phosphorus. Soc. de chim., 4th series, **27** & **28,** 507–18.

Do. The technique of the determination of phosphoric acid as barium phosphomolybdate. *Loc. cit.,* 564–8. Cited from Chem. Abstr.

Pouget, L., and Chouchak, D. 1909, 1911. Dosage colorimétrique de l'acide phosphorique. Bull. Soc. Chim. France, Series 4, **5,** 104, and **9,** 649.

Raben, E. 1916–1920. Quantitative Bestimmung der im Meerwassergelösten Phosphorsäure. Wiss. Meeresuntersuch., **18,** 1–24.

Strasburger, E. 1921. Text-book of botany. London.

Note.—Up to the end of November, 1923, the phosphate content of the fresh waters studied has been far below the 1922 values, obtained on stored samples. This indicates that the possible error from storage, mentioned on p. 140, l. 1–3, may be very considerable. The accuracy of Fig. 8 is thus impaired.

4

Reprinted from *Trans. Wisconsin Acad. Sci. Arts Letters*, **23**, 233–248 (1927)

PHOSPHORUS CONTENT OF LAKE WATERS OF NORTHEASTERN WISCONSIN

C. Juday, E. A. Birge, G. I. Kemmerer and R. J. Robinson

Notes from the Biological Laboratory of the Wisconsin Geological and Natural History Survey.* XXVIII.

Introduction

Quantitative studies of the phosphorus content of ocean waters have been made by several investigators in recent years, but relatively few determinations have been made on fresh waters. Of the very large number of analyses of river and lake waters of the United States that are given by Clarke (1924), only 27 include phosphorus determinations; 20 of these are lake waters and they represent 19 different lakes. The waters of several of these lakes are more or less alkaline and some of them contain unusually large amounts of phosphorus. In four of them, for example, it ranged from 33 mg. to 133 mg. per liter of water. On the other hand, the water of Crater Lake, Oregon, yielded only 0.032 mg. per liter and three others gave only a trace of phosphorus.

McHargue and Peter (1921) studied the phosphate content of spring and stream waters in Kentucky. They found that the quantity of phosphorus varied with the character of the geological strata; a much larger amount was found in an Ordovician area than in Mississippian and Pennsylvanian areas. Seventeen samples from the first region yielded an average of 0.253 mg. per liter of water, while the averages for the latter areas were only 0.087 and 0.061 mg. respectively.

Kemmerer (1923) reported that the phosphorus content of the waters of five lakes situated in Idaho and Washington ranged from 0.013 to 0.026 mg. per liter.

---

*This investigation was made in cooperation with the U. S. Bureau of Fisheries and the results are published with the permission of the Commissioner of Fisheries.

Atkins (1923) found that the waters of several uncontaminated streams and water supplies in England and Scotland contained less than 0.022 mg. of phosphate phosphorus per liter; one stream yielded only 0.003 mg. per liter. Atkins and Harris (1924) found a maximum of 0.055 mg. in one pond which they studied and 0.04 mg. in another. Only a small amount or none at all was found in these ponds in the summer; the available supply of phosphate phosphorus was used up by the phytoplankton. There was an increase in winter due in part to regeneration and in part to inflowing water. These authors reached the conclusion that the exhaustion of the available phosphorus in early spring set a limit to the further growth of the algae.

Fischer (1924) states that an immediate and marked increase in carp production was obtained at the Bavarian Pond Fishery Experiment Station whenever the ponds were fertilized with ground basic slag or with superphosphate. Additions of nitrogen and potassium without phosphorus did not increase the yield very much.

Investigations relating to the phosphorus content of ocean waters have been discussed by Brandt (1919) and by Atkins (1926). The former reported that the quantity of phosphorus in the surface water of the North Sea ranged from 0.029 to 0.07 mg. per liter; the amount was smallest in May and June and largest in November and February. Atkins has·found a summer minimum and a winter maximum at various stations off the coast of England; the vernal diminution was proportional to the increase in phytoplankton. Practically no phosphate phosphorus was found in the surface water from May to August, but much evidence was obtained to show that the deep water of the ocean serves as a reservoir of phosphate phosphorus since it contains from 0.022 to 0.035 mg. per liter or more.

In addition to other observations, a quantitative study of the phosphorus content of the lake waters of northeastern Wisconsin was begun in 1925 and was continued in 1926. These lakes are situated in Vilas and Oneida counties; the former county possesses 347 lakes which are large enough to be shown on the county map and the latter county has 264 lakes. All of these lakes occupy typical morainal basins and have sandy or gravelly shores for the most part. In outline these lakes vary from almost circular to elongated and very irregular bodies of water. In general the main axes of the elongated ones have approximately a northeast-southwest trend, being parallel to the line of movement of the glacial ice which was from the northeast to the southwest. In size these lakes vary from about one hectare in area up to about 1,300 hectares. The maximum depth ranges from one or two meters up to a maximum of 35 meters. A number of them possess neither an inlet nor an outlet.

Observations have been made on 88 of these lakes during this investigation and the results obtained on 15 of them are shown in table 1. So far the work has been confined to the spring and summer seasons. The ice usually disappears from these northeastern lakes during the first week in May and it has been the aim to visit them as soon as possible after the ice has dissappeared. The summer observations have been made in July and August.

In 1925 the observations on phosphorus were confined to a quantitative determination of the phosphate phosphorus which is designated as the "soluble" phosphorus in the following discussion. As the work progressed however, it became evident that a determination of the total phosphorus (soluble plus organic phosphorus) would contribute to a better understanding of the problem and a method was devised for such a determination. The soluble phosphorus was determined by the ceruleomolybdic method of Denigès and the total phosphorus was determined by the same method after the organic material in the sample had been thoroughly oxidized. The two methods are described below. Deducting the soluble phosphorus from the total gives what is called the "organic phosphorus" in this report. This organic phosphorus represents that which is contained in the plankton and other organic material that is present in the water, and a certain part of it may consist of dissolved phosphorus that is not in a pentavalent state.

SOLUTIONS AND METHODS FOR DETERMINING PHOSPHORUS

Denigès (1921) devised a ceruleomolybdic method for the quantitative determination of phosphorus in biological products. Florentin (1921) changed the acidity of the molybdate solution and used this method to determine the quantity of phosphorus in water. Atkins (1923) modified Florentin's procedure by using 100 cc. of water for a sample instead of 10 cc., thereby greatly increasing the accuracy of the method. This colorimetric method, however, shows only the quantity of phosphate phosphorus that is dissolved in the water. The additional phosphorus obtained in the determination of the total phosphorus, as described below, is present in the sample either as soluble compounds of phosphorus other than phosphates or as organic compounds of phosphorus.

The following solutions are required for this colorimetric method of determining the phosphorus content of water:—

Molybdate Reagent. Dissolve 10 grams of ammonium molybdate in 100 cc. of distilled water; add this solution to 300 cc. of cold 50 per cent (by volume) sulfuric acid. This molybdate reagent must be protected from strong light; it should be kept in the dark when not in use. The molybdate solution is not affected by light before it is added to the sulfuric acid, so that the two may be kept separate and then mixed just before using.

Stannous Chloride. Dissolve 1 gram of tin in 20 cc. of concentrated hydrochloric acid by warming and adding 2 or 3 drops of a 5 per cent solution of copper sulfate. When the tin is dissolved, dilute the solution to 100 cc. with distilled water and add a piece of tin.

This solution may be more easily prepared by dissolving 2.15 grams of stannous chloride ($SnCl_2$ $2H_2O$) in 20 cc. of concentrated hydrochloric acid. Dilute to 100 cc. and add a piece of tin.

Standard Phosphorus Solution. Dissolve 4.394 grams of potassium di-hydrogen phosphate (KH_2PO_4) which has been dried over sulfuric acid, in phosphorus free water and make up to 1 liter; 1 cc. of this solution contains 1 mg. of phosphorus. Various dilutions of this stock solution are made up for standard solutions; the most useful ones for this investigation have been 1 to 100 and 1 to 1000.

Determination of Soluble Phosphorus. Measure 100 cc. of the water to be tested into a graduated colorimeter or into a Nessler tube. Add 2 cc. of the molybdate reagent and 3 or 4 drops of the stannous chloride. The blue color is fully developed in 5 to 10 minutes; when the color is completely developed, compare with standards in similar tubes containing known amounts of phosphorus. Standards are made up by adding known amounts of the standard phosphorus solution to 100 cc. of distilled water; then add 2 cc. of the molybdate reagent and 3 or 4 drops of the stannous chloride. The blue color fades somewhat on keeping so that a fresh standard should be made after about half an hour. Atkins (1923) states that the decrease in color averages about 1 per cent per hour over a period of 25 hours.

Silica dissolved from the reagent bottles or present in the water in large quantities (49 mg. per liter or more) also produces a blue color. It is best, therefore, to keep the reagents in Pyrex glass bottles. None of these lake waters contained enough silica to interfere with the phosphorus determinations since a maximum of only 16 mg. per liter was found in them.

Determination of Total Phosphorus. The colorimetric method is used for the determination of the total phosphorus in water after acidifying, evaporating and oxidizing the sample. This liberates the phosphorus that is combined in the organic compounds and oxidizes all of the phosphorus to the pentavalent state. The important part of the procedure is the removal of the last trace of the oxidizing agent before the molybdate reagent is added. A sample of 100 cc. of water is measured into a 250 cc. Erlenmeyer flask; 0.2 cc. of concentrated sulfuric acid is added and the sample is evaporated to a volume of 5 to 10 cc. Not more than 0.2 cc. of sulfuric acid should be used or the results will be too low. Then 0.5 cc. of concentrated nitric acid is added and the evaporation is continued just to fumes of the sulfuric acid. After cooling, 10 cc. of distilled water and 3 cc. of concentrated hydrochloric acid are added and the evaporation to fumes of sulfuric anhydride is repeated. There is danger of losing phosphorus if fumes are allowed to pass out of the flask. The sample is now cooled and rinsed into a colorimeter or into a Nessler tube. It is then diluted to 100 cc. with distilled water, after which 2 cc. of the molybdate reagent and 3 or 4 drops of stannous chloride are added. Compare with standards.

The phosphorus determinations should be made as promptly as possible after the samples are taken, otherwise the ratio of the soluble to the organic phosphorus may change. In these studies the determinations were made within one to three hours after the samples were secured.

CHEMICAL RESULTS

Only 15 of the 88 lakes on which observations have been made, are included in table 1; these were selected as representatives of the various types that were found in this survey. A more detailed account will be published when the investigation is completed. Of the total number of lakes visited in July and August, only 19 have been visited in May and none later than September 28.

The results given in table 1 for fixed carbon dioxide, phosphorus and plankton are stated in milligrams per liter of water. The data for phosphorus are stated in terms of the element rather than as PO_4 or as P_2O_5. The plankton was extracted from the samples used for this determination by means of a Foerst electric centrifuge (Juday 1926) and the figures given in the table indicate the amount of organic material in these centrifuge catches.

Fixed Carbon Dioxide. The fixed carbon dioxide was determined by titrating the water with N/44 HCl, using methyl orange as an indicator. There is a wide range in the amount of fixed carbon dioxide in the waters of the northeastern lakes and the 15 lakes shown in table 1 cover the entire range. They vary from a minimum of a little less than 1 mg. in one lake to a maximum of slightly more than 44 mg. per liter in another. Out of these 88 lakes, the water of 28 of them possesses less than 5 mg. of fixed carbon dioxide per liter and 13 others show more than 20 mg. per liter, so that more than half of them fall between these two limits. As a whole these bodies of water may be grouped into what has been designated as medium and soft water lakes. The lakes having neither an inlet nor an outlet possess the smallest amount of fixed carbon dioxide.

Soluble Phosphorus. The quantity of soluble phosphorus in the surface water varied from none in two lakes (Bass and Mary) to as much as 0.015 mg. per liter in Adelaide.

In most instances the amount falls between 0.003 and 0.006 mg. per liter. Similar amounts have been found in the hard water lakes of southeastern Wisconsin where the fixed carbon dioxide ranges from 70 to 100 mg. per liter.

The observations which were made on the northeastern lakes in May, 1926, came soon after the disappearance of the ice. The ice did not completely disappear from Trout Lake until May 5 and a set of samples was obtained on May 8. The temperature of the water on the latter date was 3.9° C. at both surface and bottom, so that the vernal circulation of the water was still in progress when the samples were secured. During this period of circulation, the various substances held in solution by the water are uniformly distributed from surface to bottom. This uniform distribution is shown by the results obtained for fixed carbon dioxide and soluble phosphorus in May on Black Oak, Crystal, Trout, and Webb Lakes. On May 12, 1926, the temperature of the surface water of Black Oak Lake was 6.9° and of the bottom water 4.8°.

In Adelaide Lake there was a marked decrease in the quantity of soluble phosphorus in the surface water between July 17 and August 21. Similar decreases, though not so marked, were noted between spring and summer in the upper water of Big Arbor Vitae, Brandy, and Webb Lakes. In Brandy Lake, for example, the amount decreased from 0.005 mg. per liter on May 13 to only a trace on July 13. Decreases were noted also in the upper water of Black Oak and Crystal Lakes, but they are too small to be of any significance.

In Bass Lake, on the other hand, there was a definite increase in the soluble phosphorus in the upper water between May 10 and July 1; this lake has a maximum depth of only 7 m. so that substantially all of the water is kept in circulation during the summer and this is favorable for the regeneration of the phosphorus. The surface water of Clear Lake yielded only a trace of soluble phosphorus on May 10, but 0.005 mg. per liter was found on August 12. Clear Lake differs from Bass in that it has a maximum depth of 25 m. and is stratified during the summer, and this is unfavorable for the process of regeneration. The increase can not be attributed to a renewal of the upper stratum by inflowing water because the lake has neither an inlet nor an outlet. A similar increase was found in the upper water of Island Lake which is also deep enough to be stratified in the summer and which does not possess an inlet or an outlet.

The surface water of Plum Lake yielded 0.005 mg. of soluble phosphorus per liter in May and in July; the same amount was found in the upper water of Trout Lake in May, June, and August and in Wild Cat Lake in July and August. None was found in the upper water of Lake Mary in July and August.

The soluble phosphorus is uniformly distributed from surface to bottom during the vernal circulation, but there is no intermixture of the upper and lower water in the deeper lakes during the summer period of stratification; as a result there is an accumulation of soluble phosphorus in the lower strata of some of the stratified lakes in summer. This phenomenon is due to the decomposition of organic matter in the lower water; during this process the phosphorus which is combined with the decaying organic material is liberated in a soluble form. The most striking example of this accumulation in the lower water was noted in Lake Mary where the sample from 20 m. yielded 0.75 mg. of soluble phosphorus per liter on July 12, 1926, with none at the surface. This is a small circular body of water, approximately 122 m. (400 ft.) in diameter, but it has a maximum depth of 22 m. (72 ft.).

All of the lakes in table 1, with the exception of Bass, Brandy, and Crystal, show a larger amount of soluble phosphorus in the lower water during the summer than in the upper strata; the quantity found in the lower water in most instances is from two to ten times as much as in the upper water.

Organic Phosphorus. The organic phosphorus includes that which is combined with the organic material that is present in the water, such as the plankton and the organic debris. The quantity varied from a minimum of 0.007 mg. per liter in the surface sample of Crystal and Webb Lakes in May to a maximum of 0.12 mg. in a sample from a depth of 15 m. in Lake Mary. With the exception of the surface sample from Adelaide Lake on July 17, of the bottom sample

of Black Oak Lake on August 25 and of the lower water of
Lake Mary, the quantity of organic phosphorus equalled or
exceeded that of the soluble phosphorus at the corresponding depth. The above exceptions are the only ones that
were encountered in samples from 70 different lakes in
1926. In most instances the amount of organic phosphorus
was from two to ten times as large as that of the soluble
phosphorus, or more.

Plankton

The various lakes showed a wide range in the amount of
plankton which they possessed. Of the 96 plankton catches
listed in table 1, 27 yielded less than a milligram of organic
matter per liter of water; 50 yielded from one to two milligrams and the remaining 19 yielded more than two milligrams. Two catches in the last group contained more than
four milligrams of organic matter per liter. Minimal
amounts were found in Clear, Crystal, and Island Lakes
where the organic material in the catches from the upper
water was considerably less than one milligram per liter.
The average for the entire series of catches obtained from
Crystal Lake on June 28 was only 0.67 mg. per liter and that
from Clear Lake on August 12 was 0.88 mg.

The large catch obtained at 3 m. in Lake Mary on July 12
consisted chiefly of Zoochlorella. The maximum catch
found at 15 m. in Plum Lake on July 14 consisted chiefly of
organic debris; only a relatively small part of it was made
up of recognizable plankton organisms, but most of the
debris was undoubtedly derived from the plankton which
flourished in the upper water.

Discussion of Results

Quantitative studies of both marine and freshwater phytoplankton have shown that there is a more or less definite
seasonal periodicity in the growth of these organisms. In
several Wisconsin lakes on which regular observations have
been made, for example, there are spring and autumn
maxima which are separated by summer and winter
minima. These rhythmic changes have been attributed to
various factors, such as temperature, light, nitrogen, and
phosphorus.

16

In recent papers Atkins (1923-1926) has put forth the theory that phosphorus is the limiting factor. The phytoplankton needs an available supply of nitrogen, potassium, and phosphorus for growth just as the land plants do. Nitrogen and potassium are usually present in natural waters in sufficient quantities to supply the demands of these organisms, but phosphorus is generally present in such small amounts that it will serve as a limiting factor in the growth of the algae.

During the summer period the phytoplankton is limited to the upper stratum, or epilimnion, because it needs a certain amount of light for the process of photosynthesis; as a result it draws upon the stock of soluble phosphorus in this stratum. When the various plankton organisms die and sink into the lower water, they carry down with them a certain amount of this phosphorus. Since the soluble phosphorus is usually present in small quantities in the upper water, the supply is soon exhausted by a large growth of phytoplankton and, according to the theory, this sets a limit to the further growth of these organisms.

The chemical results show that there is only a small amount of soluble phosphorus present in the upper water of these Wisconsin lakes in spring and summer; in most instances there is only a very slight or no decrease at all in the quantity as the summer season advances. This was true of lakes which maintained a fairly large growth of plankton in the upper water, such as Black Oak, Plum, and Trout. In Black Oak and Trout the crop of phytoplankton was sufficient to produce the usual summer "bloom" in July and August. In Lake Mary on the other hand, there was no trace of soluble phosphorus in the upper three meters on July 12 and none in the upper two meters on August 21, yet an abundant growth of phytoplankton was found in these strata on both dates. A supply of soluble phosphorus was present just below these strata and enough may have diffused into the upper water to supply the demands of the phytoplankton, but diffusion takes place so slowly that it is doubtful whether much could be obtained in this way. The decomposition of organic matter in the upper water would also furnish a certain amount.

In a few lakes there was an actual increase in the amount

of soluble phosphorus in the upper water between May and July or August, such as noted in Bass, Clear, and Island Lakes. Silver Lake also, which is not included in the table, showed an increase from only a trace at all depths on May 8, 1926, to 0.005 mg. per liter from the surface to 13 m. on July 8. In Bass Lake the whole body of water is kept in circulation during the summer so that the increase may be accounted for by the decomposition which takes place at the bottom as well as that which takes place in the water. In the other three lakes, however, the water is deep enough to be stratified in summer and the increase in them can not be due to decomposition below the epilimnion. Some of the lakes, such as Adelaide, Brandy, and Webb, showed a more or less marked decrease in the stock of soluble phosphorus in the upper water as the season advanced.

This quantitative study of the soluble phosphorus was undertaken with the hope of confirming Atkins' theory, but the results obtained in 1925 and 1926 do not give any positive evidence that the soluble phosphorus is a limiting factor in the production of phytoplankton. That is, the great majority of these lakes showed little or no decrease in the amount of soluble phosphorus in the upper water as the season progressed, yet some of them supported relatively large crops of phytoplankton during this interval. Just how some of these lakes are able to support a crop of phytoplankton from May to July or August without any appreciable decrease in the soluble phosphorus of the upper water or only a slight one, is not known at present. An increase in the lower water during this time suggests that part of this increase at least, and most probably a very large part of it, comes from plankton material that sinks into the lower strata from the upper water and decomposes there. Such a transfer ought to result in a decrease of the soluble phosphorus in the upper stratum. The fact that there is none or only a very slight one may mean that regeneration of the soluble phosphorus through the decomposition which takes place in the upper water is sufficient to make good the loss that is sustained. The whole problem needs further investigation and plans have been made for further studies.

The data given in table 1 show that there is no direct correlation between the quantity of organic phosphorus and the amount of organic matter in the plankton. The phosphorus content of 23 samples of net plankton from Lake Mendota amounted to an average of 0.61 per cent of the dry weight of the material (Birge and Juday 1922). Applying this percentage to the total quantity of plankton obtained in these northeastern lakes would give a smaller amount of organic phosphorus than was actually obtained from the water. This may be accounted for by the fact that these lake waters contain a relatively large amount of organic matter, either in the particulate or dissolved form, which can not be extracted with a centrifuge. In both Bass and Turtle Lakes for example, it has been found that less than 3 per cent of the total organic matter in the water consisted of plankton (Birge and Juday 1926).

SUMMARY

1. Only a small amount of soluble phosphorus was found in the waters of 88 lakes situated in northeastern Wisconsin.

2. The quantity of soluble phosphorus is not correlated with the amount of carbonates (fixed carbon dioxide) in solution.

3. In some lakes which support a relatively large crop of plankton, there is no decrease in the amount of soluble phosphorus, or only a very slight one, in the upper water from May to July or August.

4. No definite evidence was found to indicate that soluble phosphorus is a limiting factor in the production of phytoplankton in these lakes.

5. There was no correlation between the amount of centrifuge plankton and the quantity of organic phosphorus.

6. The presence of a comparatively large amount of organic phosphorus shows that determinations of both the soluble and the organic phosphorus must be made within a few hours after the samples are taken in order to obtain reliable results concerning the relative proportions of the two.

LITERATURE CITED

Atkins, W. R. G. 1923. The phosphate content of fresh and salt waters in its relationship to the growth of algal plankton. Jour. Mar. Biol. Assoc. 13 (1): 119–150.

1925. Seasonal changes in the phosphate content of seawater in relation to the growth of algal plankton during 1923 and 1924. Jour. Mar. Biol. Assoc. 13 (3): 700-720.

1926. The phosphate content of seawater in relation to the growth of algal plankton. Jour. Mar. Biol. Assoc. 14 (2):447-467.

Atkins, W. R. G. and G. T. Harris, 1924. Seasonal changes in the water and heleoplankton of freshwater ponds. Proc. Roy. Dublin Soc. 18 (1): 1–21.

Birge, E. A. and C. Juday. 1922. The inland lakes of Wisconsin. The plankton. I. Its quantity and chemical composition. Bul. No. 64, Wis. Geol. and Nat. Hist. Survey. ix+222 pp.

1926. Organic content of lake water. Bul. Bur. Fish. 42: 185-205.

Brandt, K. 1919. Ueber den Stoffwechsel im Meere. Wissensch. Meeresuntersuch. 18 (3): 187–429.

Clarke, F. W. 1924. The composition of the river and lake waters of the United States. Prof. Paper 135, U. S. Geol. Survey. 197 pp.

Denigès, G. 1921. Détermination quantitative des plus faibles quantités de phosphates dans les produits biologiques par la méthode céruléomolybdique. Compt. Rend. Soc. Biol. Paris. 84 (17): 875-877. Also C. R. Acad. des Sci. 171: 802. 1920.

Fischer, H. 1924. The problem of increasing the production in fish ponds by the use of chemical fertilizers. Internat. Rev. Sci. and Prac. Agric. 2 (4): 822–830.

Florentin, D. 1921. The determination of phosphates in water. Ann. chim. anal. chim. appl. 3 : 295-296. Chem. Abstracts **16** : 601.

Juday, C. 1926. A third report on limnological apparatus. Trans. Wis. Acad. Sci., Arts and Let. **22** : 299-314.

Kemmerer, George, J. F. Bovard and **W. R. Boorman.** 1923. Northwestern lakes of the United States: Biological and chemical studies with reference to possibilities in production of fish. Bul. Bur. Fish. 39 : 51-140. 22 figs.

Mc Hargue, J. S. and **A. M. Peter.** 1921. The removal of mineral plant food by natural drainage water. Ky. Agric. Expt. Sta. Bul. No. 237 : 333-362.

TABLE 1. *This table shows the quantity of fixed carbon dioxide, phosphorus and plankton found in the waters of several Wisconsin lakes in 1926. The results are stated in milligrams per liter of water. The figures for plankton show only the amount of organic matter in this material. Tr. means trace.*

Lake	Date	Depth in meters	Fixed carbon dioxide	Phosphorus			Plankton
				Soluble	Organic	Total	
Adelaide..........	July 17.....	0	3.2	0.015	0.012	0.027	2.27
		5	3.5	0.015	0.015	0.030	1.75
		10	3.5	0.015	0.015	0.030	0.96
		15	3.6	0.018	0.015	0.033	1.18
		20	4.8	0.027	0.033	0.060	1.81
	August 21	0	3.2	0.006	0.022	0.028	1.76
		5	3.3	0.006	0.022	0.028	1.73
		10	3.6	0.006	0.021	0.027	0.96
		15	4.3	0.008	0.020	0.028	1.48
		20	5.5	0.018	0.029	0.047	1.43
Bass..............	May 10.....	0	1.5	0.000	0.020	0.020	....
	July 1......	0	1.7	0.006	0.014	0.020	1.57
		3	1.7	0.006	0.014	0.020	....
		5	1.9	0.006	0.016	0.022	2.80
Big Arbor Vitae....	May 13.....	0	22.8	0.006	0.024	0.030	2.03
	July 30.....	0	24.0	0.004	0.016	0.020	1.18
Black Oak..........	May 12.....	0	9.5	0.006	0.009	0.015	1.90
		22	9.5	0.006	0.011	0.017	....
	August 25	0	9.6	0.005	0.014	0.019	1.09
		5	9.9	0.005	0.012	0.017	1.10
		10	9.9	0.005	0.018	0.023	1.10
		15	9.9	0.007	0.015	0.022	1.14
		20	10.5	0.011	0.044	0.055	1.20
		23	10.7	0.045	0.040	0.085	1.34
Brandy...........	May 13.....	0	....	0.005	0.040	0.045	2.63
	July 13.....	0	16.1	Tr.	0.022	0.022	1.48
		12	21.1	Tr.	0.022	0.022	1.76
Clear.............	May 10.....	0	3.5	Tr.	0.020	0.020	....
		25	3.8	0.005	0.015	0.020	....
	August 12	0	2.8	0.005	0.014	0.019	0.62
		10	3.5	0.005	0.014	0.019	0.81
		15	3.8	0.008	0.014	0.022	0.90
		23	3.8	0.012	0.017	0.029	1.20
Crystal...........	May 9.....	0	2.2	0.005	0.007	0.012	....
		18	2.2	0.005	0.013	0.018	....
	June 26.....	0	2.0	0.004	0.011	0.015	0.27
		5	2.0	0.004	0.011	0.015	0.47
		10	2.0	0.004	0.011	0.015	0.56
		15	2.0	0.004	0.014	0.018	0.77
		18	2.0	0.005	0.015	0.020	1.29
	August 7...	0	1.6	0.005	0.010	0.015	0.81
		10	1.6	0.005	0.010	0.015	0.86
		15	1.6	0.005	0.010	0.015	1.20
		18	1.6	0.005	0.010	0.015	0.78
Island.............	May 11....	0	1.8	0.000	0.012	0.012	....
		12	1.8	0.005	0.010	0.015	....
	August 25	0	1.5	0.005	0.015	0.020	0.71
		5	1.7	0.005	0.015	0.020	0.73
		9	1.7	0.005	0.019	0.024	0.92
		11	1.7	0.007	0.022	0.029	1.14

Lake	Date	Depth	Fixed carbon dioxide	Phosphorus			Plank-ton
				Soluble	Organic	Total	
Little Arbor Vitae	May 13	0	23.0	0.007	0.029	0.036	2.35
	July 30	0	24.1	0.008	0.032	0.040	2.80
Mary	July 12	0	2.5	0.000	0.040	0.040	1.74
		3	2.8	0.000	0.045	0.045	4.06
		5	4.0	0.100	0.020	0.120	2.46
		10	4.8	0.225	0.075	0.300	2.03
		15	7.5	0.500	0.050	0.550	1.56
		20	11.8	0.750	0.000	0.750	2.24
	August 21	0	2.6	0.000	0.040	0.040	2.49
		2	2.7	0.000	0.030	0.030	2.11
		3	3.1	0.007	0.031	0.038	--------
		5	4.2	0.100	0.050	0.150	2.11
		15	7.5	0.280	0.120	0.400	1.92
		20	10.9	0.440	0.100	0.540	1.35
North Turtle	July 20	0	14.0	0.010	0.020	0.030	0.98
		6	14.0	0.010	0.020	0.030	1.09
		10	14.8	0.010	0.020	0.030	1.02
		13	15.8	0.013	0.021	0.034	0.90
Plum	May 14	0	18.6	0.005	0.025	0.030	1.76
		14	18.9	0.006	0.038	0.044	1.40
	July 14	0	17.6	0.005	0.016	0.021	1.18
		7	18.3	0.005	0.016	0.021	1.60
		10	19.0	0.005	0.022	0.027	1.60
		12	20.0	0.006	0.021	0.027	--------
		15	21.5	0.007	0.023	0.030	4.34
Trout	May 8	0	20.0	0.005	0.025	0.030	1.00
		30	20.0	0.005	0.025	0.030	1.57
	June 25	0	20.0	0.005	0.015	0.020	1.32
		10	20.0	0.005	0.015	0.020	0.98
		20	20.0	0.008	0.012	0.020	0.78
		30	20.0	0.008	0.017	0.025	0.82
	July 31	0	19.2	0.004	0.011	0.015	0.98
		10	19.2	0.004	0.011	0.015	1.12
		20	19.2	0.004	0.011	0.015	0.93
		25	19.2	0.005	0.013	0.018	0.87
		30	19.2	0.010	0.017	0.027	1.40
	August 23	0	18.7	0.005	0.017	0.022	1.14
		5	19.0	0.005	0.015	0.020	0.93
		10	19.0	0.005	0.015	0.020	1.33
		20	19.0	0.005	0.015	0.020	0.85
		25	19.5	0.005	0.025	0.030	0.91
		31	19.5	0.005	0.025	0.030	2.21
Webb	May 9	0	2.3	0.005	0.007	0.012	--------
		12	2.3	0.005	0.013	0.018	--------
	July 5	0	1.2	0.003	0.013	0.016	1.21
		8	1.2	0.003	0.013	0.016	1.08
		10	1.3	0.004	0.016	0.020	--------
		12	1.3	0.004	0.021	0.025	2.78
	August 18	0	0.8	0.004	0.013	0.017	0.98
		8	0.9	0.004	0.013	0.017	1.06
		10	1.0	0.004	0.014	0.018	1.00
		12	1.7	0.018	0.048	0.066	1.67
Wild Cat	July 11	0	30.3	0.006	0.014	0.020	1.42
		5	30.6	0.006	0.014	0.020	1.56
		8	31.5	0.006	0.014	0.020	1.14
		11	41.5	0.008	0.016	0.024	1.78
	August 24	0	30.5	0.006	0.024	0.030	1.70
		5	30.5	0.005	0.022	0.027	1.75
		8	32.8	0.007	0.028	0.035	2.31
		11	44.1	0.012	0.028	0.040	2.17

Reprinted from *James Johnstone Memorial Volume*, Liverpool University Press, Liverpool, 176–192 (1934)

ON THE PROPORTIONS OF ORGANIC DERIVATIVES IN SEA WATER AND THEIR RELATION TO THE COMPOSITION OF PLANKTON [1]

ALFRED C. REDFIELD

PROFESSOR OF PHYSIOLOGY, HARVARD UNIVERSITY, AND SENIOR BIOLOGIST, WOODS HOLE OCEANOGRAPHIC INSTITUTION

(*Received September 5, 1933*)

" Chemical analysis shows that the animal and plant body is mainly built up from the four elements, nitrogen, carbon, hydrogen, and oxygen.   Added to these are the metals, sodium, potassium and iron, and the non-metals, chlorine, sulphur and phosphorus.   Calcium or silicon are also invariably present as the bases of calcareous or siliceous skeletons.   All these, with some others, are indispensable constituents of the organic body, and in an exhaustive study of the cycle of matter from the living to the non-living phases, and *vice versa*, we should have to trace the course of each."   JAMES JOHNSTONE, " Conditions of Life in the Sea," p. 273.   1908.

IT is now well recognized that the growth of plankton in the surface layers of the sea is limited in part by the quantities of phosphate and nitrate available for their use and that the changes in the relative quantities of certain substances in sea water are determined in their relative proportions by biological activity.   When it is considered that the synthetic processes leading to the development of organic matter are limited to the surface layers of the sea in which photosynthesis can take place, it becomes evident that the chemical changes which occur in the water below this zone must arise chiefly from the disintegration of organic matter.   In so far as this disintegration goes to completion, the changes in the derived inorganic constituents of sea water must depend strictly upon the quantity and composition of the organic matter which is being decomposed.   This is true quite irrespective of the agencies of decomposition, be they bacterial action,

---

1.   Contribution No. 30, from the Woods Hole Oceanographic Institution.

176

the autolytic enzymes of the original tissues themselves, or the metabolic processes of deep sea animals which utilize the organic matter as food. Thus, in the decomposition of a given quantity of organic matter, the quantity of oxygen consumed must be determined by the quantity of carbon, nitrogen, hydrogen, sulphur, and phosphorus to be oxidized, and the relative changes in the quantity of oxygen, nitrate, phosphate, carbonate, and sulfate must depend exactly on the elementary composition of the plankton.

It has seemed possible that the chemical composition of the population of the seas may be sufficiently uniform, and the contributions of the substances in question from other sources sufficiently limited, to permit the discovery of relations between their concentration in sea water which would be serviceable in the study of oceanic problems.

An examination of the data on nitrate and phosphate in the water of various seas secured by the " Dana " in the course of the Carlsberg Foundation's oceanographical expeditions around the world in 1928-1929 proved encouraging from this point of view. Thanks to the collaboration of Dr. Norris Rakestraw a more complete set of data, showing the relative concentrations of oxygen, nitrate, phosphate, and carbonate at different depths, was obtained from a number of stations made by the " Atlantis " in the deep waters of the western Atlantic Ocean. Six stations were occupied in the Sargasso Sea between Bermuda and the southern coast of the United States. In this region water samples were obtained at successive depths from at least three distinct water masses : (1) the surface water in.intimate relation with the atmosphere and in which photosynthetic activity has reduced the nutrient substances to a minimum ; (2) the layer of intermediate depth—most clearly marked at 700 meters and characterized by minimal oxygen concentrations and high concentrations in organic derivatives ; and (3) the deep water of polar origin of relatively uniform composition below 1,500 meters, which has also been the seat of organic decomposition. Between these layers lie zones in which the water has intermediate characters due to the mixing of the primary water masses.

The characteristics of the water at the various stations occupied are unusually uniform both as regards temperature and salinity and the concentrations of the biologically significant substances. A seventh station was occupied in the western edge of the Gulf Stream outside the Gulf of Maine. Here the layer characterized by minimal oxygen concentration is nearer the surface and the deep cold water lies within 500 meters of the surface. The numbers and positions of the stations are given in Table I. The station data will be published in the Bulletin Hydrographique of the International Council for the Exploration of the Sea.

TABLE I

Ship	Station No.	Position	Date
" Atlantis "	1463	34°02′ N. 68°15′ W.	February 8, 1933
,,	1465	31°25′ N. 66°30′ W.	,, 13, 1933
,,	1467	30°30′ N. 68°30′ W.	,, 14, 1933
,,	1469	29°26′ N. 70°42′ W.	,, 15, 1933
,,	1472	27°56′ N. 73°53′ W.	,, 17, 1933
,,	1475	26°49′ N. 76°09′ W.	,, 19, 1933
,,	1677	40°53′ N. 66°21′ W.	June 25, 1933

From the data for oxygen content and temperature, the amount of oxygen which has disappeared from the water since its exposure to the atmosphere has been estimated, assuming that the water was saturated at that time and had the temperature which characterized it *in situ*. While neither of these assumptions are probably strictly correct, they provide the only definite method available to arrive at the amount of oxygen utilized at the various depths. From the data for pH as measured at the temperature of the ship's laboratory the total carbonate content, Σ_{CO_2}, of the water samples was estimated with the aid of the tables provided by Buch, Harvey, Wattenberg, and Gripenberg (1932). These, together with the direct analyses of nitrate and phosphate, give a measure of the principal products formed by the complete oxidation of the carbon, nitrogen, and phosphorus of the biological population, together with the amount of dissolved oxygen which has disappeared in the process.

FIG. 1.

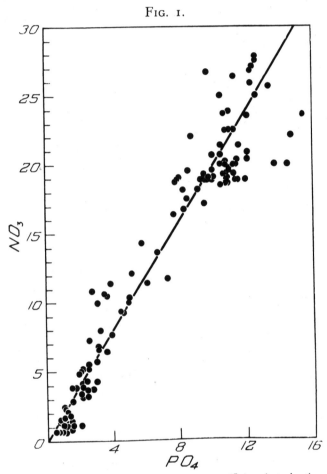

Correlation between concentrations of nitrate and phosphate in the waters of western Atlantic Ocean. Ordinate, concentration of nitrate, units 10⁻³ millimols per liter; abscissa, concentration of phosphate, units 10⁻⁴ millimols per liter. The line represents a ratio of $\triangle N : \triangle P = 20 : 1$ milligram atoms.

The data obtained for the water masses underlying the various stations are plotted in Figs. 1, 2 and 3. In these diagrams the concentration of nitrate characterizing each water sample is plotted against one of the other constituents in order to show whether the concentrations of these constituents vary in a correlated fashion. The concentrations are expressed in milligram atoms or in millimols per liter, as suggested by Cooper (1933), since this notation yields values proportional to the number of atoms of nitrogen, carbon, and phosphorus present in the constituents in question.

To consider first the relation between nitrate and phosphate concentration in the water samples, shown in Fig. 1, it is apparent that a very close correlation exists. Any gain or loss in nitrate content observed in comparing one water sample with another is accompanied by a strictly proportional gain or loss in phosphate content. The water samples in which the concentrations were low were from near the surface ; the quantities of nitrate and phosphate increase with increasing depth. The straight line drawn through the points describing the nitrate-phosphate correlation has the slope demanded if, for every three grams of nitrogen added to or subtracted from the water, one gram of PO_4 (or $\frac{3}{4}$ gram of P_2O_5) also made its appearance or disappearance. This corresponds to a ratio of 20 atoms of nitrogen to 1 atom of phosphorus. On the assumption that these changes are due solely to the decomposition or synthesis of organic matter, these ratios may be taken to reflect the proportions of nitrogen and phosphorus in the plankton community taken as a whole.

The correlation between the gain in concentration of nitrate and the loss in concentration of oxygen, as shown in Fig. 2, is almost equally good for the water samples collected from depths less than 1,000 meters in the Sargasso Sea, and 400 meters in the Gulf Stream,—depths which mark the transition between the intermediate layer of low oxygen content and the deeper layer of cold water. The data for greater depths, indicated by open circles in the figure, depart strongly from the expected correlation—an anomaly to which we will return. Above these depths, however, the variation of nitrate concentration and oxygen utilization is in the proportion of 1 millimol of nitrate to 6 millimols of oxygen.

The correlation between the concentrations of nitrate and carbonate shown in Fig. 3 is less precise owing to the greater variation in the carbonate measurements. This is due to the fact that the total variation in carbonate content is

FIG. 2.

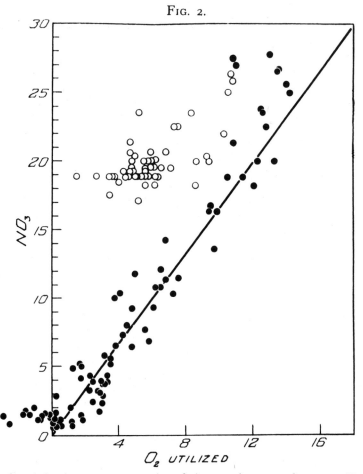

Correlation between concentration of nitrate and amount of oxygen utilized in waters of western Atlantic Ocean. Ordinate, concentration of nitrate, units 10^{-3} millimols per liter. The line represents a ratio of oxygen used : $\triangle N = 6 : 1$. Points representing stations from a depth below 400 meters in the Gulf Stream and below 1,000 meters in the Sargasso Sea are represented by open circles.

only a small part of the amount present, to the indirect nature of the method of estimating carbonate, to possible variations in buffer capacity which have been neglected in these estimations, and to exchange in CO_2 between air and the surface waters. Nevertheless, taking the data as a whole, it is clear that carbonate content is correlated with nitrate content. The envelopes drawn about the points in Fig. 3 have a slope corresponding to a ratio of 1 millimol of nitrate to 7 of carbonate. The value of this ratio is not as securely established as that relating nitrate to phosphate and to oxygen utilization.

FIG. 3.

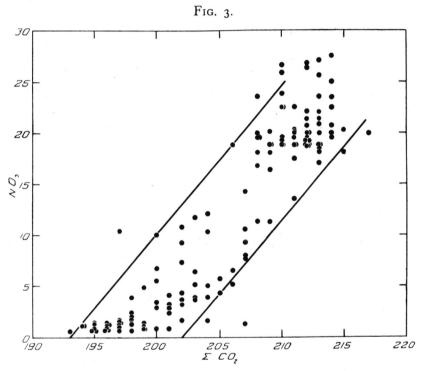

Correlation between concentration of nitrate and total carbonate in waters of western Atlantic Ocean. Ordinate, concentration of nitrate, units 10^{-3} millimols per liter; abscissa, concentration of carbonate, units 10^{-2} millimols per liter. The slope of the envelopes corresponds to a ratio of $\triangle C : \triangle N = 7 : 1$ milligram atoms.

If we take the ratio of change in nitrate to carbonate to be 7 : 1 and that of nitrate to oxygen utilization to be 6 : 1, it follows that the ratio of carbonate to oxygen is 7 : 6 or 1·17 : 1. This ratio is close to that characterizing photosynthetic processes and the decomposition of the principal classes of organic compounds. Its variation from the theoretical is not greater than the uncertainty of the carbonate measurements.

It appears from the foregoing data that, with the exception of the anomaly quoted in connection with the oxygen content of the deep water, the concentrations of nitrate, phosphate, and carbonate in samples of oceanic sea water of widely different origin vary in such a way as might be expected if the different samples contained the products of the complete disintegration and oxidation of organic material of similar composition, and that they differ only in the quantity of such material which has arrived at and remains in this degree of decomposition. Furthermore, the loss in oxygen agrees approximately with that to be expected from the quantity of carbonate gained. The proportions of carbon, nitrogen and phosphorus present in the organic material from which the carbonate, nitrate, and phosphate may be supposed to be derived are approximately 140 : 20 : 1 atoms or 100 : 16·7 : 1·85 grams.[1]

It is pertinent to inquire how these proportions agree with those actually found in various members of the plankton community. Naturally each kind of organism is found to have a characteristic composition different from that of other kinds, and unfortunately no adequate means of obtaining a truly representative sample of the entire population is available. However, by considering the composition of the plankton yielded by various kinds of haul, some idea may be obtained as to the validity of the foregoing considerations.

A number of elementary analyses of various kinds of plankton are recorded in Table II. They show that the proportions of carbon, nitrogen, and phosphorus as calculated from the carbonate, nitrate, and phosphate composition of the sea are not greatly different from those observed in

1. This comparison presupposes that nitrate nitrogen may be regarded as the sole source of nitrogen available to the plankton. Strictly speaking, this is not the case. Of other sources of nitrogen, nitrite is known to occur in oceanic waters in concentrations too small to be significant in computations of this sort. Professor Krogh has recently made observations on the nitrogen present as ammonia and as organic compounds in the deeper waters of the western Atlantic. His results indicate that ammonia is present in only small quantities. Organic compounds in the sea water contain relatively large amounts of nitrogen, but as the concentration of these does not vary greatly from surface to bottom, this source of nitrogen does not appear to be readily available for conversion into living matter.

various plankton, and on the whole the latter differ among themselves much more than their average differs from the calculated ratios.

TABLE II

Proportions of carbon, nitrogen, and phosphorus in various samples of plankton.

Sample	Parts by Weight		
	Carbon	Nitrogen	Phosphorus
Mixed copepods from Buzzards Bay ..	100	21	1·98
Centropages typicus, Gulf of Maine ..	100	25·6	1·06
Calanus finmarchicus, Gulf of Maine ..	100	13·4	2·04
Calanus finmarchicus, Gulf of Maine ..	100	15·8	2·26
Diatoms––Bay of Fundy, almost entirely *Thalassiosira nordenskiöldi*.	100	18·2	1·36
Diatoms—Off Nova Scotia coast—17 species of somewhat the same abundance.	100	15·6	2·26
Peridinians—Meyer (1914)	100	13·2	2·2
Chiefly peridinians—average of samples No. 1, 2, 3, 4, of Brandt (1898).	100	8·1	—
Chiefly diatoms—average of samples No. 6 and 7, Brandt (1898).	100	12·4	—
Chiefly copepods—average of samples No. 8 and 9, Brandt (1898).	100	15·3	—
Mixed plankton—sample no. 10, .. Brandt (1898).	100	11·3	—
Average all samples	100	15·4	1·88
Estimated from analyses of sea water ..	100	16·7	1·85

In this connection it is of interest to note that Braarud and Føyn (1930) have observed that each cell of *Chlamydomonas* removes $2·98 \times 10^{-12}$ gr. NO_3 nitrogen and $0·98 \times 10^{-12}$ gr. P_2O_5 from the sea water in which it grows. Here we see in a laboratory experiment an organism modifying the concentration of nitrate and phosphate in the medium in a ratio ($\triangle N : \triangle P = 15 : 1$) not very different from that observed in the oceans as a whole.

89

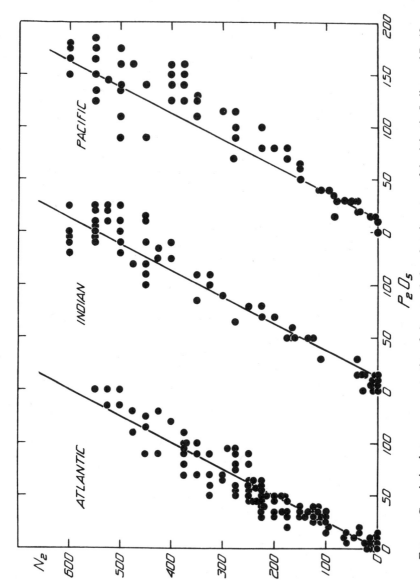

FIG. 4. Correlation between concentration of nitrate and phosphate in waters of the Atlantic, Indian, and Pacific Oceans. Data of Thomsen. Ordinate, concentration of nitrate nitrogen, units milligrams of nitrogen per cubic meter; abscissa, concentration of phosphate, units milligrams P_2O_5 per cubic meter. The lines correspond to a ratio of $\triangle N : \triangle P = 20 : 1$ milligram atoms.

The establishment of a general, if approximate, relation between the concentration of the various organic derivatives in sea water and the chemical composition of plankton would provide a valuable tool in the analysis of many oceanographic problems. Observations made at a single group of stations such as that reported cannot be considered to establish such a relation, though the results doubtless are encouraging. The data collected by the " Dana " (Thomsen, H., 1931) enable the generalization to be tested more widely—at least so far as nitrate and phosphate are concerned, for the measurements were made in the most diverse regions by the same investigators using uniform methods. In Fig. 4 I have plotted the simultaneous measurements of phosphate and nitrate obtained by the " Dana " at all stations and depths in the three oceans traversed. No attempt has been made to separate the data obtained at different stations within each ocean, for, as in the case of the observations collected by the " Atlantis," the water underlying a given station at various depths has quite different origins. It is apparent that a definite correlation exists between the quantity of nitrate and phosphate occurring in any sample. The proportions between these constituents, as indicated by the lines drawn through the points, are essentially the same in the three oceans visited by the " Dana," and in the Atlantic stations occupied by the " Atlantis ".

The stations occupied by the " Dana " and the " Atlantis " were all in temperate or tropical latitudes. In Fig. 5 are plotted data selected at random from the stations in Barents Sea, 70°-76° N, reported by Kreps and Verjbinskaya (1932) and from stations in the south Atlantic Ocean between 55° and 62° S. published by Ruud (1930). Here the proportions between phosphate and nitrate nitrogen vary in the same ratio as in the more temperate waters found at low latitudes, though the quantities of these constituents are much lower in the north than in the south.

These results indicate that the relations observed in the western Atlantic may be of general application, at least so far as the oceanic waters are concerned. By inference they

FIG. 5.

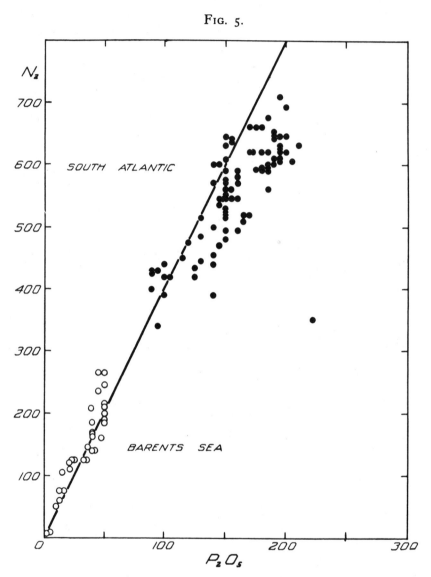

Correlation between concentration of nitrate and phosphate in waters of Barents Sea (data of Kreps and Verjbinskaya) and south Atlantic (data of Ruud, from stations between 55°-62° S., 0°-40° W.). Ordinate, concentration of nitrate nitrogen, units milligrams of nitrogen per cubic meter, abscissa, concentration of phosphate, units milligrams P_2O_5 per cubic meter. The line corresponds to a ratio of $\triangle$ N : $\triangle$ P= 20: 1 milligram atoms.

suggest a remarkable uniformity in the chemical composition
of the planktonic communities occupying the various oceans.
It is scarcely to be expected that these relations will hold in
water bodies of limited extent, where local conditions such
as the proximity of land, the inflow of rivers, and the
peculiarities of a local flora and fauna may alter the picture.
Examination of data for such regions as the Baltic (Buch, 1932),
the Norwegian fjords (Braarud and Klem, 1931), the Denmark
strait (Bohnecke Føyn, and Wattenberg, 1932), indicate that
this is the case.   In the English Channel the ratio in the
variation in nitrate and phosphate in the surface water with
season was about 2 mg. nitrate nitrogen to 1 mg. P_2O_5
($\triangle N : \triangle P = 10 : 1$ mg. atoms) according to data published
by Harvey (1928).   Cooper (1933) has shown that in this
region where active regeneration takes place, variation in the
ratio of nitrate to phosphate is attributable to the longer time
required to complete the regeneration of nitrate from
decomposing albuminous matter.   In the Gulf of Maine data
recently secured by Dr. Rakestraw indicate that the ratio of
the variation of nitrate and phosphate with depth is lower
and more variable than in the open ocean outside the Gulf.
However, the value of a generalization such as that suggested
by the data from the open oceans lies in the fact that it makes
clearer the exact nature of the anomalous conditions found
in local situations and thus defines more precisely the questions
which must be answered before the local situation is understood.   This point can be illustrated by a consideration of
the anomaly in the concentration of oxygen which is present
in the data secured by the " Atlantis ".

Below 1,000 meters there is less oxygen removed from the
water than is to be expected from the amount of nitrate
present.   This expectation, however, is based on the
assumption that the nitrate has all appeared as the result of
decomposition of organic matter after the water was removed
from contact with the atmosphere.   The observations suggest
that the deeper water of the Atlantic contained a considerable
amount of nitrate at the time it sank.   Actually one must
assume most of the samples from the greater depths to have

contained about 140 mg. nitrate nitrogen per cubic meter (equivalent to 10×10^{-3} mg. atoms per liter) at the time they became separated from the atmosphere. Now the water at these depths is thought to originate at high latitudes where nutrient substances are commonly observed at high concentrations in the surface water. Thus Kreps and Verjbinskaya (1932) observed nitrate nitrogen concentrations in the surface water of Barents Sea of 128 to 200 mg. per cubic meter in March and April, and Bohnecke, Føyn and Wattenberg (1932) record values of 125 mg. per cubic meter in the mixed water of the polar front east of Greenland. The values of nitrate nitrogen recorded by Ruud (1930) in the surface waters of the Antarctic exceed 600 mg. per cubic meter in many cases. Thus the history of the deep water suggested in explanation of the anomalous character of the ratios in which oxygen enters is confirmed by independent observations.

One fact of great general interest emerges from an examination of the data recorded in Figs. 1 and 4. It may be noted that as the quantities of nitrate and phosphate diminish, as they do as one approaches the surface layers of the sea, they diminish simultaneously in such proportions that no great excess of one element is left when the supply of the other has been exhausted. In 1926 Harvey wrote: "It is a remarkable fact that plant growth should be able to strip sea water of both nitrates and phosphates, and that in the English Channel the store of these nutrient salts, formed during autumn and winter, should be used up at about the same time." The relation noted by Harvey is thus of much wider application than he could have known at that time. It appears to mean that the relative quantities of nitrate and phosphate occurring in the oceans of the world are just those which are required for the composition of the animals and plants which live in the sea. That two compounds of such great importance in the synthesis of living matter are so exactly balanced in the marine environment is a unique fact and one which calls for some explanation, if it is not to be regarded as a mere coincidence. It is as though the seas had been created and populated with animals and plants and all of the nitrate and phosphate which the water contains had been derived from the decomposition of this original population.

Professor Huntsman has suggested to me that the correspondence between the proportions of phosphate and nitrate in the seas and the statistical composition of living matter may be due to the fact that, although different members of the plankton community have different requirements in regard to phosphorus and nitrogen, the numbers of each of these members may depend on the relative availability of the substances they most need. Thus a population requiring relatively much nitrogen may thrive until the nitrate supply has been depleted, when it will be replaced by a population requiring relatively more phosphorus. By the balancing of such communities the ratio of the elements in the plankton as a whole might come to reflect the ratio of the nutrient substances in sea water rather closely.

Another explanation of the correspondence between the proportions of nitrate and phosphate in the sea and the composition of living matter may be sought in the activities of those bacteria which form nitrogenous compounds from atmospheric nitrogen or liberate. nitrogen in the course of the decomposition of organic matter. In the case of the nitrogen-fixing bacterium, *Azotobacter*, it is known that for every unit of nitrogen fixed or assimilated and synthesized into microbial protein, about one half a unit of P_2O_5 must be available. The physiological activity of such organisms must tend to bring the relative proportions of organic nitrogen and phosphorus toward the ratio in which these substances occur in the bacterial protoplasm. Comparable studies upon the physiology of the denitrifying bacteria do not appear to have been made. It is evident, however, that the composition of these organisms must be more or less fixed in regard to their relative phosphorus and nitrogen content and that when living in an environment containing an excess of nitrate, considered in relation to phosphate, the growth and assimilation of the organisms may continually tend to bring the proportions of nitrogen and phosphorus nearer to that characteristic of their own substances. It would appear inevitable that in a world populated by organisms of these two types the relative proportion of phosphate and nitrate must tend to approach that characteristic of protoplasm in general and that, given time enough and freedom from systematic disturbing influences, a relationship between phosphate and nitrate such as that observed to occur in the sea must inevitably have arisen. On this view the quantity of nitrate in the sea may be regulated by biological agencies and its absolute value determined by the quantity of phosphate present.

These explanations are in no wise mutually exclusive; nor do they exclude other possibilities. Whatever its explanation, the correspondence between the quantities of biologically available nitrogen and phosphorus in the sea and the proportions in which they are utilized by the plankton is a phenomenon of the greatest interest.

LITERATURE CITED

BOHNECKE, G., FØYN, B., and WATTENBERG, H. 1932. Beiträge zur Ozeanographie des Oberflächenwassers in der Dänemarkstrasse und Irminger See. Teil 2. *Ann. Hydrogr. Berl.*, Bd. 60, S. 314.

BRAARUD, T., and FØYN, B. 1930. Beiträge zur Kenntnis des Stoffwechsels im Meere. *Avhandl. Norske Vidensk.-Akademi Mat. Nat. Klasse*, 1930, No. 14, pp. 1-24. Oslo.

―――― and KLEM, A. 1931. Hydrographical and Chemical Investigations in the Coastal Waters off Møre and in the Romsdalsfjord. *Hvalrådets Skrifter Norske Vidensk.-Akademi*, No. 1, pp. 1-88. Oslo.

BUCH, K. 1932. Untersuchungen über gelöste Phosphate und Stickstoffverbindungen in den Nordbaltischen Meeresgebieten. *Merentutkimuslaitoksen Julkaisu Havsforskningsinst. Skrift*, No. 86, pp. 1-30. Helsingfors.

――――, HARVEY, H. W., WATTENBERG, H., and GRIPENBERG, S. 1932. Uber das Kohlensauresystem im Meerwasser. *Rapp. Cons. Explor. Mer.*, Vol. 79, pp. 1-70. Copenhague.

COOPER, L. H. N. 1933. Chemical Constituents of Biological Importance in the English Channel, Nov. 1930-Jan. 1932. *J. Mar. Biol. Ass.*, n.s., vol. 18, pp. 617-628. Plymouth.

HARVEY, H. W. 1926. Nitrate in the Sea. *Ibid.*, vol. 14, p. 71.

——— 1928. Nitrate in the Sea. II. *Ibid.*, vol. 15, p. 183.

KREPS, E., and VERJBINSKAYA, N. 1932. The Consumption of Nutrient Salts in Barents Sea. *J. Cons. int. Explor. Mer*, vol. 7, pp. 25-46. Copenhague.

RUUD, J. T. 1930. Nitrates and Phosphates in the Southern Seas. *Ibid.*, vol. 5, pp. 347-360.

THOMSEN, H. 1931. Nitrate and Phosphate Contents of Mediterranean Water. Report on the Danish Oceanographical Expeditions 1908-1910 to the Mediterranean and Adjacent Seas. No. 10, vol. 3, pp. 1-11.

II
Control Mechanisms

Editor's Comments on Papers 6 Through 13

In the 1940s two innovations in technique appeared which had far-reaching results for the study of essential elements. One was the rather sudden availability of both radioactive and heavy isotopes of elements for use in tracer experiments. The other was the application of the culture methods of bacteriology and the rapidly expanding knowledge of biochemistry to the study of the nutritional requirements of phytoplankton.

The importance of tracer experiments to biology in general can hardly be overstated. Before the availability of isotopic tracers, the movement of materials through cells and organisms was largely inferred from experiments based on chemical analysis of the concentrations of the materials in the cells or organisms. In living systems there is a dynamic equilibrium of elements and compounds. The amount of any given element may be about the same today as it was yesterday, but the molecules that were present yesterday are in fact gone and have been replaced by an almost identical number of newly arrived molecules. If we wish to measure the rate of replacement (turnover rate) of a particular group of molecules, it is necessary to identify that particular group, apart from all other molecules of the same kind. This can be done by adding to the group some radioactive molecules of the same kind. Then, as those molecules are replaced, the radioactivity of that group, or pool, decreases, and other pools become radioactive, showing the direction as well as the rate of movement of those molecules.

Studies of the cycle of phosphorus in ecosystems quickly turned to the use of ^{32}P when it became available. Small quantities had been prepared in cyclotrons, but by the early 1950s the reactors at Oak Ridge were producing essentially unlimited quantities of it. This made possible not only laboratory experiments with microcurie quantities but field experiments with a substantial fraction of a curie to label, for example, the water in an entire small lake. The first experiment of this kind was done by Hutchinson and Bowen of Yale University, using for their experiment Linsley Pond, which was already well known as a study site for Hutchinson and his associates. The labeling of Linsley Pond had two effects on ecology. Its general effect was to set in motion a series of experiments by other investigators on whole lakes. These were followed by experiments on whole watersheds and whole landscapes. Investigators came to realize that they could treat whole ecosystems as functioning units, just as the physiolgist deals with whole organisms, and that they could do experiments at the ecosystem level. This attitude existed already in biological oceanography, where the massive size of the system made detailed approaches less appealing, but Hutchinson and Bowen brought the whole-system experiment out of the sea and pointed it toward the land.

The Linsley Pond experiments also showed that the old paradigm of seasonal cycles of phosphorus flux no longer fitted the facts. The fatal blow to that paradigm came with Rigler's tracer study of phosphate in lake waters, which revealed phosphate turnover times on the order of 1 minute in some cases. Rigler's paper gave insights into the evolution of a research project which investigators often conceal. The rapidity of turnover of phosphate in lakes had not been anitcipated fully, and the original experimental design could not cope with turnover times of minutes. Here we see the investigator altering both his paradigm and his experimental design as he goes along. Worthwhile discoveries are often made in this way, but the investigator either glosses over the moments of uncertainty in his publication or he refuses to abandon his original paradigm and leaves the opportunity for discovery to someone else.

While Rigler's clear Canadian lakes proved to have exceptionally rapid recycling of phosphate, Pomeroy showed that phosphate had a turnover time of hours, or at most days, in natural waters generally. It was no longer possible to explain the cycle of phosphorus as a purely seasonal regime of uptake and decay. Phosphorus was being recycled through the community rapidly at all times, and the seasonal changes in phosphate concentration in natural waters looked like a shifting equilibrium of a dynamic process. Recycling mechanisms within the community now had to be identified and quantified. Of course, it had long been postulated that recycling was done by bacterial consumption of organic matter and the release of chemical elements in available inorganic forms, such as PO_4^{3-}, HPO_4^{2-}, NH_3, NO_2^- and NO_3^-. In fact, there had not been a direct demonstration of these processes in nature. The association of bacteria with recycling of phosphorus was based on circumstantial evidence.

In 1937 Gardiner of the Plymouth Laboratory published an account of the production of rather large amounts of phosphate by zooplankton which had been caught in a net and held for several hours in water. At that time there were no metering plankton nets, so Gardiner could not estimate what the rate of phosphate production by zooplankton would be in nature. Pomeroy and Bush, and later Pomeroy, Mathews, and Min (1963), combined population density estimates with measurements of the rate of excretion of phosphate and total phosphorus by both planktonic and benthic populations. It became apparent that the smaller organisms, such as zooplankton, were important recyclers of phosphorus, more important than larger organisms and possibly more important than bacteria. Some of the phosphate produced by heterotrophic organisms is an excretion product, and some of it probably originates from fecal material that was never assimilated. Bacteria sometimes may produce phosphate by incomplete assimilation of substrates, but they do not excrete phosphate characteristically. Instead they store it as polyphosphate (volutin), and that phosphorus is recycled only under anaerobic conditions, when all available high-energy bonds are mobilized, or at the death of the bacteria. Johannes carried this paradigm an important step further, showing that there was indeed an inverse relationship between body size and the rate of phosphate excretion per unit of body weight. This meant that the smallest organisms, except for bacteria, should be the most important recyclers of phosphate.

The artificial production of ^{32}P started a second avalanche of research on the cycle of phosphorus as massive as the first one that was set off by Denigès (1921). Nitrogen is as important and probably as often present in limiting concentrations, but there is no radioactive isotope of nitrogen. Therefore, both analytical and tracer work with nitrogen is more difficult, and there is less of it. Moreover, the cycle of nitrogen, like that of carbon, involves gases (N_2) and volatile compounds (NH_3), adding to the difficulty and complexity of investigating it. Until recently the study of the cycle of nitrogen did not receive the emphasis it deserved. The cycle of nitrogen appears to be similar to that of phosphate in some respects. Ammonia is an abundant excretory product and it is recycled rapidly. Bacteria do not store excess nitrogen, so their role in the recycling of nitrogen may not contrast with that of protozoa. However, specialized bacteria will denitrify ammonia to biologically unavailable N_2 in anaerobic environments, and other specialized bacteria will oxidize ammonia to NO_3^- and NO_2^- in aerobic environments. Both nitrogen and phosphorus pass through a community many times in the course of a single season (Stumm, 1972).

Comparable recycling processes probably occur in soils. Waksman (1916) did some early, suggestive experiments with soil protozoans which seem to be confirmed by more recent work (a more detailed account is given in Pomeroy, 1970).

The other major advance in the 1940s was the application of bacteriological and biochemical methods to the study of phytoplankton in pure culture. The discovery of antibiotics made it much easier to start and maintain axenic (bacteria-free) cultures of algae. With axenic cultures it was possible to determine the nutritional requirements of a number of species of algae by the screening methods of bacteriology. To do this many cultures were started simultaneously from a single algal clone, using variations in the culture media. The differential responses showed both absolute and relative growth requirements. The leaders in this research were Seymour Hutner and Luigi Provasoli of the Haskins Laboratories in New York City. Hutner and Provasoli showed that the requirements of algae are many and varied, including not only the major essential elements but trace metals, vitamins, and sometimes other specific compounds such as amino acids. They found that some trace metals are more readily available and sometimes only available when they are kept in solution by organic chelating compounds. In culture, ethylenediaminetetraacetic acid and its salts were used as chelators. In nature, humates, some amino acids, and other compounds are naturally occurring chelators.

The work with axenic batch cultures in the 1940s and 1950s helped us understand the principles of uptake of essential elements by phytoplankton. Limits on our understanding of algae in nature were set by the conditions of growth in batch culture. A culture in a flask is a transitory thing, perhaps like an algal bloom in some respects, but certainly not like the normal quasi-steady-state populations we ordinarily find in nature. To understand the normal problems of algal cells in the usual low concentrations of nutrients, a different approach was necessary. Continuous culture methods provided the means to approximate more closely natural conditions for planktonic algae. In batch cultures it had been necessary to keep the concentrations of essential elements higher than one finds in nature by orders of magnitude. As a result batch cultures would bloom to high densities and then die as the nutrients were depleted. In continuous culture a new supply of essential elements could be provided at a rate that just balanced growth. The cells were harvested continuously at a rate just balancing growth, so that populations in steady states were achieved. Realistic laboratory studies could now be performed on the uptake and utilization of essential elements at the concentration found in nature. In nature the continuous supply comes from the recycling mechanisms. In a monospecies culture recycling cannot occur because the heterotrophic organisms that are involved in recycling are not present.

Most of the continuous culture technique, including the methods of graphic expression and the calculation of results, were borrowed whole from the biochemical literature. The results are sometimes confusing. Most of the early continuous culture work (as well as some later batch culture work) was based on the assumption that the uptake of a substrate follows a hyperbolic relation to its concentration (Monod, 1942). The kinetics and graphical analysis describing this were developed by Michaelis and Menten (1913) to describe an enzyme–substrate reaction at specific sites on a cell or organelle membrane. Analytical methods that were developed to deal with clean, defined chemical systems are now used to deal with dirty, undefined natural systems, as well as cultures or organisms that fall somewhere between. Even for a single species in axenic culture the situation proves to be somewhat more complex, as Caperon and Meyer have shown. Their results have been published so recently that it is difficult to say what the impact will be. The study may do a great deal to make experimental work with cultures applicable to understanding the dynamics of mixed populations in real-world communities.

Ecology is thus starting to come of age as a science. Hitherto, it has been constructed of all sorts of bits and pieces from other, better-established disciplines. The interdisciplinary nature of ecology gave it initial strength, but now it must break away from cookbook applications of the methods and the mathematics of other sciences and develop its own. In the case of the experimental culture of planktonic organisms we see recognition and acceptance of the more complex uptake kinetics of cells and populations compared with the single enzyme systems and metabolic pathways of biochemistry. As Caperon and Meyer have shown, we must also take into account the fact that today's growth kinetics are the response of the population to its environmental experience yesterday, which may not have been the same as today's.

6

Reprinted from *Ecology*, **31**(2), 194–203 (1950)

LIMNOLOGICAL STUDIES IN CONNECTICUT. IX.
A QUANTITATIVE RADIOCHEMICAL STUDY
OF THE PHOSPHORUS CYCLE IN
LINSLEY POND

G. Evelyn Hutchinson and Vaughan T. Bowen

*Osborn Zoological Laboratory, Yale University and American Museum of
Natural History, New York City*

Evidence has been presented (Hutchinson, '41) that in small stratified lakes phosphorus is continually passing from the mud to the water and at the same time is being removed from the water by incorporation into the bodies of plankton organisms which sediment when they are turned to faecal pellets or otherwise die. The persistent low concentrations of phosphorus, observed in the surface waters of eutrophic lakes during the summer appear to represent fluctuations around a steady state concentration presumably determined by the rate of liberation of phosphorus from the mud and the rate of sedimentation of sestonic phosphorus. Such a dynamic view of the phosphorus cycle, even at times of extreme thermal stratification, can also be derived from the experimental study of Einsele ('41) published almost simultaneously with that of Hutchinson, but only recently available in this country. Einsele added massive quantities of phosphate to the surface waters of the Schleinsee and found a rapid uptake and rapid sedimentation of this phosphate, so that the lake ultimately returned to its original condition.

Hutchinson and Bowen ('47) demonstrated the sedimentation of phosphorus from the epilimnion into the hypolimnion of Linsley Pond by the use of radiophosphorus. They added approximately 10 mc. of P^{32} in the surface waters of Linsley Pond on June 21, 1946; a week later 47% of the radiophosphorus present in the lake had descended below the 3 m. level and 10% below the 6 m. level, in spite of very stable thermal stratification below three meters throughout the period of the experiment. The present paper describes a second experiment on the same lake, in which the use of larger amounts of radiophosphorus and more sensitive counting equipment have permitted quantitative estimates of the rate of movement of phosphorus in the lake, and have also brought to light certain additional and unexpected features of the cycle of the element.

Description of the Experiment

The experiment was started on the morning of July 25, 1947. During the taking of the temperatures and water samples in the deepest part of the lake the wind was observed to be variable in direction. It was therefore decided to introduce the radiophosphorus in twenty-five portions while rowing in an approximately circular course between the central deep part of the lake and the margin; this arrangement is believed to have provided adequate opportunity for the mixing of the radiophosphorus with the superficial layer of the lake. The radiophosphorus was added as potassium phosphate, one-fifth of an irradiation unit of initial activity 350 mc., from the Clinton Laboratories, Oak Ridge. The quantity of radiophosphorus actually introduced corresponds, on the arbitrary scale employed in presenting the results and explained below, to 1350×10^6 counts per minute.

Temperature curves for the entire period of the experiment are given in figure 1. On August 1 the mercury column of the only available thermometer was found to be sticking badly, so that no reliable readings are available for this date. The only important temperature change observed during the entire period of the experiment was a marked superficial heating prior to and a superficial cooling subsequent to August 15. This change is

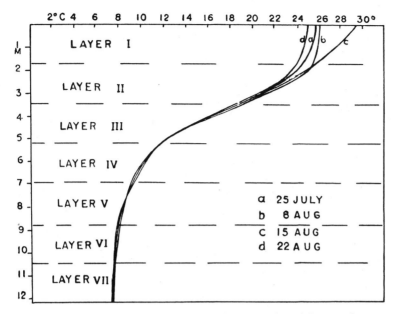

FIG. 1.   Distribution of temperatures during the period of the experiment.

probably reflected in the behavior of the phosphorus in the epilimnion in a way subsequently to be discussed.

The samples for analysis and radio-activity determinations were collected in a Nansen bottle lowered on a graduated rope. This rope had originally been marked at meter intervals and then water-proofed in melted paraffin; during the waterproofing it contracted so that it is necessary to correct the depths. This circumstance explains the choice of boundaries for the seven layers into which the lake is divided for the purpose of the experiment. Two samples were collected with the top of the bottle at the top, two at the bottom of each layer, and two at three intermediate depths in each layer. Since the bottle is 54 cm. long, there is a slight overlap between the samples from each intermediate depth and an overlap of about 54 cm. between each layer. This arrangement insures that every horizontal layer in the lake was included in the sampling. It has the slight disadvantage that in any computation based on the sum of several layers, the depth constituting the boundary is overweighted, since it contributes to the bottom of one layer and the top of the layer immediately below it. There is, however, no reason to believe that any serious error is introduced by this procedure. All the samples were taken in the central station in the deepest part of the lake, normally used in our limno-logical studies of Linsley Pond. Total phosphorus was determined in water samples taken at the time that the tempera-tures were taken (table I). Since seston

TABLE I.   *Total phosphorus* (mg. per m.³)

	July 25	Aug. 1	Aug. 8	Aug. 15	Aug. 22
0 m.	16	16	14	19	19
1.75 m.	20	21	21	21	22
3.5 m.	22	22	27	28	34
4.4 m.	—	37	—	—	—
5.3 m.	62	33	50	43	44
7.0 m.	48	53	75	88	82
8.8 m.	140	120	100	150	150
10.5 m.	60(?)	180	160	220	100(?)
12.2 m.	200	240	310	310	360

phosphorus was determined from the seston samples used for radioactivity de-terminations, it should be compared with the mean total phosphorus at the top and bottom of each layer (table II). The

TABLE II. *Total and seston phosphorus, mean concentrations in each layer: mg. per m.*[3]

Depths		Aug. 1		Aug. 8	Aug. 15		Aug. 22	
		tot.	sest.	tot.	tot.	sest.	tot.	sest.
I	0–1.75 meters	19	10	18	20	18	21	7
II	1.75–3.5 meters	22	10	24	25	16	28	13
III	3.5–5.3 meters	31	8	39	36	19	39	25
IV	5.3–7.0 meters	46	15	63	66	46	63	21
V	7.0–8.8 meters	87	36	88	119	78	116	41
VI	8.8–10.5 meters	150	50	130	190	67	125(?)	83
VII	10.5–12.2 meters	210	148	240	270	23	230(?)	117

seston phosphorus samples for August 8 were obviously grossly contaminated, possibly due to accidental use of a phosphate-containing cleaner on the glass-ware employed on that date, and have been rejected.

The water samples for radioactivity determinations, coming from five depths at which a Nansen bottle holding 1.25 liters was filled, consisted of approximately 12.5 liters. The samples were filtered through large No. 54 Whatman filter paper using suction from a Cenco Highvac pump. Several changes of paper were required to insure filtration within a reasonable period. The papers (*seston fraction*) were dried and ignited. The filtrates (*filtrate fraction*) were evaporated almost to dryness, oxidized with nitric and perchloric acids; phosphoric acid was then added as a carrier and the phosphorus precipitated as ammonium phosphomolybdate. The ash of each seston fraction and the phosphomolybdate precipitate of each filtrate fraction was spread out on the bottom of a small Petri dish for counting. The seston ash was later analyzed for phosphorus as indicated above.

The radioactivity determinations were made with a modified Scott type bell counter, filled with an argon-alcohol mixture at 10 cm. pressure, used in a thick lead shielding. This instrument gave a background count of 20 per minute, and permitted easy determination of the radioactivity of all the samples examined. All the radioactivity determinations are corrected for decay to correspond to August 8, the mid-point of the experiment. All have been corrected for variations in counter sensitivity to correspond to a sensitivity giving 150 counts per minute for an arbitrary uranium standard (2.53^{-3} g. Uranyl Acetate C. P.). The results given as counts per minute per cubic meter are set out in table III.

DISCUSSION OF THE RESULTS

Total recovery of radiophosphorus: It is immediately apparent from the table that significant amounts of radiophosphorus were recovered in all layers on all dates. The only way in which the radiophosphorus can be carried rapidly into the very stable hypolimnion of the lake is by the sedimentation of seston. The quantitative aspect of this process is of considerable interest. If the only movement occurring were vertical, the total quantity of radiophosphorus in the water under unit area of the deepest part of the lake, assuming a reasonably uniform original horizontal distribution, could not possibly be greater than the original introduction per unit area. Since the amount introduced corresponds to 1350 × 10^6 counts per minute, and the area of the lake is 94,400 m.^{2}, the upper limiting amount per square meter should be 14,300 counts per minute. The amount to be expected in a water column from 0–12.2 meters under unit area might be less than this, for the deepest layer studied is not quite the deepest layer of the lake and some radiophosphorus could have reached the bottom and could have been incorporated in the sediments. The actual values obtained by

TABLE III. *Quantities of radiophosphorus in Linsley Pond, by layers*

		August 1		August 8		August 15		August 22	
		Seston	Filtrate	Seston	Filtrate	Seston	Filtrate	Seston	Filtrate
I 0–1.75 meters	Counts per m.³ per min.	850	1600	770	1560	730	1250	750	2000
	Entire layer Millions counts per min.	128	241	116	235	110	188	113	301
II 1.75–3.5 meters	Counts per m.³ per min.	940	1410	730	1280	1020	1260	740	810
	Entire layer Millions counts per min.	117	176	91	160	128	157	93	101
III 3.5–5.3 meters	Counts per m.³ per min.	800	1490	1880	1080	1050	1620	1070	2040
	Entire layer Millions counts per min.	98	182	229	132	128	198	131	249
IV 5.3–7.0 meters	Counts per m.³ per min.	650	820	790	416	1100	1700	1210	1610
	Entire layer Millions counts per min.	64	81	78	45	109	168	120	159
V 7.0–8.8 meters	Counts per m.³ per min.	560	700	480	650	1060	150	630	1270
	Entire layer Millions counts per min.	36	45	31	42	68	10	40	81
VI 8.8–10.5 meters	Counts per m.³ per min.	550	420	260	390	1180	140	750	850
	Entire layer Millions counts per min.	22	17	10	16	47	6	30	34
VII 10.5–12.2 meters	Counts per m.³ per min.	420	510	860	460	—	710	710	1130
	Entire layer Millions counts per min.	10	12	21	11	(19)[1]	17	17	27
Sum, entire epilimnion (I, II) Millions of counts per min.		245 (662)	417	207 (602)	395	238 (583)	345	206 (608)	402
Sum, entire hypolimnion (III–VII) Millions of counts per min.		230 (567)	337	369 (615)	246	371 (770)	399	338 (888)	550
Sum, entire lake Millions of counts per min.		1229		1217		1353		1496	
% Recovery in water		91.1		90.2		100.2		109.6	

[1] Sample lost: interpolated value used in sums.

summing from table III, and the percentage of the expected values are as follows:

Aug. 1	Aug. 8	Aug. 15	Aug. 22	
20,570	20,470	22,930	27,240	counts per min. per m.^{2}
144.3%	142.05%	160.4%	190.5%	

It will be observed that on every date the recovery, computed on the hypothesis of purely vertical movement of phosphorus, is impossibly high.

In table III the entries described as "entire layer, millions of counts per minute" refer to the product of the radioactivity per cubic meter and the volume of the layer. The sum of these products for any day (sum entire lake) gives the total amount of radiophosphorus actually in solution and suspension in the lake water. This quantity may then be compared with the amount of radioactive material introduced. It will be apparent that on the first two days about 90% of the material added is accounted for by this mode of calculation, while on the third day the radioactivity of the lake corresponds almost exactly to the added material. Only on the fourth day does an excess appear to have been present and, as is indicated below, this excess probably can be explained. These computations clearly indicate that when the sedimenting seston comes to rest on the mud at small and intermediate depths the radiophosphorus is rapidly liberated and passes out into the free water, being distributed in the same sort of way as is bicarbonate (Hutchinson, '41). Computation by entire layers is obviously the correct procedure and is employed in all further quantitative discussion.

Variation in the epilimnion: The total radiophosphorus in the top two layers is found to fall during the first three weeks of the experiment, but rises in the fourth week. The excess of radiophosphorus apparently present on August 22 is largely due to this rise, which concerns only the upper layer from 0 to 1.75 meters. The most reasonable explanation of the excess

in this layer is that radiophosphorus previously held by marginal rooted vegetation was being rapidly liberated at this time and that an irregular distribution of radiophosphorus was set up, the water at the surface at the collecting station not being truly representative. It is reasonable to suppose that such a sudden liberation of phosphorus in the littoral as is implied was due to enhanced decomposition during the period of very high temperature that the surface layers had experienced. There is evidence that great superficial rises in total phosphorus may occur at the height of summer during Cyanophycean blooms (Hutchinson, '41; Einsele, '41) and are doubtless similarly determined by high superficial temperatures.

During the course of the period of the experiment the total phosphorus content of the entire epilimnion rose from 5.34 kg. to 6.66 kg. A least squares solution gives 0.32 kg. per week as the best value for the rate of increase, or 0.26 kg. per week for the period from August 1 to August 15. As it will appear from the next section, the hypolimnion was receiving at least 1.64 kg. phosphorus per week from the epilimnion during this period. It is evident therefore that the epilimnion was receiving 1.96 kg. of phosphorus per week from the littoral. Omitting the last week when a rise took place, the rate of decrease in radiophosphorus in the epilimnion, from August 1 to August 15, is about 40.10^6 counts per minute, while the hypolimnion is at the same time gaining radiophosphorus at the rate of about $101.5.10^6$ counts per minute. This implies that there must be a source of radiophosphorus other than the free water, from which the isotope can enter the epilimnion.

Samples of *Potamogeton crispus* from the north end of the lake were collected and their radioactivity and phosphorus contents determined; the results, given as counts per minute per gram and as % P, both referred to dry weight, and the specific activities as counts per minute per mg. P, are as follows:

Aug. 1	Aug. 8	Aug. 15	Aug. 22	
19.9	15.5	9.0	11.3	Counts per min. per gm.
0.216	—	0.285	0.190	% P.
9.2	—	3.2	5.9	Activity of P. Counts per min. per mg.

While the samples were small and may not have been representative of the lake as a whole, there is a clear indication of loss of radiophosphorus when the first two samples are compared with each other and with the third. In view of the slight increase on the last day, this decline may not be very significant.

At least between August 1 and August 15 radiophosphorus corresponding to 123 × 10⁶ counts per minute must have been liberated from the littoral plants or other marginal reservoirs to counterbalance the loss to the hypolimnion. If this radiophosphorus came from the plants, in which the specific activity is about 6.2 counts per minute per mg., it corresponds to 10 kg. per week. Since the known increment was only 1.96 kg. per week it is evident that either the plants contain mobile phosphorus of high and immobile phosphorus of low activity or that active phosphorus temporarily immobilized in the littoral sediments is involved. It is quite likely that the main littoral reservoir is in epiphytic diatoms and other algae with relatively short life-spans. If the amount liberated by the littoral be added to the total activity for the entire lake on August 1, it will become 1352 × 10⁶ counts per minute, or essentially the original sample put into the lake. Actually the analyses just given indicate that only about half the littoral supply was exhausted by August 15 or even by August 22, so that it is certain that any correction made to the August 1 figure, for the content of the marginal vegetation, will inevitably raise the total in the lake to an amount at least 10% in excess of the quantity introduced. There is evidently some error involved here, which is perhaps inherent in the morphometric data derived from Riley's ('39) map of the lake, or perhaps referable to some systematic instrumental error. In view of the fact that the experiment was not conducted under controlled laboratory conditions, but in nature, it is remarkable that such close agreement between the radioactive material introduced and that recovered can be obtained.

Variation in the hypolimnion as a whole: The total radiophosphorus content of the hypolimnetic layers III–VII is plotted against time in figure 2. It will be observed that the variation indicated by this plot possesses features of considerable interest. There is an initial rise of about 540 × 10⁶ counts per minute in the first week and then a slower and approximately linear rise throughout the second, third and fourth weeks of the experiment. The solid line fitted by least squares to the four points August 1, August 8, August 15, August 22 has a slope corresponding to 111.8 × 10⁶ counts per minute per week, so that the rate of increase during the first week was almost five times that in the subsequent weeks.

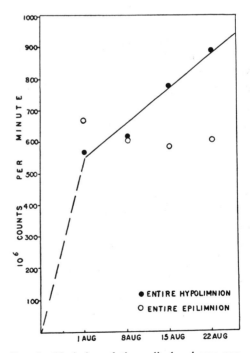

Fig. 2. Variation of the radiophosphorus contents of the epilimnion and hypolimnion.

The only reasonable explanation of these results is that initially the phytoplankton of the epilimnion took up the whole of the radiophosphorus, which, being introduced as phosphate, would be readily assimilable. At this time the sedimenting seston would presumably be much richer in P^{32} than at any later date. Later the phytoplankton may be supposed to have lost some of the phosphorus in a less assimilable, probably organic, form. Throughout the greater part of the experiment the distribution of radiophosphorus in the seston and in the free water may be regarded as representing an equilibrium between the seston, the less assimilable, and the ionic phosphate radiophosphorus. It is, therefore, obvious that only the radiochemical data of the period from August 1 to August 22 can be used in determining the rate of entry of phosphorus into the hypolimnion. The mean sestonic phosphorus content of the entire epilimnion during this period is 3.376 kg. while the mean P^{32} content of the same layer for the same date is 230×10^6 counts per minute. The rate of increase of radiophosphorus in the hypolimnion, namely 111.8×10^6 counts per minute per week, therefore, corresponds to a weekly passage of 1.64 kg. of sestonic phosphorus across the 3.5 meter plane in the middle of the thermocline. A similar computation, using only the data for the period August 1–August 15, gives a rate of increase of P^{32} of 101.5×10^6 counts per mm. per week corresponding to a weekly increment of 1.55 kg. phosphorus.

The total phosphorus of the hypolimnion increased throughout the period of the experiment until the last week. Unfortunately, two of the determinations at 10.5 meters given in table I are suspiciously low and almost certainly merely indicate incomplete reduction to the blue ceruleomolybdate used in the colorimetric determination of phosphorus. If, as seems probable, the 10.5 meter determination on August 22 is erroneous and if it

be replaced by an interpolated figure, the hypolimnetic phosphorus rises, though less rapidly than before, throughout the last week of the experimental period.

If the period from August 1 to August 15 alone be considered, the total phosphorus of the hypolimnion rises at the rate of 3.75 kg. per week. During the same period it has just been shown that the rate of increase of radiophosphorus corresponds to a movement of 1.55 kg. of phosphorus per week from the epilimnion into the hypolimnion. It would appear, therefore, that during this period rather under half the increment in the hypolimnion is derived from the epilimnion and rather over half from some other source, that can only be mud in contact with the hypolimnetic water. If the corrected figure for the August 22 total phosphorus be acceptable, the least squares solution for the mean rate of increase of total hypolimnetic phosphorus is 3.1 kg. per week, while the rate of movement from the epilimnion is 1.64 kg. per week.

The dual origin of the phosphorus of the hypolimnion is in accordance with what is generally known about the phosphorus cycle in small lakes (Einsele, '36, '38; Hutchinson, '41). It is also apparent from an examination of the events in some of the individual layers, which provide evidence that the output of phosphorus from the sediments does not proceed uniformly at all hypolimnetic depths nor equally at all times during the period of the experiment.

The radiophosphorus content of layer III (3.5–5.3 m.): The radiophosphorus in this layer may be presumed to have entered the layer solely as sedimenting seston. The variability of the specific activity of the epilimnetic sestonic phosphorus is great enough to suggest rapid variations so that a mean value is probably best used in computing the quantity of phosphorus entering the hypolimnion. Using such a figure the amount of phosphorus in layer III that originated from falling seston can

	Aug. 1	Aug. 8	Aug. 15	Aug. 22
Computed	33.7 mg. per m.³	43.7	39.2	45.8
Observed	31 mg. per m.³	39	36	39

be computed and compared with the total phosphorus in the layer.

It will be observed that although the computed values are about 10% too high they vary with the observed values. The excess of the computed over the observed values suggests that the mean conversion factor used above may be a little too high. The general parallel variation of the observed and computed values, however, strongly supports the hypothesis that the whole of the phosphorus present at any time in this layer has been derived from seston falling from the layer immediately above it and that no phosphorus enters layer III from preexisting sediments in contact with the layer. On all occasions there is evidently a rapid liberation of soluble radiophosphorus from the falling seston. The radiophosphorus content of the sestonic phosphorus in the layer appears to fall progressively. Initially it is exactly 100 counts per minute per mg., falling to 55.3 counts per minute on August 15 and to 43 counts per minute on August 22. The mean value is 57.2 counts per minute per mg., which is little less than the mean epilimnetic value of 68.3 counts per minute per mg.

The radiophosphorus content of the lower layers of the hypolimnion: Throughout most of the hypolimnion there is an uncertainty as to how much of the seston phosphorus separated on the filter was really in suspension in the lake and how much was precipitated as ferric phosphate during filtration. This circumstance makes detailed quantitative treatment of the individual layers impossible, but a good deal can be learned from the specific activities of the total phosphorus at different depths on the various days on which samples were taken.

Reference to the total phosphorus contents of table II and the specific activities of table IV indicates quite clearly that the main rise in the total phosphorus of layer IV during the week August 1 to August 8 cannot have been due primarily to sedimenting plankton. The specific activity falls during a period of marked increase in total phosphorus content and it has already been indicated that during the period August 1 to August 15 the specific activity of the seston of layer III is considerably greater than that of layer IV either on August 1 or August 8. During the week of August 15 to August 22 the total phosphorus concentration remained practically unchanged in layer IV, the specific activity rose markedly, indicating a considerable replacement of the less radioactive phosphorus derived from the sediments by more radioactive phosphorus derived from falling seston.

The rise in total phosphorus in layer V during the week August 8 to August 15, though accompanied by a very slight rise in total radioactivity, leads to an overall decrease in specific activity, again indicating entry of phosphorus from the mud. As in layer IV, there is evidence of a

TABLE IV. *Specific activities of total and sestonic phosphorus counts per minute per mg.*

		Aug. 1		Aug. 8	Aug. 15		Aug. 22	
		tot.	sest.	tot.	tot.	sest.	tot.	sest.
I	0–1.75 meters	129	85	122	94	40.5	131	107
II	1.75–3.5 meters	107	94	86	91	42.5	55.5	57
III	3.5–5.3 meters	74	100	76	74	55.1	80	42.8
IV	5.3–7.0 meters	32	47.6	19.8	42.4	24.0	45.4	57.6
V	7.0–8.8 meters	14.6	15.1	12.8	10.2	13.6	16.4	15.4
VI	8.8–10.5 meters	6.7	11.0	5.0	12.2	17.6	—	9.0
VII	10.5–12.2 meters	4.4	2.8	5.5	—	—	—	6.1

replacement of some of this phosphorus by more radioactive phosphorus in falling seston during the last week of the experiment. While the events in the lowest two layers do not lend themselves to analysis, it is evident that in layers IV and V, but not in layer III, the dual origin of the phosphorus present is well established. It is probable that the mud surface in contact with layer III was still oxidized at the time of the experiment and that any phosphate that tended to leave the mud was precipitated in the oxidized microzone as $FePO_4$. The process of regeneration of phosphorus from falling seston at the mud surface is evidently a different process from that responsible for the rise of non-radioactive phosphorus in the deeper hypolimnetic layers.

Our best thanks are due to Miss Anne Carbone for much assistance in the determination of the radioactivity of our samples.

SUMMARY

1. The movement of radiophosphorus added at the surface of Linsley Pond was studied over a four week period in summer.

2. Estimates of the total activities, corrected for decay, indicate that throughout this period phosphorus is sedimented from the epilimnion into the hypolimnion, but that such phosphorus as reaches the sediments is rather rapidly regenerated, passing again into the free water.

3. Phosphorus passes into the littoral vegetation and from such vegetation into the free water again. The specific activity of the phosphorus in material of *Potamogeton crispus* from the shallow water of the lake was always much less than that of the total epilimnetic phosphorus. It appears that an immobile fraction of low activity and a mobile fraction of high activity are present in the littoral zone. The mobile active fraction may be in part in epiphytic diatoms and similar short-lived organisms.

4. In the first week of the experiment the increment in activity in the hypolimnion below the 3.5 meter plane was five times as great as during the three subsequent weeks, during which it remained essentially constant. It is believed that immediately on addition practically the whole of the radiophosphorus enters the phytoplankton, which can easily be sedimented. During the first week phosphorus was probably lost from the phytoplankton in a form less easily assimilated than ionic phosphate. A steady state between ionic, less easily assimilable, and seston phosphorus is presumably set up.

5. There is evidence of movement of phosphorus into the hypolimnion both from the epilimnion and from the mud, but the latter process (other than regeneration at the mud surface) does not occur above 5.3 meters.

6. During the period August 1 to August 15, for which the best data are available, the following values for the rate of movement of phosphate in the lake were obtained.

Observed rate of increase in epilimnion	0.26 kg. per week
Loss to hypolimnion	1.55 kg. per week
Total movement from littoral into epilimnion	1.81 kg. per week
Observed rate of increase in hypolimnion	3.75 kg. per week
Gain from epilimnion	1.55 kg. per week
Gain from hypolimnetic sediments	2.20 kg. per week

Since the mean phosphorus content of the epilimnion is 5.82 kg. the observed rate of entry in this layer corresponds to replacement every three weeks. Since some phosphorus regeneration certainly occurs in the epilimnion the rate of plankton replacement must be somewhat greater.

REFERENCES

Einsele, W. 1936. Über die Beziehungen des Eisenkreislaufs zum Phosphatkreislauf im eutrophen See. Arch. Hydrobiol., 29: 664–686.

——. 1938. Über chemische und kolloidchemische Vorgänge in Eisen-Phosphat-Systemen unter limnochemischen und limnogeologischen Gesichtspunkten. Arch. Hydrobiol., **33**: 361–387.

———. 1941. Die Umsetzung von zugeführtem, anorganischem Phosphat im eutrophen See und ihre Rückwirkung auf seinen Gesamthaushalt. Zeit. Fischerei u. d. Hilfswissens., **39**: 407–488.

Hutchinson, G. E. 1941. Limnological studies in Connecticut. IV. The mechanisms of intermediary metabolism in stratified lakes. Ecol. Monogr., **11**: 21–60.

Hutchinson, G. E., and V. T. Bowen. 1947. A direct demonstration of the phosphorus cycle in a small lake. Proc. Nat. Acad. Sci., **33**: 148–153.

Riley, G. A. 1939. Limnological studies in Connecticut. I. General limnological survey. Ecol. Monogr., **9**: 53–66.

7

Reprinted from *Ecology*, **37**(3), 550–562 (1956)

A TRACER STUDY OF THE PHOSPHORUS CYCLE IN LAKE WATER[1]

F. H. Rigler[2]

Atomic Energy of Canada, Ltd., Chalk River, Ontario

Introduction

The phosphorus cycle in lakes has long been of interest to limnologists because available phosphorus has been thought to be a limiting factor in the production of algae and hence of fish in many lakes. Before radioactive phosphorus became available, what information there was about the circulation of phosphorus came from measurements of temporal changes or vertical differences in concentration of inorganic phosphate or organic phosphorus under natural conditions or after addition of large quantities of fertilizer containing available phosphorus. But these methods were inadequate and only with the introduction of radioactive tracer techniques did it become possible to study the rate and mechanism of phosphate circulation directly.

More definite information about the mechanism by which phosphate is removed from the trophogenic zone has now been obtained by several workers who have added radioactive phosphate

[1] A.E.C.L. No. 185.

[2] Present address: The Laboratory, Citadel Hill, Plymouth, England.

($P^{32}O_4$) to the surface of small lakes. It is known that inorganic phosphate is fairly quickly removed from solution in the epilimnion because a loss of P^{32} from solution (*i.e.* from filtered or Foerstcentrifuged water) has been demonstrated by Coffin *et al.* (1949), Hutchinson and Bowen (1950), Hayes *et al.* (1952) and Whittaker (1953). Hayes demonstrated that the loss of P^{32} from solution could be explained by postulating a turnover of inorganic phosphate with the phosphate of 'lake solids' but did not show which fraction of the solids was most active in causing this turnover. However, it appears from the work of Hutchinson and Bowen, and Whittaker that plankton is primarily responsible. Following the uptake of inorganic phosphate by plankton there is a transfer of phosphorus to the hypolimnion, presumably by sedimentation of plankton, and a subsequent release of soluble, inorganic phosphate in the hypolimnion (Hutchinson and Bowen 1950).

There is still some doubt as to whether there is a loss of P^{32} from the open water. The only figures on the amount of P^{32} remaining in the open water indicated that there was almost no loss (Hutchinson and Bowen 1950), but these workers as well as Coffin *et al.* (1949) found P^{32} in rooted aquatic plants and Hayes *et al.* (1952) found an uptake of P^{32} in bottom deposits.

Although tracer studies have shown that plankton take up P^{32} rapidly, no attempt had been made to fractionate the plankton to show which organisms are most active in this process. That planktonic algae take up and utilize inorganic phosphate for growth is beyond question. However, it is also known that when natural waters are stored, there is a large increase in the bacterial population and a loss of dissolved phosphate (Waksman *et al.* 1937; Renn 1937). Although it may be inferred from these results that bacteria might also take up phosphate under natural conditions, this possibility has received little attention, probably because the rapid multiplication of planktonic bacteria has been thought to be peculiar to stored water.

The present paper describes the results of a tracer experiment which give additional information about the circulation of phosphate in a small lake, particularly the removal of phosphorus from the open water by littoral organisms and bottom deposits. It is also shown that the turnover time of dissolved inorganic phosphate is much shorter than had been previously supposed (minutes instead of days) and that bacteria are largely responsible for causing this turnover.

The work reported in this paper was carried out under the auspices of Atomic Energy of Canada Limited. Financial assistance was also received from the National Research Council of Canada. The author wishes to acknowledge the assistance given by Dr. R. R. Langford of the University of Toronto and by the late Dr. A. J. Cipriani together with other members of the staff of A.E.C.L.

METHODS

Radioactive phosphorus (P^{32}) was used as a tracer to demonstrate the movements of naturally occurring phosphorus. Carrier-free P^{32} was obtained from Atomic Energy of Canada Limited as phosphate dissolved in .005 N hydrochloric acid. Radioactivity was measured with a shielded, end-window, Geiger-Müller tube which had a background of 15-20 c.p.m (counts per minute) and a counting efficiency of approximately 15% for P^{32} under the conditions used. Whenever possible, the time for 5,000-10,000 counts was measured. All measurements were corrected for radioactive decay and for variations in counter efficiency but corrections were not applied for self-absorption of radiations. Samples in which radioactivity was to

be determined were spread evenly over the bottom of flat aluminum trays, one and one-eighth inch in diameter with a rim one-eighth of an inch high. In order to prevent concentration around the rim and to ensure even distribution of the dry residue from liquid samples, a disc of filter paper slightly smaller than the tray diameter was placed in the tray prior to addition of the sample.

Inorganic phosphate was determined by the ammonium molybdate-stannous chloride method. Four milliliters of 2.5% ammonium molybdate in 10 N sulphuric acid and 0.5 ml of a 0.2% solution of potassium chlorate were added to duplicate, 100-ml samples. Then 0.3 ml of a 1.3% solution of stannous chloride in 0.3 N hydrochloric acid was added and the color which developed in 25 min. was measured in a Lumetron colorimeter made by the Photovolt Corporation, N. Y. A 650 mμ filter and a 150-mm light path were used. All glassware was soaked in dilute nitric acid between analyses. Total phosphorus was measured by the method of O'Reilly and Papson (Manuscript).

The method of distributing P^{32} over Toussaint Lake was essentially the same as that used by earlier workers. The sample of radioactive phosphorus (200 millicuries) was diluted with 20 liters of distilled water containing five grams of potassium dihydrogen phosphate which was added to prevent adsorption of P^{32} on the glass container and was slowly siphoned into the water from a moving boat. The addition of tracer took over three hours, during which time the lake was circled and crossed many times in order that the tracer would be distributed as evenly as possible.

Samples of water for measurement of radioactivity and for chemical determinations were taken at meter intervals with a one-liter Friedinger water bottle. Two stations were sampled (Fig. 1), one at the deepest part of the lake and the other near weed beds at the east end where the depth was five meters. Water temperatures were measured with a thermistor thermometer similar to that described by Mortimer and Moore (1953) and dissolved oxygen was measured by the unmodified Winkler method. Bottom sediments were collected with a sampler built at the Ont. Fish. Res. Lab. The sampler is essentially a transparent, plastic cylinder one inch in diameter and one foot long with a hinged valve at the top end. The valve allows water to leave the cylinder while it is being lowered into the mud and prevents water from entering the cylinder while it is being pulled out of the mud. As a precaution against loss of mud, the bottom end is closed by means of a spring-loaded cup as soon as the sampler is withdrawn from the mud. Only the samples in which

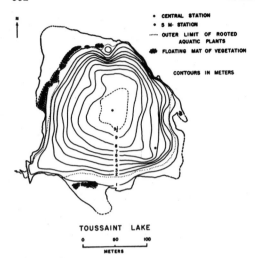

• CENTRAL STATION
• 5 M· STATION
····· OUTER LIMIT OF ROOTED
        AQUATIC PLANTS
⬤ FLOATING MAT OF VEGETATION

CONTOURS IN METERS

TOUSSAINT LAKE

0   50   100
        METERS

FIG. 1. Contour map of Toussaint Lake showing sampling stations used in tracer experiment.

the surface mud was undisturbed and the water above it perfectly clear were used for determination of radioactivity.

Radioactivity of water and plankton was measured separately. The plankton was separated from one liter of water by means of a Foerst centrifuge, plated directly on a counting tray and dried under a heat lamp. Water samples were acidified in beakers with a few drops of nitric acid, concentrated to about four milliliters and transferred to counting trays for final drying. Beakers were covered with watch glasses and, after the sample was removed, were fumed out with a few ml of dilute nitric acid which was then added to the sample. This method of concentrating the water samples gave better than 95% recovery of P^{32}.

The top quarter-inch of mud was used for measurement of P^{32}. Mud samples from deeper water were plated directly on a counting tray and dried. Mud samples from shallow water contained a large proportion of sand and were therefore extracted for one hour with boiling nitric acid (70%), diluted and filtered. The filtrate was concentrated, plated on a counting tray and dried. This method extracted 90-95% of the P^{32} and considerably reduced the self absorption of the sample.

Bacteria for laboratory experiments were isolated from plate cultures of filtered, freshly collected or stored Toussaint lake water. No attempt was made to identify the bacteria although it was observed that all species were Gram-negative rods; the different types were separated on the basis of colonial characteristics alone. Alchem 299, an agar medium produced by Alchem Ltd., Burling-

ton, Ontario for culturing freshwater bacteria, was used for isolation and measurements of growth. Standard bacteriological techniques were used for dilution of bacteria in preparation for measurement of populations. One-tenth-milliliter samples from the dilution tubes were spread evenly over the surface of agar plates with a glass spreader. Three to six plates were prepared for each dilution. Plates were incubated at room temperature for 24-72 hours.

Bacteria for the inoculum were grown in nutrient broth or sterile lake water. When nutrient broth was used, bacteria were centrifuged, washed with distilled water, recentrifuged and resuspended in distilled water. When lake water was used, bacteria were introduced directly into experimental vessels. Experiments were carried out in continuously agitated, Pyrex Erlenmeyer flasks stoppered with cotton wool. Results from contaminated flasks were discarded.

In the first experiments the amount of P^{32} in bacteria was measured by difference. Bacteria were separated from the water by centrifuging five milliliters of the culture at 11,000 r.p.m. for 15 min. Then the radioactivity of two milliliters of the supernatant was measured. In later experiments the amount of P^{32} in bacteria was measured directly. Complete separation of bacteria was accomplished by filtering the culture through Millipore bacterial filters purchased from the Lovell Chemical Company, Watertown, Mass. The filters, which were of a slightly smaller diameter than the counting trays, were dried and tested for radioactivity.

LAKE EXPERIMENT

Description of Lake

Toussaint Lake is a bog lake situated on the property of Atomic Energy of Canada Limited near Chalk River, Ontario. The surrounding terrain is Precambrian granite overlain with a thin layer of postglacial sand. The lake drains an area of approximately 200 hectares, has a surface area of 4.7 hectares and a maximum depth of 9.8 meters. It has a small permanent outlet and two temporary inlets. The water is deep brown. At each end there is a large area of rooted aquatic plants representing the following genera: *Brasenia, Chamaedaphne, Myrica, Myriophyllum, Nuphar, Nymphoides, Potamogeton, Sphagnum* and *Utricularia*.

During the summer there is a stable thermal stratification and by late July the hypolimnion becomes almost devoid of dissolved oxygen (Fig. 2). The summer stratification of inorganic and organic phosphorus (Fig. 2) is similar to that reported by

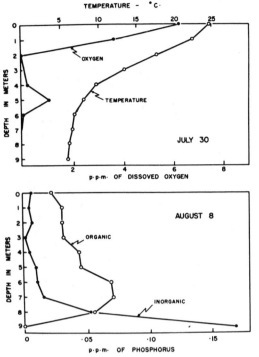

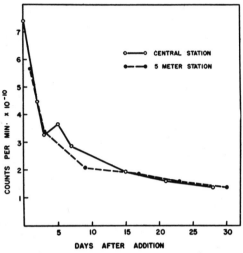

method was applicable to Toussaint Lake and therefore the total suspended P³² was calculated from radioactivity at the central station. "Suspended P³²" is used to mean all P³² in the open water, both in solution and in plankton. Results are given in Table I. There was a rapid loss of P³² from suspension during the first seven days followed by a slower loss during the next 21 days; at the end of four weeks only 23% of the P³² remained in suspension.

Fig. 2. Typical series of measurements showing variation in temperature, dissolved oxygen, soluble inorganic phosphate and organic phosphorus with depth in Toussaint Lake during tracer experiment.

Fig. 3. Total radioactivity in the upper five meters calculated from radioactivity of vertical series of water samples taken at the central station and at the five-meter station.

Hutchinson (1941) in Linsley Pond. From the surface to a depth of seven meters the organic phosphorus concentration was, on the average, eight times as high as the inorganic phosphate, but below seven meters the concentration of organic phosphorus decreased and inorganic phosphate increased sharply with depth. These were the conditions found in the lake when P³² was added on July 25, 1951.

Loss of P³² from Suspension

To calculate the total P³² in suspension or solution the earlier workers assumed that horizontal mixing of P³² was complete at all depths. To determine whether this were true in Toussaint Lake, samples for measurement of radioactivity were collected at two widely separated stations. Total suspended P³² in the top five meters was calculated from the concentration of P³² at each station by the method proposed by Coffin *et al.* (1949) and is shown in Figure 3. Since there was no appreciable difference between the total radioactivity calculated from the concentrations of P³² at the two stations, it was concluded that this

TABLE I. Amount of P³² remaining in suspension in Toussaint Lake after addition of 200 mc. (7 x 10¹⁰ c.p.m.) on July 25, 1951.

Days after Addition	c.p.m. x 10⁻¹⁰	% of Total
0	7.40	105
2	4.51	64
5	3.74	53
7	2.92	42
15	2.03	29
21	1.96	28
28	1.59	23

Mechanism of Loss from Suspension

There are several ways in which tracer phosphorus could have been removed from the open water; it could have been lost through the outlet, carried down into the bottom mud with sedimenting seston, or utilized by organisms inhabiting the littoral region. Loss through the outlet was calculated (with an error possibly as great as 50%) from volume of flow and radioactivity of outlet water. A total of 1.2% of the added P^{32} was lost through the outlet during the first week and only 0.5% during the last three weeks. Obviously this was not a major factor in the removal of P^{32} from suspension. Loss to bottom sediments was calculated from the radioactivity of 21 mud samples which were taken as described above at depths of 2 to 9.5 meters in a straight line across the lake from east to west. The average radioactivity of mud was $(50 \pm 20) \times 10^3$ c.p.m. per.m² on August 6, $(30 \pm 3) \times 10^3$ c.p.m. on August 23 and $(45 \pm 4) \times 10^3$ on September 5. No trend is shown by these figures, but if the one extremely high reading from the August 6th series is deleted, the average for that date becomes $(30 \pm 5) \times 10^3$ c.p.m. There is then an indication of a slight increase of radioactivity with time. Nevertheless, the radioactivity of all samples was averaged to give $(42 \pm 8) \times 10^3$ c.p.m. per m² and used as an approximation of maximum radioactivity of mud below the two-meter contour. The total surface area of mud below the two-meter contour was estimated to be 5.15×10^4 m² (110% of surface area of the lake). From these figures the amount of P^{32} lost to the mud below the two-meter contour was calculated to be 2.2×10^9 c.p.m. or 3% of the total P^{32} added to the lake. Since a total of less than five per cent of the radiophosphorus was lost through the outlet and to the bottom sediments in four weeks it appears that transfer of phosphorus to the littoral region was largely responsible for the rapid decrease of suspended P^{32}.

A better understanding of the fate of P^{32} which was lost from the open water of Toussaint Lake can be obtained by considering the P^{32} content of the epilimnion and of the deeper water separately. For the purpose of this calculation, a depth of two meters will be taken as the lower limit of the epilimnion. Actually, measurements of temperature and dissolved oxygen (Fig. 2) showed that the thermocline extended to the surface at times, and measurements of vertical distribution of P^{32} indicated that it mixed rapidly only to a depth of one meter, but the choice of two meters ensured that all rooted aquatic plants and associated organisms would be included within the "epilimnion."

The changes, during the experiment, of total suspended P^{32} above two meters and of P^{32} both in suspension and in the mud below two meters are shown in Figure 4. P^{32} was rapidly lost from the upper two meters but little passed into the water and mud below two meters or through the outlet. At the end of four weeks 88% of the total P^{32} had been lost from the epilimnion but only 14% had passed below the two-meter level or been lost through the outlet. The remaining 74% had been

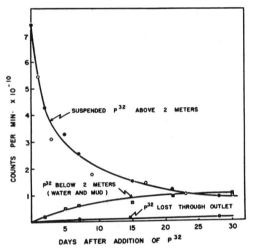

FIG. 4. Loss of P^{32} from the upper two meters of Toussaint Lake (open circles calculated from radioactivity at the five-meter station and closed circles from radioactivity at the central station) and increase in water and mud below two meters. Cumulative loss of P^{32} through the outlet is also shown.

lost elsewhere, probably to the littoral organisms in the upper two meters.

The loss of P^{32} from the epilimnion of Toussaint Lake by no means indicates that there was a depletion of phosphorus during the experiment. Table II shows that there was actually an increase of inorganic phosphate and that organic phosphorus remained relatively constant. Therefore, as well as a loss of phosphorus to the littoral region, there must have been a return of phosphorus from the littoral region to the epilimnion. This situation can be best described as a turnover of the "mobile" phosphorus of the epilimnion with respect to the following exchange:

Mobile P of water and plankton in epilimnion $\rightleftharpoons$ P of littoral solids in contact with epilimnion

The phosphorus participating in this exchange is described as mobile because it is not known which fraction of the total phosphorus is involved. Probably the inorganic and part of the plankton phosphorus are involved but the soluble organic phosphorus may also enter into the exchange to a limited extent.

When tracer ions or molecules have been added to one phase of a system such as the one described above, the rate of disappearance of tracer can be used to give a measure of the turnover time of the naturally occurring ions or molecules in that phase (Zilversmit et al. 1943). The application of this calculation to a system similar to the one described

TABLE II. Weight of phosphorus in the upper two meters of Toussaint Lake during tracer experiment, 1951

Date	KILOGRAMS OF PHOSPHORUS	
	Inorganic	Organic
July 23...............	.072	2.09
Aug. 1...............	.220	3.08
Aug. 8...............	.485	2.09
Aug. 15..............	.628	2.77
Aug. 22..............	.789	2.76

above has been adequately discussed by Hayes *et al.* (1952) and will not be treated here. From the rate of disappearance of P^{32} from the epilimnion of Toussaint Lake the turnover time of mobile phosphorus was calculated to be 3.6 days. The similarity between this value and the values of 5.4 and 7.6 days obtained by Hayes *et al.* is fortuitous since they calculated the turnover time of inorganic phosphate of the whole lake from the rate of disappearance of P^{32} from filtered water, and did not imply that all or even a large fraction of the P^{32} lost from solution was lost from the open water.

Uptake of P^{32} by Plankton

The percentage of the total suspended P^{32} in the 0-2m and 8-9 m strata which was in the plankton removed by a Foerst centrifuge is shown in Figure 5. At the surface (0-2 m) the percentage of P^{32} in plakton increased rapidly to a maximum of approximately 50% in three days, and remained relatively constant until the end of the experiment although the total amount of P^{32} in this layer decreased continuously. Since P^{32} was lost from the open water to the littoral region, there must have been a rapid equilibration between planktonic

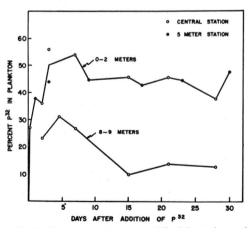

FIG. 5. Percentage of suspended P^{32} of the surface and deep water of Toussaint Lake in the fraction of the plankton removed by Foerst centrifugation.

phosphorus and inorganic phosphate so that the ratio of P^{32} in solution to that in plankton remained constant.

In the deepest water studied (8-9 m layer) the percentage of P^{32} in solids was consistently lower than at the surface. Since the P^{32} almost certainly reached this depth with sedimenting plankton, it is apparent that there was a rapid regeneration of inorganic phosphate in the hypolimnion. The regeneration may have been more complete than is indicated by these results since there was probably some precipitation of ferric phosphate in the centrifuge.

UPTAKE OF PHOSPHATE BY BACTERIA IN THE ABSENCE OF OTHER PLANKTON

Although it was demonstrated that phosphate is rapidly taken up by plankton in the epilimnion of Toussaint Lake and that there is an exchange of phosphorus between the open water and the littoral region, it is not possible to calculate the actual quantities of phosphorus involved because the specific activities of the various phosphorus fractions are not known. Therefore, it appeared that the value of further lake experiments would be limited until more was known about certain transformations of phosphorus such as the uptake and loss of inorganic phosphate by plankton.

Bacteria and algae are the two groups of planktonic organisms known to take up phosphate from solution in significant amounts. Bacteria were chosen for further study because they were known to take up phosphate rapidly in stored lake water, although their position in the phosphorus economy of lakes was uncertain.

The uptake of phosphate by aquatic bacteria has received relatively little attention because their rapid multiplication has been thought to be peculiar to stored water. Consequently, interest has centered on the cause of bacterial multiplication in such water. Zobell (1943) suggested that growth in stored water occurred on the walls and was stimulated by adsorption and concentration of organic matter on walls of storage vessels. However, Taylor and Collins (1949) were not able to confirm Zobell's hypothesis and proposed that the walls simply acted as a convenient site for attachment and multiplication of bacteria. In spite of their difference of opinion, these workers agreed that increase of bacterial population was due to multiplication of attached bacteria. This is not necessarily the only explanation for the increase of bacterial numbers in stored water, for an earlier suggestion by Waksman and Carey (1935) might be equally valid. They suggested that the bacterial populations of natural waters are held in check by organisms which feed on bacteria and that these predators die off when the water is stored. If this is true, it is possible that there is a rapid turnover of the bacterial populations of natural waters and that bacteria take up considerable amounts of inorganic phosphate in some lakes.

The growth of bacteria in stored lake water was therefore studied to determine if attachment to the walls of storage vessels was essential for multiplication of the species responsible for the uptake of inorganic phosphate.

Preliminary experiments showed that when water from a number of lakes was enriched with inorganic phosphate and stored, there was an initial decrease in the concentration of inorganic phosphate during the first 48 hours, followed by a slow regeneration. This was in accord with previous observations of Waksman and Stokes (1937) and Renn (1937) who also observed an increase in bacterial population coincident with the loss of inorganic phosphate and inferred that the phosphate was utilized by growing bacteria. If, as suggested by Zobell, and Taylor and Collins, some or all bacterial multiplication takes place on the walls of storage vessels, it would be expected that some of the phosphate lost from solution would go to the walls.

Experiment 1

The first experiment was carried out to determine whether phosphate lost from solution remained in suspension or was lost to the walls. Fifteen liters of water was collected from the surface of Toussaint Lake on July 18, 1952 and filtered through a Whatman No. 12 filter paper to remove larger plankton. On the following day the phosphate concentration was increased to .011 p.p.m. by addition of potassium dihydrogen phosphate and P^{32} was added to give approximately 250 c.p.m. per ml. The water was maintained at 15°C and stirred continuously. Samples were taken for measurement of inorganic phosphate, total suspended P^{32} and P^{32} in solution.

Results are shown in Figure 6. Inorganic phosphate and P^{32} in solution decreased rapidly during the first ten hours and then more slowly until, at the end of 32 hours, more than 85% had been lost from solution. Since the loss of P^{32} from solution paralleled that of inorganic phosphate, it appears that there was a direct removal of phosphate from solution and not a turnover of phosphate.

At no time during the experiment was there a significant loss of P^{32} from suspension. Therefore, it is obvious that phosphate was removed from solution by living or non-living, suspended, particulate matter of a size which would pass through filter paper and was not lost to the walls of the storage vessels.

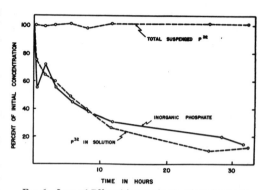

FIG. 6. Loss of P^{32} and inorganic phosphate from solution and total suspended P^{32} in filtered, Toussaint Lake water stored in a Pyrex container. Initial concentration of inorganic phosphate was .011 p.p.m.

Experiment 2

To demonstrate that bacteria could cause the loss of inorganic phosphate from solution observed in the previous experiment, the uptake of phosphate by pure cultures of aquatic bacteria was measured. Uptake of P^{32} was used as a measure of uptake of inorganic phosphate and, as in the previous experiment, loss of P^{32} from suspension as a measure of attachment of bacteria to the walls. Of six species isolated and tested, five grew in autoclaved lake water and utilized inorganic phosphate during their growth phase. The species used differed in the extent to which they attached to the walls during growth: the two extreme results are shown in Figure 7. In culture 1 the P^{32} taken up was equally distributed between bacteria in suspension and bacteria attached to the walls during the growth phase. A decrease in the percentage of P^{32} on the walls and an increase in suspended P^{32} after 20 hours' incubation suggested that as the culture aged, the cells broke away from

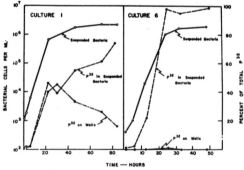

FIG. 7. Uptake of P^{32} by suspended and sessile cells of two species of bacteria during growth in stored Toussaint Lake water.

the walls and remained in suspension. In culture 6 there was no significant loss of P^{32} to the walls during or on completion of bacterial growth. It appears that all the cells of this species remained in suspension during and on completion of growth.

In control flasks containing sterile lake water there was almost no loss of P^{32} from solution. The average loss from controls of all experiments was 1.8%.

Conclusions

Since there was almost no loss of P^{32} from solution when sterile lake water was stored in Pyrex vessels, adsorption of phosphate on walls or suspended inanimate material was of no importance in causing the decrease of inorganic phosphate. Pure cultures of aquatic bacteria are capable of causing this decrease, presumably by utilizing inorganic phosphate for their growth. The multiplication of some species which utilize inorganic phosphate may take place on the walls of storage vessels but other species grow well in suspension. The species which multiply in unsterilized Toussaint Lake water do not need to adhere to the walls in order to grow. Therefore, it appeared that the rapid multiplication of aquatic bacteria might not be peculiar to stored water. If this is true the problem has an important bearing on the circulation of phosphate in lakes because studies by Birge and Juday (1934), Kusnetzow and Karzinkin (1931) and Bere (1933) have shown that the bacterial population of lake water is of the order of 10^4 to 10^6 cells per ml. The multiplication of populations of this magnitude must be an important factor in the removal of phosphate from solution.

UPTAKE OF PHOSPHATE BY BACTERIA IN THE PRESENCE OF OTHER PLANKTON

In the preceding laboratory experiments the uptake of phosphate by bacteria was studied in the absence of other plankton: in the following experiments, it was studied in water in which all of the naturally occurring plankton was present. Radioactive phosphorus was added to unfiltered, unsterilized water which was later filtered to fractionate the plankton.

In order to study uptake of P^{32} under completely natural conditions, tracer would have to be added to the whole lake but this is not practicable. Since the results of the first experiment in Toussaint Lake indicated that uptake of P^{32} by plankton might not be complete for several days (Fig. 5) it did not seem advisable to carry out experiments in glass containers in the laboratory where effects of storage would be pronounced. Therefore, in order to make experimental conditions approach natural conditions as closely as possible, the first experiment was carried out in a large polyethylene bag which was suspended in the lake. This experiment showed that uptake of P^{32} was complete in the time required to bring to the laboratory and filter the first samples. Consequently, the remaining experiments were carried out in the laboratory, but owing to the short period of time involved in these experiments, it was considered that the results would be comparable with those obtained from water under natural conditions.

Polyethylene Bag Experiment

Methods. The first experiment was carried out on August 28, 1953 in a cylindrical polyethylene bag two meters in diameter and one and one-half meters deep. The bag was suspended at the surface by a circular polyethylene float attached to the rim and fastened to a wooden frame anchored near the center of Toussaint Lake. After the bag had been filled with surface water, 0.4 mc. of P^{32} was added and mixed thoroughly with an oar. One-liter samples of water were taken at different depths within the bag at one, three and five hours after addition of P^{32} and immediately taken to the laboratory where the whole sample was filtered through No. 10 silk bolting cloth to remove large zooplankton. Then 50 ml was filtered through No. 20 silk bolting cloth to remove smaller zooplankton and larger phytoplankton, through No. 2 Whatman filter paper to remove smaller phytoplankton and finally through a Millipore filter to remove bacteria. All filters which were used had been cut to fit the standard counting trays so that they could be dried and counted directly. A 10-ml sample of the filtrate was evaporated and counted.

Results. Since the results from samples taken at different depths were essentially the same, they were averaged for presentation in Table III. The uptake of P^{32} was much more rapid and complete than was expected on the basis of the lake experiment. When the first samples were filtered, one and one-half hours after the P^{32} was added, over 97% of the P^{32} had been taken up by plankton. After this time the percentage of P^{32} in the plank-

TABLE III. Per cent of total P^{32} in fractions of plankton and in solution at intervals after addition of tracer phosphate to water inside a polyethylene bag August 28, 1953

Fraction	1.5 hr.	4.5 hr.	6 hr.
No. 10 Bolting cloth....	0.9	1.3	3.4
No. 20 Bolting cloth....	1.9	1.8	1.5
No. 2 Whatman Paper...	33.8	28.9	21.3
Millipore Filter.........	60.4	63.2	68.4
Filtrate................	2.9	4.8	5.2

ton decreased slightly. At the end of six hours when 95% of the P^{32} was in the plankton only five per cent of the total P^{32} had been taken up by zooplankton and larger phytoplankton; by far the largest amount (68.4%) had been taken up by the fraction of the plankton which passed through filter paper and was removed on the Millipore filter.

The concentration of inorganic phosphate within the bag changed only slightly during the experiment: it was .0020 p.p.m. at the beginning and .0018 p.p.m. at the end. But P^{32}, and hence phosphate, was taken up by plankton. Therefore, phosphate must have been returned to the water at the same rate as it was removed. These results are best explained by postulating a turnover of inorganic phosphate caused by the simultaneous uptake and release of phosphate by plankton. Since the uptake of P^{32} was complete when the first samples were analysed this experiment gave no measure of the rate of turnover.

Laboratory Experiments

In the following experiments the turnover of inorganic phosphate and the uptake of P^{32} by plankton was studied in water which had been brought into the laboratory. Since all experiments were of short duration, it was assumed that water and plankton would not be changed appreciably by storage in small containers.

Methods. Surface water was collected from the center of Toussaint Lake, immediately transferred to the laboratory and poured into experimental vessels. A ten-liter sample was used in the first experiment and a one-liter sample in the second. Radioactive phosphorus was added to give 100-200 c.p.m. per ml and the first samples were taken for filtration as soon as possible. Filtration took 10-30 sec.; the mid-time of filtration was recorded. Water was stirred after the addition of P^{32} and before each sampling. During these experiments the water was illuminated by skylight and by a 15-W daylight fluorescent lamp.

To determine whether algae, as well as bacteria, passed through the filter paper, the filtrate was examined microscopically. Surface water for this purpose was collected from Toussaint Lake on September 10, 1953 and October 27, 1953 and again on September 28, 1954 when a third measurement of turnover rate of inorganic phosphate was also made. The plankton in unfiltered water, in water filtered through No. 20 bolting cloth and No. 2 Whatman filter paper was concentrated by means of Sedgwick-Rafter funnels on September 10, 1953, a Foerst centrifuge on October 27, 1953 and an International No. 1 centrifuge running at 3,000 r.p.m. on September 28, 1954. All algae and algal fragments in the concentrates were counted.

Results. Loss of P^{32} from solution and uptake by the two plankton fractions is shown in Figure 8. Since all experiments to measure turnover gave essentially the same results, only the series made on Sept. 18, 1953 is presented. There was a rapid loss of P^{32} from solution—in 20 minutes when equilibrium had been attained, approximately 93% of the added P^{32} had been taken up by plankton. Again the fraction retained by the Millipore filter took up more than half of the added P^{32}. Phosphate concentration was approximately .0008 p.p.m. and did not change during the experiment. Therefore, the uptake of P^{32} represents a turnover, not a direct uptake, of phosphate. Turnovertime was calculated from the semi-log plot of loss of P^{32} from solution and was found to be 3.6 min. on September 11, 1953 and 5.4 min. on September 18, 1953. This is more rapid than was expected on the basis of the work of Hayes *et al.* (1952) and Whittaker (1953) and from results of the 1951 experiment in Toussaint Lake. It is believed that earlier work did not give a true measure of the rate of turnover of phosphate because the methods of separating plankton did not completely remove the smallest plankton which were shown above to be most important in causing turnover of inorganic phosphate.

Results of the algal counts are given in Table IV. The bolting cloth removed an average of 66% of the algal cells. The largest algae such as *Dinobryon* and *Asterionella* were almost completely removed by this filter. An average of 98% of the total algal cells were removed by the filter

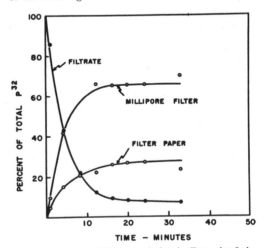

Fig. 8. Loss of P^{32} from solution in Toussaint Lake water and uptake of P^{32} by the plankton removed on a Whatman #2 filter paper and by the fraction of the plankton which passed through the filter paper and was removed on a Millipore filter, Sept. 18, 1953.

paper. Therefore, since a very small percentage of the algal plankton passed through the filter paper to the Millipore filter and since on each of the three dates the bolting cloth and filter paper fractions took up much less P[32] than the Millipore filter fraction, it was concluded that bacteria were the organisms which caused the turnover of inorganic phosphate. Also, it is possible, although it was not demonstrated, that a large part of the radioactivity of the filter paper fraction was caused by bacteria which were retained by the filter paper.

TABLE IV. Total number of algal cells in unfiltered and filtered Toussaint Lake water, as determined by microscopic counts

| | ALGAL CELLS PER ML | | | Average per cent removed |
	Sept. 10 1953	Oct. 27 1953	Sept. 28 1954	
Unfiltered water......	900	3200	860	–
Filtrate of No. 20 bolting cloth.......	230	1400	–	66
Filtrate of No. 2 Whatman paper....	13	17	35	·98

Disappearance of Added Phosphate

It was shown in the previous experiments that, although there was a rapid turnover of inorganic phosphate in Toussaint Lake water, the concentration of phosphate remained constant for as long as six hours. Obviously the rate of uptake of phosphate was exactly balanced by the rate of loss in these experiments. In order to determine the effect on this equilibrium of changing the concentration of phosphate in the water, the concentration was measured at intervals after addition of small quantities of inorganic phosphate to Toussaint Lake water.

Surface water was collected from the center of Toussaint Lake and divided into three two-liter samples. The phosphate concentration of the first sample was unaltered, the second was increased by .001 p.p.m. and the third by .005 p.p.m. The samples were placed in a water bath and maintained at the same temperature as when collected. Phosphate analyses were carried out as often as possible over a six hour period.

The experiment was repeated three times and the results of all tests were essentially the same. The results of the most complete series are shown in Figure 9. As was expected, the phosphate concentration in the unaltered sample remained almost constant, but in the samples to which phosphate had been added it decreased rapidly until it reached approximately the original concentration. It appears that under natural conditions the concentration of inorganic phosphate is maintained at a level where the rate of uptake by plankton is balanced by the rate of loss from plankton and that no increase of plankton phosphorous is possible unless the water receives an increment of phosphate from the inflowing water or from the littoral region. When the inorganic phosphate is added to the water the rate of loss by plankton remains constant but the rate of uptake increases until the concentration of phosphate in solution drops back to its original value and equilibrium between uptake and loss by plankton is re-established.

The initial increased rate of uptake, averaged from all tests was between 2×10^{-5} and 3×10^{-5} µg/ml/min. This increased rate of uptake is quite small in comparison with the turnover rate which was 17×10^{-5} µg/ml/min. It might be expected that the rate of uptake should be higher when .005 p.p.m. of phosphate was added than when .001 p.p.m. was added. If this is true the difference was small and was not detected in the limited number of tests carried out.

Comparison between Turnover of Phosphate in Toussaint Lake and in Other Lakes

Although it was shown that the turnover time of phosphate in Toussaint Lake was extremely short and that planktonic bacteria were largely responsible for causing this turnover, it is possible that Toussaint Lake might be exceptional in these respects. Therefore, the turnover of phosphate was measured in several other lakes for comparison with Toussaint. Water was tested from Oiseau Lake, a small, colorless, oligotrophic lake; Maskinonge Lake, a large, slightly coloured oligotrophic lake, and from the Ottawa River which, in the vicinity of Chalk River, more closely resembles a large lake than a river.

The results are summarized in Table V. It can be seen from column 3 that the turnover time of phosphate in Toussaint Lake was not exceptionally short since Oiseau and Maskinonge Lakes had turnover times for inorganic phosphate of 3.6 and 26 minutes respectively. However, the turnover time in Ottawa River water (30 hours) was very much longer than in any of the lakes. Since turnover time is a function of phosphate concentration as well as of plankton metabolism, it is perhaps preferable to compare the rate of turnover of phosphate in the different waters. In this respect, also, the three lakes were similar, while the turnover rate in Ottawa River water was much slower (Table V). Although it is not wise to make generalizations from so few results it does appear that the rapid turnover of inorganic phosphate is not peculiar to Toussaint Lake but is common to lakes

TABLE V. Comparison of phosphate turnover in several lakes in the vicinity of Chalk River, Ontario

Lake	Area in hectares (1)	PO₄ conc. p.p.m. (2)	Turnover time (3)	Turnover rate (μg/ml/ min.) (4)	Ratio of max. P³² in M.F. fraction to P³² in F.P. fraction (5)
Oiseau 28/IX/54	3.6	.0003	3.6 min.	0.8×10^{-4}	3.8
Toussaint 11/IX/53 & 18/IX/53*	4.7	.0008	4.5 min.	1.8×10^{-4}	2.5
Maskinonge 17/IX/54	160	.0012	26 min.	0.5×10^{-4}	1.8
Ottawa River 14/IX/54	—	.0049	30 hr.	0.03×10^{-4}	2.2

*Average for the two dates is tabulated.

in this region, whether oligotrophic or acid-bog, that have a low concentration of inorganic phosphate.

To give an indication of the importance of the Millipore filter fraction of the plankton in causing turnover of inorganic phosphate in the different lakes, the ratio of the maximum radioactivity in the Millipore filter fraction to the maximum radioactivity in all larger plankton was measured and is given in Table V. It is apparent that in each of the waters tested, the Millipore filter fraction took up appreciably more P³² than the larger plankton organisms.

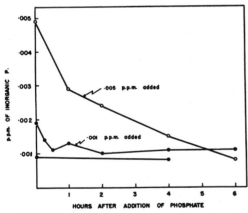

FIG. 9. Concentration of inorganic phosphate in water from the surface of Toussaint Lake measured at intervals after the addition of (i) .005 p.p.m. of inorganic phosphate, (ii) .001 p.p.m. of inorganic phosphate and (iii) no phosphate.

DISCUSSION

The results presented in this paper emphasize more strongly the transient nature of inorganic phosphate in the epilimnion of small lakes, already demonstrated by the earlier tracer studies of Hutchinson and Bowen 1950), Coffin et al. (1949)

and Hayes et al. (1952). It is shown that inorganic phospate remains constant at a low concentration for periods as long as several hours, not because plankton are unable to take up phosphate, but because the rate of uptake does not exceed the rate of metabolic loss, and that the inorganic phosphate in solution in the water is replaced many times in an hour.

The turnover times of inorganic phosphate in the small lakes studied were quite different from the estimates of turnover times (5-8 days) in two Nova Scotia lakes presented by Hayes et al. (1952). No doubt the extreme rapidity of turnover observed in lakes in the Chalk River district was related to the low concentrations of inorganic phosphate in these waters: in lakes with higher concentrations the turnover time might be expected to be much longer. Nevertheless it is believed that Hayes' values may be high because his methods of separating plankton from the water did not remove all bacteria which have been shown in the recent investigation to be most active in causing turnover of inorganic phosphate. In future studies of phosphate metabolism in lake water or of uptake of phosphate by plankton, the activity of aquatic bacteria must be considered and care should be taken to distinguish between uptake by algal plankton and uptake by bacteria.

The uptake of phosphate by bacteria in lake water certainly requires more study because planktonic bacteria, if they take up small increments of phosphorus received from inflowing water or from marginal vegetation, may compete with algae for this essential element. Whether this would be advantageous in terms of productivity is an interesting point. Bacteria are not generally autotrophic; they primarily transform existing organic matter, either particulate or dissolved, into bacterial protoplasm, a form that can be utilized by zooplankton. If, in this process, they utilize inorganic phosphate, they would reduce the amount of phosphate available to algae and thus reduce the amount of organic matter produced by algae. Since our present knowledge of the utilization of either bacteria or algae by zooplankton is incomplete, it is not possible to say what effect the utilization of phosphate by bacteria would have on productivity.

In addition to the rapid exchange of inorganic phosphate between water and plankton, a slower exchange of phosphorus between the open water of the epilimnion and littoral organisms was demonstrated. In Toussaint Lake the turnover time of "mobile" phosphorus in the epilimnion during the last week of July was found to be 3.5 days. In larger lakes and in lakes with fewer littoral organisms the turnover of phosphorus between open water and the margin would be much slower and of little or no importance in determining the amount of phosphorus in the epilimnion, but in Toussaint Lake a slightly greater uptake or loss by the littoral organisms could greatly decrease or increase the total phosphorus of the epilimnion within a few days.

A rapid turnover of phosphorus of the epilimnion has also been demonstrated in Linsley Pond by Hutchinson and Bowen (1950). But it differed from the turnover in Toussaint Lake in that phosphorus, although it entered the epilimnion from the littoral region, was lost almost exclusively to the hypolimnion. In Toussaint Lake there was also a loss to the hypolimnion but it was insignificant in comparison with loss to the littoral region.

Finally, little is known about the position of soluble organic phosphorus compounds in the phosphorus cycle of natural waters. This is a serious gap in our knowledge since organic phosphorus is usually present in much higher concentration than inorganic phosphorus. It is generally assumed that organic phosphorus compounds are produced by decomposition of plant and animal tissue, but nothing is known about the rate of production or the possible utilization of these compounds. Until the position of organic phosphorus in the phosphorus cycle is established, conclusions based on tracer studies with inorganic phosphate will tell an incomplete story and may even be misleading. For example, in the calculation of turnover time of inorganic phosphate, it has been assumed that all P^{32} in solution was in the form of inorganic phosphate. But if some of the soluble organic phosphorus compounds in water are produced by plankton, P^{32} would soon be incorporated into these compounds. Calculations of the specific activity of inorganic phosphate based on the amount of P^{32} in solution would then be erroneous. Therefore, a study of organic phosphorus compounds in the lake water is essential for complete interpretation of tracer experiments. It is to be hoped that with more widespread application of the tracer technique in limnological research the position of soluble organic phosphorus in the phosphorus metabolism of lakes will be clarified.

SUMMARY

The phosphorus cycle in a small, acid-bog lake was studied by the use of radioactive phosphorus. Of the total P^{32} added to the surface, 77% was lost from the water and plankton in four weeks but only two per cent was lost through the outlet and three per cent to the mud. It was concluded that there was a turnover of "mobile" phosphorus of the epilimnion with phosphorus of the littoral organisms. The turnover time of phosphorus of the epilimnion with respect to this exchange was found to be 3.5 days. The similarity between this value and the turnover times of inorganic phosphate calculated by Hayes et al. (1952) is considered to be fortuitous.

Several days after the addition of P^{32} to the lake, 50% had passed into the fraction of plankton removed by a Foerst centrifuge. This was similar to the results of Hutchinson and Bowen (1950) and Whittaker (1953) who used comparable methods for the removal of plankton. Laboratory experiments showed that neither a Foerst centrifuge nor filter paper removed the organisms which took up P^{32} most rapidly. When complete removal of plankton was achieved by filtering water through a Millipore bacterial filter it was shown that over 95% of the added P^{32} was taken up by plankton within 20 minutes. The turnover time of inorganic phosphate in the surface water of Toussaint Lake was approximately five minutes. Measurements of turnover made in two other small lakes suggested that a rapid turnover of inorganic phosphate is not peculiar to Toussaint Lake.

In laboratory experiments with mixed and pure cultures of aquatic bacteria, loss of P^{32} from solution and loss from suspension were used to measure uptake of phosphate by bacteria and extent of attachment of bacteria to the walls of glass containers respectively. Growing bacteria rapidly utilized inorganic phosphate from lake water but when sterile water was stored there was no loss of phosphate from solution. Attachment to the walls was not essential for the growth of most species of bacteria which multiplied and utilized phosphate during the first few days of storage.

The turnover of phosphate under natural conditions appeared to be caused primarily by bacteria. It was concluded that aquatic bacteria might compete with algae for inorganic phosphate.

REFERENCES

Bere, A. 1933. Numbers of bacteria in inland lakes of Wisconsin as shown by the direct count method. Internat. Rev. Hydrobiol. Hydrogr., 29: 248-263.

Birge, E. A., and C. Juday. 1934. Particulate and dissolved organic matter in inland lakes. Ecol. Monogr., 4: 440-474.

Coffin, C. C., F. R. Hayes, L. H. Jodrey, and S. G. Whiteway. 1949. Exchange of materials in a lake as studied by the addition of radioactive phosphorus. Can. J. Research, Section C, 27: 207-222.

Hayes, F. R., J. A. McCarter, M. L. Cameron, and D. A. Livingstone. 1952. On the kinetics of phosphorus exchange in lakes. Jour. Ecol., 40: 202-216.

Hutchinson, G. E. 1941. Limnological studies in Connecticut: IV. Mechanism of intermediary metabolism in stratified lakes. Ecol. Monogr., 11: 21-60.

Ecology, Vol. 37, No. 3

———, and V. T. Bowen. 1950. A quantitative radio-chemical study of the phosphorus cycle in Linsley pond. Ecology, 31: 194-203.

Kusnetzow, S. I., and G. S. Karzinkin. 1931. Direct method for the quantitative study of bacteria in water and some considerations on the causes which produce a zone of oxygen minimum in lake Glubokoje. Centralbl. Bakt., 83: 169-174.

Mortimer, C. H., and W. H. Moore. 1953. The use of Thermistors for the measurement of lake temperatures. Commun. Internat. Assoc. Theoret. Appl. Limn., No. 2.

O'Reilly, J. D. and A. Papson. Manuscript, O.F.R.L., Dept. of Zoology, U. of Toronto. The estimation of phosphorus in fresh waters by means of a potassium chlorate-stannous chloride-molybdate reagent in the presence of perchloric acid: II. Total phosphorus.

Renn, C. E. 1937. Bacteria and the phosphorus cycle in the sea. Biol. Bull., 72: 190-195.

Taylor, C. B., and V. G. Collins. 1949. Development of bacteria in waters stored in glass containers. J. Gen. Microbiology 3: 33-42.

Waksman, S. A., and C. L. Carey. 1935. Decomposition of organic matter in sea water by bacteria: II. Influence of addition of organic substances upon bacterial activities. J. Bact., 29: 545-561.

Waksman, S. A., J. L. Stokes, and M. R. Butler. 1937. Relation of bacteria to diatoms in sea water. J. Mar. Biol. Assoc., 22: 359-373.

Whittaker, R. H. 1953. Removal of radiophosphorus contaminant from the water in an aquarium community. A.E.C. Document HW-28636.

Zilversmit, D. B., C. Entenman, and M. C. Fishler. 1943. On the calculation of "turnover time" and "turnover rate" from experiments involving the use of labeling agents. J. Gen. Physiol., 26: 325-331.

Zobell, C. E. 1943. The effect of solid surfaces on bacterial activity. J. Bact., 45: 39-56.

8

Reprinted from *Science*, **131**(3415), 1731–1732 (1960)

Residence Time of Dissolved Phosphate in Natural Waters

Abstract. Residence time varies from approximately 0.05 to 200 hours. Short residence times are indicative of depleted phosphate, active metabolic activity, or both. The turnover rate of phosphate is between 0.1 and 1.0 mg of phosphorus per cubic meter, per hour, regardless of phosphate concentration, except in biologically active systems where it is 1.0 to 20. The turnover rate of phosphate may be more important than the phosphate concentration in maintaining highly productive systems.

The use of $P^{32}O_4$ in studies of lakes has revealed dynamic equilibria between phosphate dissolved in the water and phosphorus in the plankton, benthic organisms, bacteria, sediments, and dissolved organic materials (*1–5*). The residence time of dissolved phosphate proved to be a matter of minutes (*3*). Estimates of the residence time of dissolved phosphate in the sea and in estuaries (Table 1) reveal a wide range of conditions under which phosphate equilibria may be established, with the residence time varying over several orders of magnitude.

Residence time and turnover rate of phosphate in freshly collected water were estimated in the laboratory. Aliquots of 1.5 liters of water were placed in 7-liter rolling bottles in a constant-temperature room, and the water was kept within ±2°C of its temperature at collection. The water was illuminated by flourescent lights with an intensity of 350 ft-ca. Sterile, carrier-free $P^{32}O_4$ was added to the water, and

changes in radioactivity were measured at intervals until equilibrium was reached. Uptake of P^{32} by bacteria and plankton was measured by filtering aliquots of water through Millipore filters of 0.45-μ porosity and counting the activity of the dried filters. The residence time of phosphate was calculated by the method used in studies of lakes (*2–4*), but only the observations of Rigler (*3*) are comparable in the method of filtration.

Turnover rate is considerably less variable than residence time, and exceeds the range of 0.1 to 1.0 mg/m³ per hour only in biologically active systems, such as plankton blooms, salt marshes, and small lakes. Residence time is influenced both by the turnover rate and the concentration of dissolved phosphate. Therefore, a system having a short phosphate residence time may be impoverished in phosphate, as in the sea, or it may be unusually active biologically, as in algal blooms. When both conditions occur together, as in small lakes, the residence time becomes vanishingly short.

Measurements of the concentration of dissolved phosphate in natural waters give a very limited indication of phosphate availability. Much or virtually all the phosphorus in the system may be inside living organisms at any given time, yet it may be overturning every hour with the result that there will be a constant supply of phosphate for organisms able to concentrate it from a very dilute solution. Such systems may remain stable biologically and chemically for considerable periods in the apparent absence of available phos-

phate. This suggests how it is possible for phytoplankton blooms to persist in water containing only a few hours' supply of dissolved phosphate. The observations presented here suggest that a rapid flux of phosphate is typical of highly productive systems, such as blooms, and that the flux rate is more important than the concentration of dissolved phosphate in maintaining high rates of organic production.

It would be of interest to learn what factors tend to stabilize the flux of phosphate over a wide range of phosphate concentrations and what factors induce a more rapid flux in certain circumstances (*6*).

LAWRENCE R. POMEROY
*University of Georgia Marine
Institute, Sapelo Island*

References and Notes

1. G. E. Hutchinson and V. T. Bowen, *Proc. Natl. Acad. Sci. U.S.* 33, 148 (1947); G. E. Hutchinson and V. T. Bowen, *Ecology* 31, 194 (1950); C. C. Coffin, F. R. Hayes, L. H. Jodrey, S. G. Whiteway, *Can. J. Research* D27, 207 (1949).
2. F. R. Hayes, J. A. McCarter, M. L. Cameron, D. A. Livingstone, *J. Ecol.* 40, 202 (1952).
3. F. H. Rigler, *Ecology* 37, 550 (1956).
4. E. Harris, *Can. J. Zool.* 35, 769 (1957).
5. F. R. Hayes and J. E. Phillips, *Limnol. Oceanog.* 3, 459 (1958).
6. Contribution No. 17 from the University of Georgia Marine Institute. This work was supported by grant G-2506 from the National Science Foundation and by grants from the Sapelo Island Research Foundation.

25 January 1960

Table 1. Residence time, concentration, and turnover rate of dissolved phosphate in natural waters

Date	Location	Type of system	Res. time (hr)	Concn. (mg atom P/m³)	Turnover (mg P/m³ per hr)	T (°C)
9/28/54 9/11 and	Oiseau Lake, Ontario*	Small lake	0.06	0.003	1.6	
9/18/53	Toussaint Lake, Ontario*	Small bog lake	0.08	0.009	3.6	
9/17/54	Maskinonge Lake, Ontario*	Lake	0.4	0.012	1.0	
7/29/58	Salt-marsh creek, Georgia	*Kryptoperidinium* bloom	1.0	0.6	19	29
5/14/59	Altamaha River, Georgia	Nostocaceae bloom	1.0	0.1	3	25
7/18/58	30°53′N, 80°28′W	Continental shelf water	5	0.1	0.6	29
7/18/58	30°58′N, 80°01′W	Gulf Stream, surface	4	0.1	0.8	29
7/18/58	30°58′N, 80°01′W	Gulf Stream, 60 m	12	0.1	0.3	29
4/ 2/59	31°25′N, 81°05′W	Coastal sea water, surface	34	0.1	0.1	18
10/15/59	31°19′N, 81°10′W	Coastal sea water, surface	155	0.5	0.1	34
10/15/59	31°20′N, 81°13′W	Coastal sea water, surface	63	0.8	0.4	33
11/19/59	31°20′N, 81°13′W	Coastal sea water, surface	50	0.3	0.2	15
11/19/59	31°19′N, 81°11′W	Coastal sea water, surface	46	0.1	0.1	16
4/20/59	31°23′N, 81°17′W	Doboy Inlet	4	0.2	1.5	22
11/12/59	31°23′N, 81°17′W	Doboy Inlet	66	1.0	0.5	16
11/19/59	31°23′N, 81°17′W	Doboy Inlet	111	0.9	0.2	15
6/26/58	31°25′N, 81°18′W	Doboy Sound	37			
2/28/59	31°25′N, 81°18′W	Doboy Sound	50	0.6	0.5	12
7/17/59	31°25′N, 81°18′W	Doboy Sound	56	1.0	0.5	30
11/12/59	31°25′N, 81°18′W	Doboy Sound	39	1.0	0.8	15
11/19/59	31°25′N, 81°18′W	Doboy Sound	30	1.0	1.0	14
1/30/59	31°29′N, 81°16′W	Salt marsh (low tide)	49	3.0	2.0	15
10/12/59	31°29′N, 81°16′W	Salt marsh (low tide)	40	5.5	4.0	27
11/12/59	31°29′N, 81°16′W	Salt marsh (high tide)	169	1.1	0.2	15
11/12/59	Altamaha River, Georgia		13	0.2	0.5	14
9/14/54	Ottawa River, Ontario*		30	0.05	0.2	

* From Rigler (*3*). Rigler's observations have been converted from other units of measurement for convenient comparison.

Reprinted from *Intern. Oceanog. Congr. Preprints*, 893–894 (1959)

Regeneration of Phosphate by Marine Animals

LAWRENCE R. POMEROY and FRANCIS M. BUSH

While much emphasis has been given to the role of microorganisms in the regeneration of phosphate, relatively few attempts have been made to evaluate the importance of animals in phosphate production. Gardiner's (1937) data suggest that zooplankton produce considerable amounts of phosphate. Margalef (1950) reports an enzymatic production of phosphate from dissolved organic material in the presence of cladocerans. Animal excretions contain phosphate, and excretion is one way in which phosphate is produced from organic materials without the action of microorganisms. In this report an attempt is made to evaluate the importance of animal excretions in the phosphorus dynamics of an estuary.

The area studied, Doboy Sound, has a high-tide area of 125 km^2, of which 88 km^2 are intertidal marshes. Phosphorus utilization by all plant populations in the area is estimated to be 5×10^6 g P/day. Of this, 3.7×10^6 g P/day are used by Spartina alterniflora in the marsh.

The bottom invertebrate populations were estimated by a Petersen-grab survey involving ten hauls of the 0.25 m^2 grab at each of 26 stations selected from a numbered grid by means of a table of random numbers. Various invertebrate populations in the marsh were estimated by J. M. Teal, A. E. Smalley, and E. J. Kuenzler, who have kindly supplied these data. Population estimates are not yet available on several animal populations, including zooplankton, marsh and bottom microfauna, oysters, fishes, birds, and small mammals.

The rate of phosphate production by the more abundant animals in Doboy Sound was estimated in the laboratory, using freshly-collected specimens. The phosphorus content of the animals was estimated by analysis for total phosphorus or by reference to Vinogradov's (1953) data.

The data (Table 1) suggest that animal excretion may produce all the phosphate required by plant populations other than Spartina. There is some suggestion of a pyramid of phosphorus regeneration, similar to the pyramid of numbers or of biomass often found in natural ecosystems. The present data suggest no consistent relationship between metabolic rate and phosphorus excretion or between biomass and phosphorus excretion. This may be because there is considerable variation in the ability of various animals to reabsorb phosphate during the process of excretion.

Table 1

Biomass, phosphorus content, and rates of phosphate production of several populations of animals in Doboy Sound, Georgia. Notes: (1) Assuming 33% efficiency of the Petersen grab. (2) Based on a commercial catch estimate. (3) Based on 50 dolphins eating 23 kg of fish per day.

Population	Biomass, 10^9 g dry wt	Biomass P 10^6 g P	Regeneration 10^6 g P/day
Benthic invertebrates	0.9[1]	5.0	0.5
Marsh crabs (Uca, Sesarma, Eurytium)	1.8	10.4	0.4
Littorina irrorata	0.2	0.8	0.25
Modiolus demissus	0.4	2.6	0.1
Penaeid shrimp	0.03[2]	0.3	0.1
Tursiops truncatus			0.005 [3]

129

10

Reprinted from *Science*, **146**(3646), 923–924 (1964)

Phosphorus Excretion and Body Size in Marine Animals: Microzooplankton and Nutrient Regeneration

R. E. JOHANNES

Abstract. *In marine animals the rate of excretion of dissolved phosphorus per unit weight increases as body weight decreases. As a consequence microzooplankton may play a major role in planktonic nutrient regeneration.*

Animal excretions are a major source of plant nutrients in the sea (*1*). The animals most often studied in this regard are those captured in plankton nets. The smaller species which are not retained by plankton nets, the microzooplankton, are often overlooked by marine ecologists, and their role in nutrient cycles has never been evaluated.

Examinations of unfiltered water samples reveal that microzooplankton may often constitute a considerable fraction of the total animal biomass. Lohmann (*2*), for example, found that protozoa and very small metazoa, including *Rotifera*, copepod eggs, and certain invertebrate larvae, constituted an annual average of over 50 percent of the total zooplankton biomass in the waters off Kiel. A number of investigators have described the abundance of colorless flagellates (*3*) and ciliates (*4*) in the plankton.

It is well known that the smaller the animal the greater the metabolic rate per unit weight. The rate of nutrient excretion of microzooplankton should therefore be higher than that of net zooplankton per unit weight. Accordingly, it seemed worthwhile to attempt to evaluate the relative importance of animals of different sizes in the production of dissolved phosphorus.

For marine animals larger than 1 mg (dry weight), phosphorus excretion rates were determined by measuring spectrophotometrically the total phosphorus content of the animals and their soluble excretions (*5*). Excretion rates of animals weighing less than 1 mg were determined with the radioisotope P^{32} (*6*). Excretion rates are reported here as the time it takes an animal to release an amount of dissolved phosphorus equal to its total phosphorus content. This will be referred to as the body-equivalent excretion time (BEET).

A marked decrease in the body-equivalent excretion time (Fig. 1) is associated with decreasing animal size. The method of least squares leads to the following linear regression equations: For the upper line, log BEET (hours) is equal to 0.67 log dry weight (g) plus 3.2 ($r = 0.96$). For the lower line, log BEET = 0.33 log dry weight plus 2.6 ($r = 0.98$). Both correlations and the difference between the two slopes are significant at the 1 percent level. An analysis of the phosphorus excretion rates of 24 lots of differently sized (0.05 to 0.1 g without shells) mussels, *Modiolus demissus* (*7*), produced the regression equation: log BEET = 0.51 log dry weight plus 3.5 ($r = 0.53$). There was no significant difference between the slope of this regression line and the upper line in Fig. 1.

Whereas a 12-g lamellibranch released an amount of phosphorus equal to its total phosphorus content every 438 days, the body-equivalent excretion time of a 0.6-mg amphipod was 31 hours, and that of an 0.4×10^{-5} μg ciliate was 14 minutes. Excretion-time of an animal the size of a 1-μ^3 phagotrophic microflagellate is estimated to be about 2 minutes (*7*), based on extrapolation of the lower regression line in Fig. 1 (*8*). It is difficult, if not impossible, to determine the excretion rates of these fragile forms directly.

When the data used in Fig. 1 were compared with data on oxygen consumption as related to body weight in marine animals (*9*) it was found that the ratio of oxygen consumed to phosphorus excreted decreases markedly with decreasing animal size. An animal weighing 1 μg releases approximately 50 times as much phosphorus per unit weight as a 100-mg animal, while the smaller animal consumes only 5 to 8 times as much oxygen per unit weight. Two other workers (*10*) have likewise noted that the "O/P ratio" was significantly lower for mixed zooplankton species than for several larger benthic invertebrates. The explanation of the marked lack of parallelism between these two physiological processes deserves investigation.

A recent example serves to demonstrate the importance of considering size distribution of fauna when computing excretion rates of faunal communities. Rigler (*11*) calculated the rate of release of dissolved phosphorus for lake zooplankton, assuming that rotifers excrete the same amount of phosphorus per unit weight as cladocerans 1000 times heavier. He acknowledged the possibility of error arising from this assumption. My results suggest that this error is indeed significant. On the basis of the lower regression line in Fig. 1 the rotifers would be expected to excrete dissolved phosphorus about ten times as fast as the cladocerans per unit weight.

Figure 1 can be used to demonstrate the relative importance of microzooplankton and macrozooplankton for

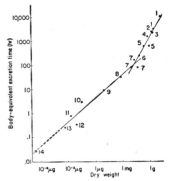

Fig. 1. Relation between body-equivalent excretion time of dissolved phosphorus and body weight of marine animals. 1, *Tridacna crocea*; 2, *Penaeus setiferus*; 3, *Crassostrea virginica*; 4, *Modiolus demissus*; 5, *Uca pugnax*; 6, *Salpa fusiformis*; 7, *Littorina irrorata*; 8, *Lembos intermedius*; 9, *Artemia salina* (nauplii); 10, *Euplotes crassus*; 11, *Euplotes trisulcatus*; 12, *Euplotes vannus*; 13, *Uronema* sp.(?); 14, hypothetical 1-μ^3 (2.5×10^{-7} μg dry wt.) phagotrophic flagellate.

phosphorus regeneration in a given population. For example: if we take a hypothetical zooplankton population composed of 10 percent by weight of ciliates weighing 1×10^{-2} μg each and 90 percent crustaceans the size of *Calanus finmarchicus*, weighing about 0.3 mg each (*12*), the ciliates would contribute over 70 percent of the total dissolved phosphorus released by the population. The substitution of phagotrophic flagellates for ciliates would shift the balance further in favor of the domination of nannozooplankton in nutrient release processes, while the substitution of smaller crustaceans, such as *Acartia clausii*, for *Calanus finmarchicus* would increase the dissolved phosphorus contribution of the net zooplankton.

Thus the contribution of microzooplankton to phosphorus regeneration is much greater than their contribution to total zooplankton biomass. Their participation in other ecologically im-

portant processes (for example, regeneration of other nutrients, food consumption) must likewise exceed the importance of their biomass.

References and Notes

1. E. Harris, *Bull. Bingham Oceanog. Coll.* **17**, 31 (1959); B. H. Ketchum, *Rapp. Procès-Verbaux Réunions Conseil Perm. Intern. Exploration Mer* **153**, 142 (1962); L. R. Pomeroy, H. M. Mathews, H. S. Min, *Limnol. Oceanog.* **8**, 50 (1963).
2. H. Lohmann, *Wiss. Meeresunters., Abt. Kiel* **10**, 129 (1908).
3. E. J. F. Wood, in *Marine Microbiology*, C. H. Oppenheimer, Ed. (Thomas, Springfield, Ill., 1963), p. 236; F. Bernard, *ibid.*, p. 215; C. A. Kofoid and O. Swezy, *Mem. Univ. Calif.* **5**, (1921).
4. E. J. F. Wood, in *Marine Microbiology*, C. H. Oppenheimer, Ed. (Thomas, Springfield, Ill., 1963), p. 28; K. Banse, *Rapp. Procès-Verbaux Réunions Conseil Perm. Intern. Exploration Mer* **153**, 47 (1962).
5. Data used to calculate *Modiolus* BEET were supplied by E. J. Kuenzler. L. R. Pomeroy, F. M. Bush, H. M. Mathews, and H. S. Min determined the BEET of all other animals weighing more than 1 mg (these data were partially published in L. R. Pomeroy and F. M. Bush, *Internat. Oceanog. Congr. Preprints,* AAAS, Washington, 1959, p. 893).
6. R. E. Johannes, *Limnol. Oceanog.* **9**, 235 (1964).
7. These data were generously supplied by Dr. E. J. Kuenzler.
8. Despite the high degree of linearity found in both regression lines in Fig. 1, this plot might be challenged as being arbitrary. A simple curvilinear function could be substituted. Within the range of animals studied this would make little difference to the discussion, but it makes some difference in extrapolating to micro-flagellates. Curvilinear functions fitted by eye to the data and extrapolated to 1-μ^3 flagellate result in an estimated BEET as high as 5 minutes.
9. E. Zeuthen, *Compt. Rend. Lab. Carlsberg. Ser. Chim.* **26** (3), (1947).
10. E. Harris, *Bull. Bingham Oceanog. Coll.* **17**, 31 (1959); M. Satomi, thesis, University of Georgia (1964).
11. F. H. Rigler, *Can. Fish Culturist* **32**, 3 (1964).
12. S. M. Marshall and A. P. Orr, *Biology of a Marine Copepod* (Oliver and Boyd, Edinburgh, 1955).
13. I thank Drs. L. R. Pomeroy and K. L. Webb who critically read this manuscript, Mrs. I. B. Webb who prepared the figure and performed the statistical analyses, and Dr. A. C. Borror who identified three species of ciliates. Supported by NSF GB-1040 and by grants from the Sapelo Island Research Foundation. Contribution No. 68 from the University of Georgia Marine Institute.

11 September 1964

Reprinted from *Trans. N.Y. Acad. Sci.*, **10**, 136–141 (1948)

ESSENTIALITY OF CONSTITUENTS OF SEA WATER FOR GROWTH OF A MARINE DIATOM

By S. H. HUTNER *

The growth requirements of marine organisms higher than bacteria are poorly understood. Knowledge of the ionic relations of marine forms may assist in testing the theory that the electrolyte balance necessary for the functioning of mammalian tissue is a reflection of the pattern imposed on protoplasm by its origin in the ancient less saline sea—a hypothesis fascinatingly sketched by Baldwin.[1] The inshore diatom *Nitzschia closterium,* isolated in England nearly forty years ago, has been studied intensively, and was therefore chosen as the starting point of this investigation. The relation of this older work to general oceanographic problems has been summarized by Harvey.[2]

Only guide-post data are presented here, but as the *methods* being developed to evaluate the role of each constituent of sea water appear widely applicable, emphasis will be placed on the theoretical foundations of the procedures, and experimental results are cited mainly to illustrate these methods. The conclusions drawn from these results are highly tentative. Much information has accumulated, but more quantitative data are needed before clearcut principles can be formulated.

Methods. A bacteria-free subculture of the classical strain was obtained from Professor E. G. Pringsheim. Stock cultures were maintained on 1 to 2 per cent sea salt (a convenient substitute for diluted sea water)—peptone agar slants, pH 7.1 to 7.7, in screw-capped tubes. Peptones of the tryptic digest of casein type (trypticase or tryptone) allowed heavy growth at the 0.1 to 0.5 per cent levels. Cultures remained viable for at least a month when

* Haskins Laboratories, New York, N. Y. This paper, illustrated by lantern slides, was presented at the meeting of the Section on January 12, 1948.

stored in darkness at 6° C. Experimental cultures were grown in 25-ml. Pyrex Erlenmeyer flasks, plugged with Pyrex glass wool. Illumination and the proper temperature (18° to 20° C.—this is close to the upper limit) were secured by placing the flasks on glass trays in a refrigerating incubator (Central Scientific Company) equipped with two 20-watt "soft white" fluorescent lamps. Growth was rapid under these conditions, and was measured after 25 to 35 days. Otherwise the procedures were similar to those previously described.[3]

Results with Citrate. Work with enriched synthetic sea water media is hindered by the formation of precipitates. These, by absorption, remove essential elements, as indicated by the repeated observations of other workers that filtered solutions do not support growth. In devising substitutes for sea water, it early became apparent that calcium, magnesium, and iron were required in rather high concentration, each inducing precipitates at pH 7.0 and higher. The use of citrate overcame this difficulty. Citrate forms soluble chelate complexes with many elements, but in so doing removes them from the field of action. The formation of the calcium citrate complex is illustrated in FIGURE 1. As the citrate concentration is increased, greater proportions of certain essential elements are bound, and it becomes necessary to adjust upwards the concentrations of these elements. By supplying an excess of all essential elements save one, one may then plot the increment of this element required to compensate for each increment of citrate. Then, by extrapolation to zero citrate, it is possible to arrive at a value closer to the *absolute* requirement for a particular element. This procedure applied to calcium and magnesium is illustrated in FIGURE 2. It is impractical to set up experiments with no citrate because of the formation of precipitates already

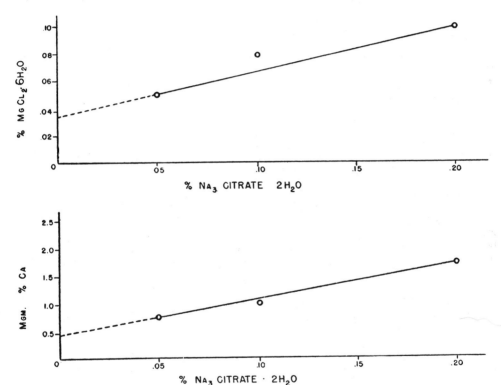

FIGURE 1. Possible manner of formation of chelate rings in the calcium citrate complex. One calcium ion may coordinate with more than one molecule of citrate. In this and the subsequent figures, only a limited effort is made to depict interatomic bond distances accurately.

FIGURE 2. Calcium and magnesium requirements as a function of citrate concentration.

mentioned. The points plotted represent the amount of metal (or salt) that must be added to restore half maximal growth.

Before accepting these "absolute" values, it is necessary to define the experimental conditions rigorously. One way to test the accuracy of these values, obtained by what may be termed the citrate extinction technique, is to determine whether other complex-forming reagents, yielding curves different in slope, yield the same intercept as the citrate curve when extrapolated to zero reagent. For this purpose, the tartrate, oxalate, and other complexes are under investigation. Additional requirements for rigorous accuracy can be posed:

different non-toxic salts of the element in question must furnish the same intercept; furthermore, when a salt of the element is purified by at least two independent methods, both must furnish the same intercept. Meeting these conditions entails much tedious work but appears to be necessary to keep impurities in commercial "chemically pure" substances from seriously influencing the results. The satisfaction of these requirements for accuracy appears well on the way to fulfillment for calcium and magnesium, but is more remote for iron and manganese. Here, the situation is complicated by simultaneously induced deficiencies in molybdenum and other trace elements. Methods are being devised for each of the constituents of the culture medium, and it is likely that these purified materials will lead to reproducible results. The citrate extinction technique is an extension of the procedure employed by Hopkins and Wann[4] for determining the iron requirement of the green alga *Chlorella*. TABLE 1 shows the type of medium used for obtaining "extinction" data. It allows very heavy growth. The concentrations of nutrients are adjusted to the level of citrate shown in the table.

TABLE 1

SYNTHETIC MEDIUM FOR *Nitzschia closterium* *

K_2HPO_4	0.04%
NH_4NO_3	0.05%
Na_3 citrate.$2H_2O$	0.1%
Na metasilicate.$9H_2O$	0.005%
NaCl	0.2%
$MgSO_4.7H_2O$	0.25%
Ca (as $CaCO_3$ + HCl)	3.5 mg.%
Fe (as Fe + HCl + HNO_3)	0.5 mg.%
Boron (as H_3BO_3)	0.05 mg.%
Zn	0.005 mg.%
Mo	0.005 mg.%
Cu	0.005 mg.%
Mn	0.00005 mg.%

* pH 7.2-7.5. Because of toxicity, Mn, while essential, must be kept low.

On consideration, the citrate extinction technique is recognized as a special case of the method of competitive inhibition now widely exploited as a tool for finding new antimicrobial agents. But here the interference is with essential inorganic metabolites rather than with organic metabolites, and the mechanism is one of interception of the essential element rather than "jamming the lock" based on structural homology. This method provides a means of minimizing the importance of chemical contamination of the environment, by exaggeration of the requirement for a specific element or group of elements. Analogously, the discovery of the ubiquitous and highly potent growth factor, *p*-aminobenzoic acid, would have been much delayed if it had not been brought into the open as the most powerful antagonist of the sulfonamides. If there are essential trace elements yet to be identified, this method might facilitate their discovery.

The Sodium Chloride Requirement. Like most marine microorganisms, *N. closterium* poorly tolerates solutions more hypertonic than sea water. As the citrate concentration was increased, it was possible (and necessary) to decrease the NaCl to low levels, and even virtually eliminate chloride. It was not possible to obtain growth with concentrations of sodium below 0.02 per cent or so, a finding in harmony with some older conclusions.[5] A medium developed for this series of experiments is shown in TABLE 2. Comparison of it with the medium of TABLE 1 indicates the liberties permissible in abandoning sea water media.

TABLE 2

SYNTHETIC MEDIUM FOR DETERMINATION OF SODIUM CHLORIDE REQUIREMENTS *

K_3 citrate.H_2O	0.15%
NH_4NO_3	0.04%
$(NH_4)_2HPO_4$	0.03%
$MgSO_4.7H_2O$	0.3%
ethyl silicate [$(C_2H_5O)_4Si$]	0.004%
Fe (as $FeSO_4.7H_2O$)	0.5 mg.%
B (as H_3BO_3)	0.1 mg.%
Mo [as $(NH_4)_6Mo_7O_{24}.4H_2O$]	0.001 mg.%
Zn (as $ZnSO_4.7H_2O$)	0.001 mg.%
Mn (as $MnSO_4.H_2O$)	0.001 mg.%
Co (as $CoSO_4.7H_2O$)	0.01 mg.%
Ca (as gluconate)	3.0 mg.%

* pH 7.4. The addition of Na_2SO_4, 0.06%, allows heavy growth.

Preliminary Conclusions as to Ionic Balances. The values in FIGURE 1 for the calcium and magnesium requirements are much higher than those usually found for non-marine organisms. The calcium requirement is of the order of that of the land plants, and may prove deeply significant in view of the importance assigned to calcium in controlling the permeability of all cell membranes.

Differential Permeability and Ion Accumulation. Many organic compounds bind metals tightly by formation of chelate rings. The metal is held in a ring by forces similar to those stabilizing aromatic rings.[6] Citrate is comparatively weak in this respect; greater avidity and specificity in binding elements, particularly those of the transitional series, is possessed by compounds imparting a high degree of resonance to the metal-containing ring. Truly extraordinary biological compounds of this type are the porphyrins, where iron is bound in heme compounds and mag-

nesium is bound in chlorophyll. It is increasingly recognized that the maintenance of differential permeability and the accumulation of ions against a concentration gradient is closely linked to respiratory activity. The suggestion has been made that this respiratory activity is required for the formation of ion-binders of the chelate type.[7] Following this line of thought, the surface of the cell may be regarded as studded with receptor patches, constantly regenerated by energy derived from respiration. The *number* of these receptor sites would help determine the concentration of a particular ion that must be present in the medium to keep the cell alive and growing; and the *avidity* or *"stickiness"* of these sites would determine the ability of the cell to compete with various types of metal-binding compounds for the available supply of essential ions. These receptor sites must be present to abstract such elements as iron and molybdenum from sea water where they occur

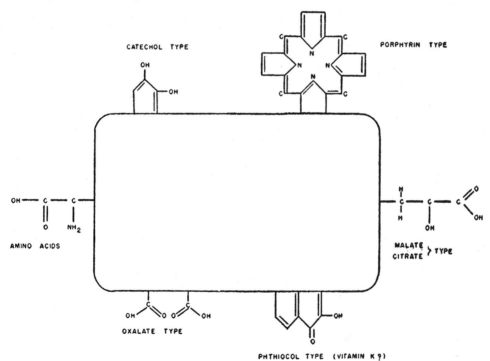

CATECHOL TYPE

PORPHYRIN TYPE

AMINO ACIDS

MALATE
CITRATE } TYPE

OXALATE TYPE

PHTHIOCOL TYPE (VITAMIN K ?)

FIGURE 3. Cell with hypothetical metal receptors on the surface.

in almost immeasurably low concentration. The remarkable sensitivity and specificity which make certain chelating reagents so useful in the analytical chemistry of trace amounts of many metals, hint at the employment by the cell of this type of compound for similar purposes. Many components of the cell other than porphyrins and citrate are more or less effective metal binders: *e.g.*, α-amino and α-hydroxy carboxylic acids, glycerol, and other compounds with *ortho* configurations of potentially chelating substituents. This concept of the nature of the receptor sites for metals on the cell surface is diagrammed in FIGURE 3. Immunologists may recognize in these ideas an invocation of the shade of Paul Ehrlich.

The competitive action of the bulky hydrophilic citrate radical likely is exercised primarily at the cell surface. Other competitive reagents may be employed for a systematic probing of these hypothesized receptors. Each such reagent may be viewed as competing with its structural counterpart (or, perhaps, more accurately,

its electronic counterparts) embedded in the cell surface. Thus o-phenanthroline (FIGURE 4) binds iron and the adjoining transitional element of the periodic table in virtually the same kind of linkage characteristic of porphyrins. The question sometimes arises as to whether this competitive action, as distinguished from irreversible poisoning, takes place entirely at the cell surface, or whether there is also an appreciable penetration of the cell, with consequent interference with metal-catalyzed reactions within the cell. The answer may be sought in a comparison of the activities of pairs of chelators, alike in metal-binding capacity, but widely different in ability to penetrate the cell. Such a pair is 8-hydroxyquinoline and 8-hydroxyquinoline-5-sulfonic acid (FIGURE 5).

8 - HYDROXY QUINOLINE

8 - HYDROXYQUINOLINE - 5 - SULFONIC ACID

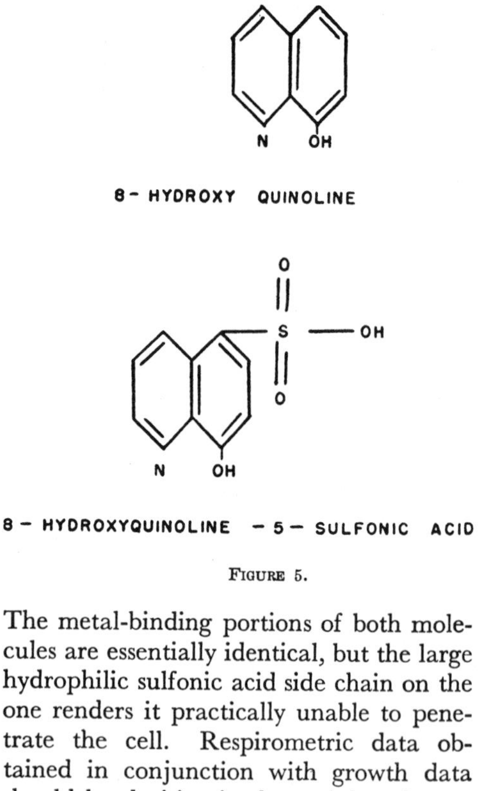

FIGURE 5.

The metal-binding portions of both molecules are essentially identical, but the large hydrophilic sulfonic acid side chain on the one renders it practically unable to penetrate the cell. Respirometric data obtained in conjunction with growth data should be decisive in determining the extent of cellular penetration.

Little can be said at this time about the data obtained by these methods. Explor-

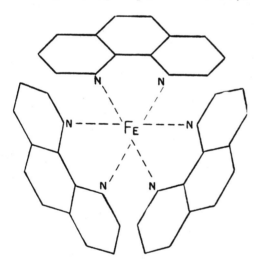

O - PHENANTHROLINE

FIGURE 4. The o-phenanthroline-ferrous complex. Double bonds in the o-phenanthroline are not shown.

atory experiments with faster growing organisms such as *Chlorella* and photosynthetic and non-photosynthetic bacteria, indicate that these methods will eventually lead to a quantitative assessment of the requirements of *N. closterium* for iron, manganese, molybdenum, and, very likely, cobalt, and will even allow some well-informed guesswork as to the avidity of its receptors for various elements as compared with other microorganisms. The power and delicacy of this method is borne out by the provocative findings of Albert et al.,[8] who noted that cobalt overcame the inhibitory action of 8-hydroxyquinoline on gram-positive bacteria, but that iron plus zinc was required instead to antagonize the action of 8-hydroxyquinoline on gram-negative bacteria.

Diatoms and Other Biological Problems. There are additional objectives in work on *N. closterium* that can be touched on only briefly. Pelagic diatoms require for growth a factor in natural sea water that has, as yet, defied analysis. It is reasonable to suppose that information on *N. closterium,* which does not require this factor,[9] will facilitate investigations of the requirements of these more exacting diatoms. The diatoms belong to a great array of brown-pigmented algae whose energy metabolism has scarcely begun to be studied. Valuable sidelights on such problems as the nature of the energy stockpile built up by photosynthesis might be gained by a thorough-going search for substances permitting growth of this form in darkness. That this substitute for photosynthesis has to be unusually efficient is suggested by the finding that the mere maintenance of turgor in a marine diatom entailed the expenditure of considerable photosynthesis-gained energy.[10]

Viewed broadly, the object of this work is to contribute to the long series of experiments in other laboratories fashioning *Nitzschia closterium* into an effective tool for a wide range of investigations into cell physiology and comparative biochemistry. Because of the ever-present temptation felt by physiologists to confine their experiments to a few well-established laboratory organisms, there is always the danger that far-reaching conclusions may be based on too narrow a choice of living material. The manner in which a fundamental broadening of the theories of photosynthesis and hydrogen transfer waited upon studies of photosynthetic bacteria, complementing the time-honored green forms, is a striking example of the use of new organisms to clarify old problems. It is reasonable to suppose that the diatoms will be an increasingly valuable addition to the armamentarium of the cell physiologist, and thereby enhance the intrinsic importance they already possess as the equivalent of grass in the economy of the oceans.

BIBLIOGRAPHY

1. **Baldwin, E.** 1937. An Introduction to Comparative Biochemistry. University Press, Cambridge.
2. **Harvey, H. W.** 1945. Recent Advances in the Chemistry & Biology of Sea Water. University Press, Cambridge.
3. **Hutner, S. H.** 1946. Organic growth essentials of the aerobic nonsulfur photosynthetic bacteria. J. Bact. **52**: 213-221.
4. **Hopkins, E. F., & F. B. Wann.** 1927. Iron requirement for *Chlorella*. Bot. Gaz. **84**: 407-427.
5. **Richter, O.** 1909. Reinkulturen von Diatomeen. III. Über die Notwendigkeit des Natriums für braune Meeresdiatomeen. Sitzb. k. Akad. Wiss., Math.-naturw. Kl. **118** (Abt. 1): 1337-1344.
6. **Calvin, M., & K. W. Wilson.** 1945. Stability of chelate compounds. J. Am. Chem. Soc. **67**: 2003-2007.
7. **Jacobson, L., & R. Overstreet.** 1947. A study of the mechanism of ion absorption by plant roots using radioactive elements. Am. J. Bot. **34**: 415-420.
8. **Albert, A., S. D. Rubbo, R. J. Goldacre, & B. G. Balfour.** 1947. The influence of chemical constitution on antibacterial activity. Part III: A study of 8-hydroxyquinoline (oxine) and related compounds. Brit. J. Exp. Path. **28**: 69-87.
9. **Harvey, H. W.** 1947. Manganese and the growth of phytoplankton. J. Mar. Biol. Assoc. U. K. **26**: 562-579.
10. **Gross, F.** 1940. The osmotic relations of the plankton diatom *Ditylum Brightwelli* (West.). J. Mar. Biol. Assoc. U. K. **24**: 381-415.

12

Copyright © 1953 by the New York Academy of Sciences

Reprinted from *Ann. N.Y. Acad. Sci.*, **56**, 839–851 (1953)

ECOLOGICAL IMPLICATIONS OF *IN VITRO* NUTRITIONAL REQUIREMENTS OF ALGAL FLAGELLATES

By Luigi Provasoli and Irma J. Pintner

Haskins Laboratories, New York, N. Y.

Introduction

The productivity of fresh-water bodies and the ocean rests on the algal photo-trophs at the base of the food pyramid. To predict the fertility of natural waters it is obviously necessary to learn what governs the proliferation of these algae.

Lucas[20] in 1947, reviewing the evidence supporting his broad concept of "non-predatory" relationships among organisms living in water, emphasized the importance of external metabolites produced by water organisms and pointed out that through these metabolites one kind of organism might affect the growth of others in several ways: (1) by producing necessary nutrients; (2) by removing inhibitory compounds; (3) by excreting inhibitory substances. The present paper describes the growth requirements of several algae in pure culture and, in so doing, documents some aspects of these "non-predatory" relationships.

Certain centric diatoms important in the pelagic phytoplankton were postulated by Harvey[8] to need one or more growth factors. This heightens interest in the observation that the one strain of *Amphora perpusilla*, a neritic pennate diatom, studied by us, requires cobalamin.[12] When coupled with other similar findings detailed in this paper, it brings into sharper relief the problem of the relative incidence of auxotrophy in coastal and oceanic phytoplankton species.

In the course of work with some fresh-water planktonic protists, a remarkable parallelism was observed between *in vitro* results on the effective concentration ranges of nutrients and the composition of natural waters established by chemical analysis, suggesting that the chemical data of the hydrobiologist are a reliable guide to pure culture studies. Conversely, nutritional studies, in revealing that many algae from a variety of environments require vitamins and trace elements, focus attention on these two classes of nutrients which have often been neglected in classical hydrobiological analyses. It would appear, accordingly, that coöperation between hydrobiologist and nutritionist should increase understanding of the factors underlying algal successions and blooms.

Cultural Techniques

General Methods. These generally were the same as those described for *Ochromonas.*[13] Distilled water was passed through a mixed-bed ion exchanger ("Deeminizer"); the purity being controlled by conductivity measurements. Volatile preservative[13] was immediately added to the water, which was stored in polyethylene carboys. The culture vessels were screw-capped tubes with the liners of the caps removed. The cork liners retained objectionable quantities of the detergent used for cleaning, and so induced precipitates in the poorly buffered very dilute media needed for the more delicate fresh-water planktonts.

Technique of Isolation. In attempting to isolate and cultivate an organism belonging to a group never previously grown aseptically *in vitro*, the main problem is to find a medium as complete as possible yet non-inhibitory. Motility is an excellent index of lack of inhibition. Loss of photosynthetic pigments serves as indicator of inhibition for non-motile forms. Before washing or plating (see Pringsheim[24]), preliminary trials of the suitability of media are made by fishing 10 to 20 individuals from the natural waters and putting them into a variety of media and observing their behavior after 6 to 24 hours. For *Synura,* one of the most delicate forms, a peat-mineral medium was developed

TABLE 1
ISOLATION MEDIUM FOR *Synura* (MG. PER CENT)

Ca (as NO_3)	0.65	K_2HPO_4	0.4
$MgSO_4 \cdot 7H_2O$	1.0	KCl	0.6
Ca (as Cl)	0.4	Fe (as SO_4)	0.06
$CaSO_4 \cdot 2H_2O$	1.0	Peat extract*	13.0 ml./100 ml.

pH 5.0–5.5; determined with the glass electrode.
The Ca salts were prepared by dissolving $CaCO_3$ in HCl or HNO_3 and driving off excess acid. Since the chloride and nitrate are hygroscopic, the concentrations shown refer to the metal, not the salt, in the final medium.
* Prepared by autoclaving (118–121°C.) black peat with an equal weight of water and filtering, then diluting to the color of "light beer" (see v. Wettstein's directions).[38]

TABLE 2
ORGANIC MEDIA FOR ISOLATION AND CONSERVATION OF FRESH-WATER ALGAE

A. Medium for *Chlorophyceae* and *Cyanophora* (mg. %)		B. Medium for *Peridinium* and *Ochromonas* (mg. %)	
Na_3 citrate $\cdot 2H_2O$	2.0	$Ca(NO_3)_2 \cdot 4H_2O$	10.0
$MgSO_4 \cdot 7H_2O$	2.0	$MgSO_4 \cdot 7H_2O$	2.0
K_2HPO_4	2.0	K_2HPO_4	2.0
Fe (as SO_4)	0.2	Glucose	50.0
Thiopeptone (Wilson)	60.0	Trypticase (B.B.L.)*	50.0
Trypticase (B.B.L.)*	16.0	pH 6.5	
Yeast extract (Difco)	5.0		
pH 6.5			

* Baltimore Biological Laboratory

(TABLE 1) with the motility response as guide (the medium is basically a modified v. Wettstein medium).[38] Other media which have proved useful for isolation of unicellular and colonial Chlorophyceae, for *Ochromonas, Cyanophora,* and for a fresh-water *Peridinium* are given in TABLE 2. These media are employed also as conservation media, either as such or slightly agarized (0.2 g per cent).

Rich media favor the growth of bacteria and molds and therefore have not been used for isolation of algae by the plating method. Mineral media have been used almost exclusively from Beijerinch on. Mineral media in effect selected for autotrophs and created the impression that algae, like most land plants, were completely autotrophic.

The designations of nutritional types employed here follow current usage. *Auxotrophs* require growth factors. *Autotrophs* derive all their carbon from CO_2. Van Niel[37] pointed out that there is a continuous quantitative series between complete obligate autotrophs and heterotrophs, and that the reliance on fixed carbon entailed by auxotrophy, while vital to the life of the organism, was only a slight quantitative impairment of autotrophy.   To select an extreme example, a photo-autotroph, auxotrophic only in respect to cobalamin, has lost less than a billionth of its autotrophic capacity.

Rich media can be used in the washing technique since accompanying organisms are eliminated mechanically before the wanted organism is inoculated in the rich isolation medium.   The advent of the washing technique permitted the discovery of auxotrophic algae.   The restriction of the plating technique to mineral media can be partially lifted by adding appropriate antibiotics.   As is to be expected, antibiotics proved to be more effective in combination. Mixtures of penicillin and chloramphenicol or streptomycin were useful.   *Amphora perpusilla* and other diatoms, for example, were isolated from organic agar media containing penicillin 1,000 units/ml. + chloramphenicol 25 μg./ml. Streptomycin, however, is especially poisonous to blue-green algae[5, 22, 27] (1 unit or less may be poisonous) and to a lesser degree to Chlorophyta (poisonous range 1–50 units/ml.).   Certain *Euglenas* are not inhibited by high concentrations of streptomycin but may become permanently colorless.[26]

Determination of Growth Requirements.   Once the organism is obtained in pure culture, much information can be gained in one step by inoculation of a variety of media ranging from mineral to increasingly complex, and at different dilutions.   If complex media are needed (peptones, soil extract, yeast autolysate, *etc.*) the next step is to identify the active constituents.   This is done by stepwise substitution of natural materials with a mixture of better-defined preparations (acid-hydrolyzed gelatin and casein; alkali-hydrolyzed yeast nucleic acid, *etc.*) and pure compounds in the form of a mixture of known vitamins and likely carbon sources.   This screening procedure is illustrated in other papers from this laboratory.[3, 34]

One of the frequent difficulties encountered in applying screening procedures to planktonic organisms is that they tolerate very low overall concentrations of media.   The ever-present problem is to supply materials in adequate yet non-toxic concentrations.   Special care has to be exercised not to overlook mineral deficiencies induced in the process of replacing crude materials with pure compounds.   To avoid precipitates of calcium, iron, and phosphates, it is helpful to add a solubilizing metal-complexer.   Experiments are easier to interpret if the complex-former is unmetabolizable and does not penetrate the organism.   In using such metal-buffering agents, essential trace elements ordinarily adequately supplied as impurities in the "chemically pure" minerals, become inadequate.   The need then becomes strikingly evident, and the proper concentrations of these trace elements have to be determined.[28]   Occasionally, growth factors can be required which may be new to biochemistry.

Chemically-Defined Media.   Media developed for determination of the growth factor requirements of *Peridinium* sp., *Gyrodinium* sp. *Cyanophora paradoxa*, and *Synura* sp. are given in TABLE 3.   The *Gyrodinium* is marine and therefore its medium is a substitute for enriched seawater.   Its isolation medium consisted of seawater enriched with NO_3, PO_4, Fe, and soil extract.

The identification of requirements for growth factors and trace elements, and the recognition of optimal ratios and concentration boundaries for major inorganic and organic nutrients are treated here as separate topics for the sake of arriving at a general picture of nutritional specializations in algae. Obviously they are some of the interrelated factors to be considered in devising good artificial media. The identification of the nutritional requirements in all these categories for *Peridinium* and *Cyanophora* is advanced enough to allow

TABLE 3

CHEMICALLY DEFINED MEDIA FOR IDENTIFICATION OF VITAMINS (MG. PER CENT)

	Cyanophora paradoxa	Peridinium sp.	Synura sp.	Gyrodinium sp.
Ethylenediamine tetraacetic acid (EDTA)		30.0	5.0	20.0
Citric acid·H₂O	30.0			
NaNO₃			2.0	
NH₄NO₃	3.0			
KNO₃		5.0		10.0
KH₂PO₄	10.0	2.0	1.4	2.0
MgSO₄·7H₂O	5.0	10.0	2.0	30.0
*MgCl₂·6H₂O				300.0
NaCl				2400.0
CaCO₃	10.0	10.0		
Ca (as Cl)		4.0	1.3	5.0
KCl			1.0	30.0
Fe (as NO₃ + Cl)	0.3	0.2	0.07	0.3
Zn (as Cl)	0.01	0.08	1.0	0.4
Mn (as Cl)	0.015	0.08	0.2	1.0
Mo (as Na salt)	0.01	0.05	0.001	0.05
Co (as Cl)	0.0004	0.001	0.003	0.003
Cu (as Cl)	0.0001	0.0001	0.0005	0.0003
†"S1 metals"				0.5 ml./100 ml.
Boron (as H₃BO₃)				0.2
Na H glutamate	15.0	100.0	10.0	2.0
Glycine		70.0		
DL-Alanine		70.0·		
Na acetate·3H₂O	10.0		4.0	
Glucose		50.0		
DL-Lysine HCl				1.0
DL-Leucine				0.2
	pH 6.3	pH 6.2	pH 5.5	pH 7.5

* Concentration in stock solution standardized by chloride determination.
† "S1 metals": 1.0 ml. of the mixture contains the following metals: Sr (as Cl) 1.3 mg.; Al (as Cl) 0.05 mg.; Rb (as Cl) 0.02 mg.; Li (as Cl) 0.01 mg.; I (as KI) 0.005 mg.; also Br (as NaBr) 6.5 mg.
Vitamin concentrations needed for growth: Cobalamin is added for *Gyrodinium* and *Cyanophora* at the 0.1 μg.% level; for *Peridinium*, 0.06 μg. per cent; and for *Synura*, 0.15 μg.per cent. The need for other vitamins is under study.
Sources of metals. The Zn was prepared by dissolving Johnson-Matthey purified zinc rod in distilled HCl. The Co and Mn were Johnson-Matthey "spec-pure" solutions of the chlorides. The Fe solution was prepared by dissolving "spec-pure" metal in aqua regia compounded of distilled HCl and HNO₃; excess acid was then evaporated off.

reproducible macroscopic growth in a month. The other organisms are, as yet, unsatisfactorily known in one or more categories. It has not been determined whether the medium for *Synura* will support indefinitely subculturable growth.

Growth-Factor Requirements

The findings to date for the flagellated and non-flagellated algae are summarized in TABLE 4. Many chlorophyceans, not included in this table, are maintained in algal collections on mineral agar. The great majority of these are probably non-auxotrophic, but a more rigorous avoidance of chemical contaminations (thiamine, cobalamin, *etc.*) from agar, cotton, and distilled water[31] might reveal auxotrophy in a few.   Several *Euglenas* are omitted, since it is not clear whether or not a sufficiently rigorous technique was applied.   It is not unlikely that all euglenids require cobalamin and thiamine.

Four general conclusions may be drawn from TABLE 4:

TABLE 4

INCIDENCE OF GROWTH-FACTOR REQUIREMENTS IN ALGAE
(Numbers in parentheses are references)

Species	Vitamins needed	Thiamine	Cobalamin	Other
CHLOROPHYTA				
Chlamydomonas agloëformis	0 (7)*			
Chlamydomonas chlamydogama	+ (23)		+	Histidine
Chlamydomonas moewusii	0 (11)			
Chlamydomonas sp. ("marine")	0 (23)			
Chlorogonium elongatum	0 (7)			
Chlorogonium euchlorum	0 (7)			
Coelastrum (*morus* ?)	+ (15)	+		
Haematococcus pluvialis	0 (7)			
Lobomonas pyriformis	0 (7)			
Lobomonas rostrata	+ (15)		+	
Polytoma caudatum	+ (7, 18)	+		
Polytoma obtusum	0 (7, 18)			
Polytoma ocellatum	+ (7, 18)	+		
Polytoma uvella	0 (7, 18)			
Polytomella caeca	+ (7, 18)	+		
Prototheca zopfii	+ (1)	+		
Selenastrum minutum (E. A. George strains)	0 (15)			
Selanastrum (*minutum* ?)	+ (15)	+		
CHRYSOPHYTA				
Amphora perpusilla	+ (12)		+	Uracil ?
Nitzschia closterium f. *minutissima*	0 (10)			
Nitzschia putrida	0 (12)			
Ochromonas malhamensis (3 strains)	+ (13)	+	+	Biotin + histidine
Poteriochromonas stipitata	+ (13)	+	+	Biotin + histidine
Synura sp.	+	?	+	?
Syracosphaera carterae	+	?	?	?
EUGLENOPHYTA				
Euglena gracilis vars. *typica, bacillaris, urophora*	+ (14)	+	+	
Astasia longa (= *klebsii* Von Dach)	+ (36)	+	+	
E. gracilis, streptomycin-bleached	+ (31)	+	+	
E. pisciformis	+ (7, 18)	+	?	
E. viridis, E. stellata	+ (11)	+	+	
PYRROPHYTA				
Chilomonas paramoecium	+ (7, 18)	+		
Cryptomonas ovata var. *palustris*	+	?	+	?
Cyanophora paradoxa	+		+	
Gymnodinium splendens	+ (35)†		+	
Gyrodinium sp. (marine)	+	?	+	?
Peridinium sp.	+	+	+	?

* Most of the earlier literature on auxotrophy in algal flagellates is reviewed in reference (7). Where the reference is not given, the data refer to the present paper.
† Personal communication from Doctor Beatrice M. Sweeney.

(1) Algae, like autotrophic bacteria, may be divided into auxotrophs and non-auxotrophs.

(2) Thiamine and cobalamin (vitamin B_{12}) are needed by the majority of the auxotrophs. The auxotrophic forms occur in all groups of algal flagellates and auxotrophy does not correlate with any particular environment: Most species of *Polytoma*, *Astasia*, and *Euglena* are found in environments rich in organic matter (polysaprobic); *Synura* and *Cryptomonas* live in waters poor in organic matter (oligosaprobic); *Gyrodinium*, *Nitzschia* (2 species, both non-auxotrophic, one, the colorless *N. putrida*, completely heterotrophic), and *Amphora* are marine. Considering the multiplicity of metabolites in nature, this pattern of dependency on cobalamin and thiamine is remarkably stereotyped and leads one to wonder whether this expresses a deep-seated evolutionary tendency in their physiology.

(3) Other vitamins may be needed besides the seemingly predominant thiamine and cobalamin; *e.g.*, biotin is required by *Ochromonas malhamensis* and *Poteriochromonas stipitata*. It might be expected that additional vitamins will be found to be required as the number of species of algae in pure culture increases.

(4) Auxotrophy is present in organisms having different degrees of heterotrophy. It may be the only heterotrophic requirement, as it appears to be in *Synura*, or it might be only one element of a many-sided heterotrophy: *e.g.*; *Euglena gracilis* utilizes various substrates, and *Ochromonas* and *Poteriochromonas* are photosynthetic phagotrophs exemplifying the coexistence of plant and animal nutrition, phagotrophy being the principal mode of heterotrophy in animals.

Auxotrophy in algae was first detected by Lwoff and Dusi[19] in several genera of colorless chlamydomonads, in a colorless cryptomonad, and in *Euglena pisciformis*. References to the earlier literature are given by Lwoff[18] and Hall.[7]

The question arises as to the extent to which auxotrophy is a sign of a heterotrophic tendency in algae. A better understanding may be obtained by reviewing the evolutionary trends in algae leading to the development of the animal forms, *i.e.*, the Protozoa. The new data on the physiology of *Euglena* and *Ochromonas* permit elaboration of the physiological implications that Lwoff[17] attached to the morphological evolutionary tendencies recognized and systematized by Pascher and by Fritsch.[6] These are outlined in TABLE 5. It will be noted that each algal group exhibits certain conspicuous morphological tendencies. These, in so far as they affect the photosynthetic apparatus, are bound to be reflected in a changed nutritional pattern. Obviously, when photosynthesis is lost, survival demands a compensatory heterotrophic ability. In different algal groups heterotrophy may assume either of two forms: In phytomonads, osmotrophy; in euglenids and all other groups, osmotrophy and phagotrophy. The clearest index of an adequate heterotrophy is the ability to grow in darkness, as is shown by *Chlamydomonas agloëformis*, *Chlorogonium euchlorum*, *Euglena gracilis*, *Ochromonas malhamensis*, *Poteriochromonas stipitata, etc.* Many algae, however, utilize fixed carbon without being able to grow in darkness. Their heterotrophy is less efficient than is that of the dark-growing species.   The existence of facultatively heterotrophic, photosynthetic organisms clearly indicates that facultative heterotrophy precedes loss of photosynthetic power and can be considered as a sign of an evolutionary tendency toward animal nutrition.   In nature, organisms with dual capacities—phototrophy and heterotrophy—are abundant.   Depending on external conditions, one or the other mode of nutrition may predominate.   It does not necessarily follow that pigmented organisms with an efficient heterotrophy have a weakened photosynthesis.   Cramer and Myers[4] have shown that *Euglena gracilis* var. *bacillaris*, when provided with sufficient light and CO_2, photosynthesizes at least as vigorously as *Chlorella*.   Furthermore, autotrophic growth under these conditions equals the best heterotrophic growth.   The relative importance of autotrophy, osmotrophy, and phagotrophy in *Ochromonas* and *Poteriochromonas* remains to be determined.   In view of the tendency already noted of all algal flagellates to lose photosynthetic pigments, the widespread

TABLE 5

GROUP TENDENCIES IN ALGAL FLAGELLATES*

	Tendency to loss of:		Tendency to phagotrophy
	chlorophyll	plastids	
Colonial Volvocales	none	none	none
Chlamydomonadaceae	pronounced	none	very rare
Polyblepharidaceae	pronounced	none	none
Euglenoids	pronounced	none	doubtful
Peranemids	total	widespread	total
Cryptomonads	rare	rare	very rare
Chrysomonads	widespread	widespread	widespread
Dinoflagellates	pronounced	pronounced	widespread

* Revised from Lwoff,[17] p. 216.

occurrence of heterotrophy and auxotrophy in pigmented forms is not surprising.   The minute heterotrophy represented by auxotrophy in algae is nevertheless all-important in determining which algae will multiply under a given set of circumstances.   It is also tempting to look upon auxotrophy as one of the forerunners of a more pronounced heterotrophy.

Presence of Growth Factors in Natural Waters.   It is a reasonable assumption that if an organism requires a growth factor *in vitro*, then this metabolite or its physiological equivalent should be found in significant amount in the environment.   Only the actual determination of the seasonal variation of these vitamins in nature will enable detection of the times when these vitamins actually become the limiting factors for growth.   A promising start has been made. Robbins *et al.*[29] charted the fluctuation of cobalamin content of a small pond which had shown *Euglena* blooms.   The amounts of cobalamin found were high enough to satisfy the cobalamin requirement of *Euglena*.   Hutchinson[9] earlier showed that a thiamine cycle existed in a pond but did not attempt to correlate this with any particular organism.

The question arises as to the identity of the principal vitamin-producers in nature.   Robbins *et al.*[29] found that soil extract contains appreciable amounts of cobalamin and that 40 per cent of the bacteria from mud isolated at random excreted cobalamin. Since cycles of metabolites in soil parallel those in water, Lochhead and Thexton's[16] demonstration is apropos that 70–84 per cent of the bacteria isolated at random from soil are cobalamin producers and that 14 per cent require soil extract. It is interesting to note that half the bacteria initially requiring soil extract were satisfied by cobalamin.

The data of TABLE 4 provide examples of the importance of external metabolites—of "non-predatory" relationships—existing in the water environment. Only a small part of the cobalamin cycle is known. The literature indicates that many other organisms participate in it. As with bacteria, algae include cobalamin-dependent organisms and cobalamin producers. Robbins *et al.* found that an extract of 7 fresh-water blue-green algae grown in mineral media contained relatively high amounts of cobalamin. The uninoculated medium showed no activity. Extracts of several red and brown algae contained cobalamin, especially *Ceramium rubrum*. Since the algae were obtained directly from the shore there is some doubt as to whether all the cobalamin was theirs. Oysters and clams were found by Robbins *et al.*[30] to contain large amounts of cobalamin, suggesting that algal flagellates, an important item in their diet, are either producers or accumulators of cobalamin.

Requirements for Trace Metals

Requirements of Synura. This organism, often abundant in unpolluted waters such as reservoirs, has long challenged the nutritionist as well as the hydrobiologist. It is selected here to exemplify the problems faced in culturing a planktont requiring very dilute media. It was cultured by Mainx[21] in a highly diluted soil extract. Rodhe[32] grew it in a very dilute mineral medium (his "medium VIII"; see TABLE 6) supplemented with an extract of soil or lake sediment. Ashed extracts were ineffective. Our medium (TABLE 3) was developed by replacing peat extract with a mixture of known vitamins. Since the medium is prone to precipitate, EDTA, an efficient solubilizing chelating agent, was added. This procedure revealed, as expected, deficiencies in trace metals which were satisfied by finding the suitable concentration of a series of different trace metal mixtures. Then, keeping constant the levels of trace elements, the major elements (Ca, Mg, K, NO_3, PO_4, and Fe) were adjusted to levels permitting better growth. Using, in turn, this information, the suitable concentration of each of the trace metals, previously added as a mixture, was determined. Inoculation of the experiments, at first, was from stock cultures maintained in the isolation medium, later from preceding experiments. Growth in defined media was better than in the original peat medium but decreased after repeated transfers. This indicated that the defined medium is still incomplete and that carry-over of factors from peat may play a role. In an attempt to identify other substances needed for growth, experiments are under way to identify the effective vitamins in the dilute vitamin mixture used. Cobalamin is required; but additional vitamins may be needed. The effect of exogenous carbon sources was also explored. In unagitated cultures, not supplied with extra CO_2, growth was unaffected by single additions of glutamate, acetate, fumarate, glucose, glycine, and alanine. It should be noted that Zn is required at a very high concentration as compared with Ca, Mg, and K, which are usually designated as major nutrients. *Euglena gracilis* var. *bacillaris*, supplied with a 10-fold higher concentration of EDTA, does not require for heavy growth more than 2.0 mg. per cent of Zn at pH 5.0.

Some acute problems are presented by dilute media; *e.g.*, adequate pH buffering is difficult to achieve because of the low tolerance to phosphate and osmotic pressure. Aconitic, trimesic, and succinic acids show promise as buffers in the acid region. Dilute media are especially difficult to reproduce, since environmental chemical contaminations from glassware, distilled water, and especially from "c. p." chemicals, may be of the same magnitude as the nutrients added intentionally. Media built around EDTA are an advance toward reproducibility. EDTA's metal-sequestering abilities make it, at the same time, a solubilizing agent preventing precipitation and a metal-buffering agent raising the threshold of metal availability and toxicity. Hit-or-miss dependence on "chemically pure" salts as a source of essential trace elements is thus minimized; now the trace elements have to be added as such.

Comparison of artificial media and lake waters. Chu and Rodhe were the pioneers in devising artificial dilute media for many planktonts. Chu[2] developed 16 different media of which "No. 10" permitted growth of several diatoms, a blue-green alga, a desmid, and *Botryococcus*. His other media were developed for the various single species. Rodhe's "No. VIII" served for cultivation of several Chlorococcales, Volvocales, Heterokontae, and desmids. Chu's and Rodhe's media were developed through determination of the concentration range for each constituent. The species cultivated by them are almost certainly complete autotrophs since growth factors were not supplied. As mentioned earlier in the discussion of growth factors, auxotrophy cannot be excluded until a rigorous technique is employed. However, Rodhe's failure to grow *Synura*, which requires cobalamin, indicates that his mineral media were not significantly contaminated with cobalamin, one of the most powerful and ubiquitous of growth factors. Our media, developed independently, ended by being as dilute as theirs. Our experiments with *Cryptomonas ovata* var. *palustris* indicate that, as a representative planktont, it has a very low osmotic tolerance. As little as 30 mg. per cent NaCl was inhibitory. These results also agreed with tolerances to neutral salts of organic acids.

TABLE 6 shows the striking similarity of the three mineral media and the similarity existing between them and Rodhe's standard mineral composition of lake waters having a similar content in total solids. This parallelism between laboratory results and analyses of natural waters validates the idea that laboratory findings in respect to growth factors and mineral requirements are directly relevant to ecological problems. As already mentioned, use of a chelator reveals sharply the indispensability of trace elements. The finding of Chu that addition of trace elements was unnecessary implies that the impurities present in his major chemicals were adequate. Rodhe made an extensive study of the precipitation—*i.e.*, unavailability—of Fe and found that addition of a Fe citrate-citric acid combination prevented precipitation. Although citric acid is only a moderately strong metal-binder and was used in very small amount, the need for Mn became evident. The use of chelators in the laboratory parallels the situation in natural waters where humic acids are among the most important, almost unmetabolizable, metal-buffering agents. Since planktonic organisms such as *Synura, Peridinium, Cryptomonas,* and *Trachelomonas* have been shown to require trace metals, more importance should be assigned to trace metals as ecological factors. A recent paper by Rodhe[33] signalizes the increasing interest in the trace element content of natural waters.

Phosphate requirements. An *in vitro* result that cannot yet be correlated with ecological data is the requirement for phosphorus. In fact, the concentrations needed in synthetic media far exceed those normally present in nature. For instance, Chu and Rodhe[32] found that *Asterionella formosa* is indifferent to

TABLE 6

COMPARISON OF LAKE WATER OF STANDARD COMPOSITION AND ARTIFICIAL MEDIA
(Mg. per liter)

Elements	Rodhe's Standard Composition*	Chu No. 10	Rodhe No. VIII	Synura Medium†
N	0.5 –0.6‡	7.0	10.2	3.3
P	0.03–0.05‡	1.8–9.0	0.89	3.1
Na	4.6	19.0	7.5	5.4
K	1.7	2.2–4.4	2.2	9.0
Mg	2.7	2.5	1.0	2.0
Ca	16.3	10.0	14.7	13.0
Fe	0.4–1.2‡	0.27	0.18	0.7
Mn	0.02–0.14‡		0.01	†
Total Solids	98.0	115–120	181	102–179**
Conductivity ($\psi.10^{-6}$)	120.0			

* Values from TABLE 2 of Rodhe, W., 1949. The ionic composition of lake waters. Proc. Intern. Assoc· Limnol. **10**: 377–386.
† See TABLE 3 for concentrations of EDTA and minor elements; 0.2 ml/100 of a vitamin supplement was added (formula in Cowperthwaite *et al.*[3]).
‡ Variation intervals, calculated from Rodhe's[33] tables, for a conductivity of 120.
** In the higher value the sodium salts of glutamic and acetic acids are not included (*Synura* can grow without this addition).
In the lower value glutamate, acetate, trace metals and EDTA are not included.

10–20 µg. P/liter. Maximal growth was obtained at about 1,000 µg. P/liter. Rodhe found that, in Lake Erken, where this organism is common, the P content scarcely attains 10 µg./liter. When his artificial inorganic basal medium was replaced by lake water (sterilized by filtration), maximum growth was obtained with the extremely small addition of 4–10 µg. P/liter. This striking discrepancy in utilization of P appears to support Rodhe's assumption that lake waters contain one or more factors facilitating the use by *Asterionella* of P at very low concentrations. A peculiarity in P utilization by *Cryptomonas ovata* may be relevant. In one experiment, different PO_4 optima were observed on varying the concentration of magnesium sulfate. For $MgSO_4 \cdot 7H_2O$ at 2.0 mg. per cent, the optimal level of KH_2PO_4 was 1.0 mg. per cent; while for $MgSO_4 \cdot 7H_2O$ 0.5 mg. per cent, the optimal level of KH_2PO_4 was 0.4 mg. per cent. If these preliminary results prove repeatable, it will be determined whether this varying response to KH_2PO_4 depends on the ratio of Mg to K. In any event, Rodhe's experiment should stimulate a search for factors in natural waters permitting growth at low P concentrations. P is undoubtedly one of the important natural limiting factors for most algae. Rodhe found, on the other hand, that blooms of *Dinobryon* and *Uroglena* appeared only when the P level was below 5 μg./liter and, experimentally, their growth was inhibited upon addition of 5–10 μg. P/liter. Our results with *Peridinium*, *Gyrodinium*, *Synura*, *Cryptomonas*, and *Trachelomonas* further document the well-established fact that the need for PO_4 varies among algae, and indicate, moreover, that some plankton organisms have a surprisingly narrow optimal PO_4 range. This last point may be important in ecology.

Discussion

The water environment is the one in which metabolites are interchanged most efficiently. It is to be expected that the interdependent growth of the different groups of water organisms should sensitively reflect the excretion and consumption of metabolites. Undaunted by new intricacies, we should envisage all the possibilities in these relationships, and not hesitate to follow Lucas's lead in constructing theoretical frameworks upon which to hang data. In the present paper, only a few aspects of the nutrition of photosynthetic forms are considered. It is possible, nevertheless, to state more definitely some of the interdependencies based upon "external metabolites": (1) the interchange of growth factors; (2) the lowering of inhibitory concentrations of several major mineral nutrients, especially PO_4; and (3) the preferential utilization of minerals, including trace metals, may condition waters, bringing their concentrations into the optimal zones for succeeding forms. The practical aim—to predict algal successions and blooms—may be achieved through a comprehensive knowledge of vitamin cycles as well as mineral cycles. An immediate problem is to trace the thiamine and cobalamin cycles. In fresh-waters, these can be worked out by applying techniques such as those used by Hutchinson and by Robbins *et al.*, but tracing these vitamins in the ocean requires the exploitation of new assay organisms and concentration techniques.

The role of other organic materials as ecological factors remains problematic at the moment. It is likely that, in waters high in organic content, such as sewage, organic compounds may serve as substrates for photosynthetic organisms endowed with heterotrophic abilities. A familiar example of this utilization of substrates is presented by the algae which form blooms in sewage lagoons and similar polysaprobic environments. The algae, *e.g.*, *Euglena gracilis* and *Chlorella*, isolated from these habitats are well endowed with heterotrophic abilities.

The coastal waters of well-vegetated, well-populated land masses should be comparatively rich in organic matter. Many phytoplanktonts appear to be coastal rather than pelagic—an indication that the land may be a significant source of essential metabolites.

Summary

Chemically defined media were devised for *Peridinium* sp., *Cyanophora paradoxa*, *Synura* sp., and a marine *Gyrodinium*. All require cobalamin (vitamin B_{12}). Possible additional growth-factor requirements are under investigation. The occurrence of auxotrophy in algae is reviewed. All algal groups and environments contain auxotrophic forms. Auxotrophy is more prevalent in algae than previously recognized. Planktonic algae often require media very low in solutes. These media correspond in composition to natural waters. The need for several trace elements, notably Zn, Mn, Cu, and Co became evident upon use of a metal-complexer (EDTA) as a solubilizing metal-buffer.

The need *in vitro* of growth factors and trace elements, and the narrow optimal ranges of phosphorus, suggest that all these are significant ecological factors. From the mapping of the cycles of growth factors and trace elements, there would accrue a better understanding of the occurrence of blooms and the succession of species in nature.

Acknowledgments

This investigation was supported (in part) by a research grant (G 3216) from the National Microbiological Institute of the National Institutes of Health, Public Health Service; the Lederle Laboratories Division of the American Cyanamid Co.; and the Rockefeller Foundation. We are indebted to Doctor C. Mervin Palmer of the Environmental Health Center, U. S. Public Health Service, for many stimulating discussions. John F. Howell of the U. S. Fish and Wildlife Service and John J. A. McLaughlin participated in the work on *Gyrodinium* and other marine organisms. We are indebted to Doctor Beatrice M. Sweeney of the Scripps Institute of Oceanography for sending us the original *Gyrodinium* material and for unpublished information on the cobalamin requirement of *Gymnodinium splendens*. Doctor Daniel L. Lilly kindly sent the original water sample containing *Synura*. Professor E. G. Pringsheim kindly sent us the *Cyanophora* and *Peridinium* material and the pure culture of *Cryptomonas ovata* along with information on their maintenance. Doctor Trygve Braarud sent us an impure culture of *Syracosphaera carterae*.

References

1. ANDERSON, E. H. 1945. Nature of the growth factors for the colorless alga *Prototheca zopfii*. J. Gen. Physiol. **28**: 287–296.
2. CHU, S. P. 1942. The influence of the mineral composition of the medium on the growth of planktonic algae. J. Ecol. **30**: 284–325.
3. COWPERTHWAITE, J., M. M. WEBER, L. PACKER, & S. H. HUTNER. 1953. Nutrition of *Herpetomonas* (*Strigomonas*) *culicidarum*. Ann. N. Y. Acad. Sci. **56**(5) : 972–981.
4. CRAMER, M. & J. MYERS. 1952. Growth and photosynthetic characteristics of *Euglena gracilis*. Arch. Mikrobiol **17**: 384–402.
5. FOTER, M. J., C. M. PALMER, & T. E. MALONEY. 1953. Antialgal properties of various antibiotics. Antibiotics & Chemotherapy. **3**: 505–508.
6. FRITSCH, F. E. 1935. The Structure and Reproduction of the Algae. Cambridge Univ. Press.
7. HALL, R. P. 1943. Growth-factors for protozoa. Vitamins and Hormones. **1**: 249–268.
8. HARVEY, H. W. 1945. Recent Advances in the Chemistry and Biology of Sea Water. Cambridge Univ. Press.
9. HUTCHINSON, E. G. 1943. Thiamin in lake waters and aquatic organisms. Arch. Biochem. **2**: 143–150.
10. HUTNER, S. H. 1948. Essentiality of constituents of sea water for growth of a marine diatom. Trans. N. Y. Acad. Sci. **10**: 136–141.
11. HUTNER, S. H. & L. PROVASOLI. 1951. The phytoflagellates. Biochemistry and Physiology of Protozoa: 27–128. A. Lwoff, Ed. Academic Press. N. Y.
12. HUTNER, S. H. & L. PROVASOLI. 1953. A pigmented marine diatom requiring vitamin B$_{12}$ and uracil. News Bull. Phycological Soc. Am. **6**: 7–8.

13. HUTNER, S. H., L. PROVASOLI, & J. FILFUS. 1953. Nutrition of some phagotrophic fresh-water chrysomonads. Ann. N. Y. Acad. Sci. 56(5): 852-862.

14. HUTNER, S. H., L. PROVASOLI, E. L. R. STOKSTAD, C. E. HOFFMANN, M. BELT, A. L. FRANKLIN, & T. H. JUKES. 1949. Assay of anti-pernicious anemia factor with Euglena. Proc. Soc. Exptl. Biol. Med. 70: 118–120.

15. LEWIN, R. A. 1952. Vitamin requirements in the Chlorococcales. News Bull. Phycological Soc. Am. 5: 21–22.

16. LOCHHEAD, A. G. & R. H. THEXTON. 1951. Vitamin B_{12} as a growth factor for soil bacteria. Nature. 167: 1034–1035.

17. LWOFF, A. 1943. L'Évolution physiologique. Actualités sci. et ind.: 970. H. HERMANN et Cie. Paris.

18. LWOFF, A. 1947. Some aspects of the problem of growth factors for protozoa. Ann. Rev. Microbiol. 1: 101–114.

19. LWOFF, A. & H. DUSI. 1938. Culture de divers flagellés leucophytes en milieu synthétique. Compt. rend. soc. biol. 127: 53–55.

20. LUCAS, C. E. 1947. The ecological effects of external metabolites. Biol. Revs. Cambridge Phil. Soc. 22: 270–295.

21. MAINX, F. 1929. Untersuchungen über den Einfluss von Aussenfaktoren auf die phototaktische Stimmung. Arch. Protistenk. 68: 105–176.

22. PALMER, C. M., M. J. FOTER, & T. E. MALONEY. Effect of streptomycin on cultures of algae. Paper in preparation.

23. PINTNER, I. J., L. PROVASOLI, & S. H. HUTNER. Unpublished.

24. PRINGSHEIM, E. G. 1946. Pure Cultures of Algae. Cambridge Univ. Press.

25. PRINGSHEIM, E. G. & O. PRINGSHEIM. 1952. Experimental elimination of chromatophores and eye-spot in Euglena gracilis. New Phytologist. 51: 65–76.

26. PROVASOLI, L., S. H. HUTNER, & I. J. PINTNER. 1951. Destruction of chloroplasts by streptomycin. Cold Spring Harbor Symposia Quant. Biol. 16: 113–120.

27. PROVASOLI, L., I. J. PINTNER, & L. PACKER. 1951. Use of antibiotics in obtaining pure cultures of algae and protozoa. Proc. Soc. Protozool. 2: 6. Detailed paper in preparation.

28. REISCHER, H. S. 1951. Growth of Saprolegniaceae in synthetic media. I. Inorganic nutrition. Mycologia. 43: 142–155.

29. ROBBINS, W. J., A. HERVEY, & M. E. STEBBINS. 1950. Studies on Euglena and vitamin B_{12}. Bull. Torrey Botan. Club. 77: 423–441.

30. ROBBINS, W. J., A. HERVEY, & M. E. STEBBINS. 1951. Further observations on Euglena and B_{12}. Bull. Torrey Botan. Club. 78: 363–375.

31. ROBBINS, W. J., A. HERVEY, & M. E. STEBBINS. 1953. Euglena and vitamin B_{12}. V. Ann. N. Y. Acad. Sci. 56(5): 818–830.

32. RODHE, W. 1948. Environmental Requirements of fresh-water plankton algae. Symbolae Botan. Upsaliensis. 10: 1–149.

33. RODHE, W. 1951. Minor constituents in lake waters. Proc. Intern. Assoc. Theoretical Appl. Limnol. 11: 317–323.

34. STORM, J. & S. H. HUTNER. 1953. Nutrition of Peranema. Ann. N. Y. Acad. Sci. 56(5): 901–909.

35. SWEENEY, B. M. 1952. Growth of the dinoflagellate Gymnodinium in culture. Abstract. Intern. Seaweed Symposium. Edinburgh.

35a. SWEENEY, B. M. Gymnodinium splendens, a dinoflagellate requiring vitamin B_{12}. In preparation.

36. THAYER, P. S. 1949. Studies on the nutrition of Astasia klebsii in synthetic medium. Thesis. Amherst College. Amherst, Mass. Cited in G. W. Kidder. 1951. Ann. Rev. Microbiol. 5: 144.

37. VAN NIEL, C. B. 1943. Biochemical problems of the chemo-autotrophic bacteria. Physiol. Revs. 23: 338–354.

38. v. WETTSTEIN, F. 1921. Zur Bedeutung und Technik der Reinkultur für Systematik und Floristik der Algen. Österr. Botan. Z. 70: 23–29.

13

Reprinted from *Deep-Sea Research*, **19**, 601–618 (1972)

Nitrogen-limited growth of marine phytoplankton—I.
Changes in population characteristics
with steady-state growth rate*

JOHN CAPERON† and JUDITH MEYER†

(*Received* 14 *December* 1971; *in revised form* 6 *June* 1972; *accepted* 9 *June* 1972)

Abstract—Steady-state growth of *Coccochloris stagnina*, *Cyclotella nana*, *Monochrysis lutheri* and *Dunaliella tertiolecta* in nitrate limiting medium and of *Dunaliella* and *Monochrysis* in ammonium limiting medium is examined in continuous culture experiments. Growth rate can not be directly related to observed nutrient concentration in a chemostat environment except perhaps in a long-term average sense. Nitrate-limited growth rate is related to internal nitrogen per unit population by a hyperbolic expression. These results are consistent with an internal reservoir nutritional mechanism. Steady-state ammonium limited growth takes place without evidence of an internal reservoir, and nitrogen per cell remains constant over all growth rates studied. The relative variability of population carbon, nitrogen, chlorophyll-*a*, cell volume and cell concentration with steady-state growth rate is examined in the context of using these indicators of the physiological state of the population, rather than environmental parameters, to determine steady-state nutrient-limited growth rate.

INTRODUCTION

SVERDRUP (1955) presented a remarkably accurate map of primary productivity in the world oceans exclusively from considerations of the supply of nutrient-rich deep water to the euphotic zone. The supply of nutrients thus appears to exercise primary control over plant growth in the marine environment. It seems likely that among the nutrients essential to phytoplankton growth, some form of fixed nitrogen is often in shortest supply (RYTHER and DUNSTAN, 1971; THOMAS, 1970; THOMAS and OWEN, 1971).

Oceanographic observations frequently include determinations of nutrient concentrations as an indicator of biological activity, and there is a rather large volume of observations on the concentration of various forms of fixed nitrogen. Rarely are these data accompanied by phytoplankton growth rate or substrate utilization rate measurements. If the relation between phytoplankton growth rate and the concentration of fixed nitrogen in the environment can be quantitatively established, then all of these data will indeed serve our objective of understanding biological activity in the ocean. Efforts to predict phytoplankton growth from environmental factors have included consideration of nutrient concentration (STEELE, 1959; RILEY, 1963). In this work the effect of nutrient concentration on growth was considered linear. This is quite satisfactory over a small nutrient concentration range and a useful approximation over more extensive ranges. A non-linear treatment of this relationship exists (DUGDALE, 1967), but this model is restricted in its present form to applications where cell nitrogen is a satisfactory measure of the phytoplankton population. The present work is directed toward a quantitative description of growth rate limited by fixed nitrogen using several other measures of population size.

Following MONOD'S (1942) demonstration that the Michaelis–Menten kinetics adequately describe nutrient-limited growth of a bacterial population, several efforts

---

*Hawaii Institute of Marine Biology Contribution Number 387.

†Hawaii Institute of Marine Biology, P.O. Box 1067, Kaneohe, Hawaii 96744, U.S.A.

were made to extend these results to phytoplankton (CAPERON, 1965; DROOP, 1961; WILLIAMS, 1965). Growth rate in phytoplankton does not depend on current environmental nutrient concentrations; rather, it depends upon the nutrient experience during a period of time preceding the time at which the growth rate determination is made (CAPERON, 1969; WILLIAMS, 1971). For this reason, growth rates obtained from batch culture experiments where the nutrient experience of the population is continually changing are difficult to interpret.

Continuous culture techniques circumvent these difficulties by making growth rate determinations at constant nutrient conditions possible. However, continuous culture experiments with phytoplankton species have uniformly failed to measure significant differences in the nutrient concentration in the growth chamber over the range of growth rates studied (CAPERON, 1968; DROOP, 1968; FUHS, 1969; WILLIAMS, 1971). WILLIAMS (1971) reports nutrient concentration to be 20 to 100 times lower than that predicted by basic theory of growth dynamics. Clearly, there are problems with this theory as applied to phytoplankton in chemostats. The continuous culture data have been used to provide data on the yield coefficient, q (the amount of limiting nutrient per unit population). This has been shown to vary with steady-state growth rate and is related to growth rate by a Michaelis–Menten type hyperbola (CAPERON, 1968; DROOP, 1968). WILLIAMS (1965), DROOP, (1968), and CAPERON (1968) have independently hypothesized an intracellular nutrient reservoir as the most likely explanation of these results. This leaves the growth rate still unrelated to the external nutrient concentration, which must be the controlling variable.

At steady state the uptake rate, V, must be proportional to the growth rate, μ. The proportionality constant is the variable yield coefficient, q. DROOP (1968) used this relationship for *Monochrysis lutheri* growing on limiting vitamin B-12 medium to combine continuous culture growth rate data with batch culture uptake rate data. The uptake rate was related to environmental nutrient concentration by a Michaelis–Menten hyperbola. The combined result provides growth rate as a function of environmental limiting nutrient. We use essentially the same approach for fixed nitrogen limited growth.

It has been pointed out that if q does not vary with growth, then uptake kinetics by themselves are sufficient to describe steady-state growth rate (EPPLEY and THOMAS, 1969). In the absence of any data suggesting that q is constant, it seems best for the present to accept the abundant data on its variability (CAPERON, 1968; DAVIES, 1970; DROOP, 1968; FUHS, 1969).

Determinations of growth rate as a function of initial nutrient concentrations in batch culture experiments give reasonable fits to Michaelis–Menten hyperbolas (EPPLEY and THOMAS, 1969; THOMAS, 1970). In these experiments, growth rate is determined from logarithmic plots of population size vs. time over a period of time (generally a few days). These growth rate values are then related to the initial nutrient concentration in the growth chamber. The nutrient and growth rate data are not contemporaneous. How much error this introduces is unknown, but it is known that maximum uptake rate is from 2 to 20 times larger than maximum growth rate (DROOP, 1968; EPPLEY and THOMAS, 1969; SYRETT, 1962). At low initial nutrient concentration, the nutrient is nearly exhausted from batch cultures by the time growth rates can be determined. THOMAS and OWEN (1971) compare ^{14}C growth rate observations with growth rates predicted from ammonium ion observations. The predictions are made on the basis of Michaelis–Menten hyperbolic equations using half-saturation constants, K_s, from uptake kinetics (MACISAAC and DUGDALE, 1969) and from batch culture growth data (THOMAS, 1970). The K_s values from uptake data (0·10, 0·55, and 0·62 with a mean of 0·42 μg at/l.) overestimated the growth rate, while the K_s values from batch culture growth data (1·68 and 1·47 with a mean of 1·57 μg at/l.) underestimated the growth rate. Only by averaging the distinctly different data sets were growth values like the ^{14}C observations obtained. It appears that continuous culture data are required, both to overcome the inherent difficulties in relating batch culture growth rate to environmental nutrient concentration and to quantify the relationship between environmental concentration of fixed nitrogen and the two rates, uptake and growth.

MATERIALS AND METHODS

All experiments were conducted in chemostat culture systems. A Pyrex, five-liter, double-walled reaction flask served as a growth chamber. Water circulating through the outer shell of the spherical reaction flask maintained the culture at 25°C. Light at an intensity of 5·8 g cal cm^{-2} hr^{-1} was supplied by constant, 24 hour, illumination from a bank of fluorescent lamps. The cultures were stirred with a teflon-coated magnetic stirring bar and, for the nitrate runs, with bubbles from the aeration tube. The air was passed through a saturated solution of $ZnCl_2$ to remove NH_3. Medium was supplied to the growth chamber through a silicone rubber tube installed in a variable rate peristaltic pump. Slow passage of ammonium-limiting medium through three or four feet of silicone rubber tubing resulted in the loss of most of the ammonium ion in the medium, most likely due to the permeability of the silicone tubing to gaseous ammonia. For ammonium-limited runs the smallest internal diameter and shortest possible length of tubing was used in the pump section. Outside the pump section the nutrient supply line was glass. Aeration in the growth chamber for the ammonium runs was reduced to about one bubble per 5 seconds—just sufficient to provide good functioning of the overflow system. Concentration of limiting ammonium ion in the incoming medium was determined from samples taken down-stream of the pump just before entry into the growth chamber.

Dunaliella tertiolecta (clone Dunal 2, Food Chain Research Group Culture Collection, Institute of Marine Resources, San Diego), *Monochrysis lutheri* (clone Mono-L, FCRG Culture Collection), *Cyclotella nana* (clone 13-1, FCRG Culture Collection), and a cyanophyte, *Coccochloris stagnina* (isolated in this lab from Fanning Island) were grown in nitrate-limited medium. *Dunaliella* and *Monochrysis* were also grown in ammonium-limited medium. The medium otherwise was identical for all experiments. It was made up with four parts nitrogen-poor sea water from waters off Oahu and one part deionized water. The medium was enriched with NaH_2PO_4 (6·9 mg/l.), Na_2SiO_3 (84 mg/l admixed with 0·6 ml 1 N HCl), thiamine HCl (500 μg/l., vitamin B-12 (2 μg/l.), biotin (1 μg/l.), and the chelated trace metal mix used in the FCRG Culture Collection at approximately one-tenth the strength specified for batch culture. Limiting nutrient concentrations varied somewhat from experiment to experiment, but were generally near 5 μg at/l. in the incoming medium. The medium was buffered with bicarbonate ion (40 mg NaHCO$_3$/l) to a pH of 7·9 and sterile filtered into autoclaved carboys that served as nutrient reservoirs. All cultures contained very small background bacteria populations during at least part of the experiments. The sterile techniques employed were directed primarily toward ensuring maintenance of a unialgal culture.

The cadmium–copper reduction method (WOOD, ARMSTRONG and RICHARDS, 1967) was used for nitrate analysis. The ammonium ion concentrations were determined by the method developed by SOLÓRZANO (1969). Cell carbon and nitrogen were determined from the particulate material retained on 47 mm diameter Selas Flotronics 1·2 μ pore size silver filters from 100 ml samples from the growth chamber. These filters were analyzed in an F & M Model 185 CHN Analyzer according to the method of GORDON (1969). Cell counts and volume analysis were made with a Celloscope electronic particle counter. Cell chlorophyll-a was determined from extracted pigments from the particulate material retained on Whatman GF/C glass fiber filters according to the method described by STRICKLAND and PARSONS (1968) using the Parsons–Strickland equations to calculate chlorophyll-a concentration.

RESULTS AND ANALYSIS

Cell concentration was the most convenient measure of population density and was used to monitor the experimental populations routinely. When changes in cell concentrations were relatively small and without trend for three or four days, the population was considered to be in steady-state growth, and the population characteristics (Table 1) were determined. The data (Table 1), limiting nutrient concentration in the growth chamber, population carbon and population nitrogen, represent four to eight samples from the growth chamber collected on the same day, spaced according to the rate at which the particular pumping rate would refill the growth chamber after sample withdrawal. Population nitrogen determined by the difference between incoming and growth chamber nutrient concentration is also presented. The chlorophyll-a values represent a single 500 to 1300 ml sample taken after other steady-state parameters had been determined.

There is no obviously appropriate single measure of phytoplankton population size. The convenience of chlorophyll-a determinations and their usefulness in separating plant from other particulate organic material (LORENZEN, 1966) make this measure the more frequent choice in field studies. Field measurements of growth rate (oxygen evolution or carbon-14 methods) involve, at least implicitly, organic carbon as the population measure. Cell numbers/ml or cell volume/ml is a convenient measure in the laboratory where one is dealing with a single species population, but counts for natural mixed populations are difficult to interpret ecologically and anything but convenient. For populations growing under limiting nitrogen conditions, the internal nitrogen is a useful measure of population size. The data in Table 1 provides some indication of the relative variation of these five measures of population over a range of steady-state nutrient-limited growth rates. The ratios, cell numbers, cell volume, carbon, nitrogen or chlorophyll-a per unit population can be expressed with any of the five serving as the denominator. No clearly preferable choice of population measure is evident. Table 2 gives the results of linear regressions of the ten possible ratios among these variables on steady state growth rate, μ for all populations studied.

Of the 42 linear regressions for which we have data, 10 involve only 3 points. Nine of these 10 have correlation coefficients greater than 0·943. Most of these are not significant at the 10% level, but this is probably due to the limited amount of data. Of the remaining 32, all but eleven are significant at the 10% level. Thus, most of these population measures show considerable growth rate dependent variability. Among the eleven exceptional instances, seven are ratios involving cell volume of *Cyclotella*,

Coccochloris and *Dunaliella*. Cell volume for these three organisms shows insignificant growth rate variability (a result consistent with an earlier observation with *Isochrysis galbana*, CAPERON, 1968). *Monochrysis* volume/cell vs. μ gives a correlation coefficient of 0·967 significant at the 1 % level. Two others of these eleven exceptional cases involve N/cell for the ammonium-limited experiments with *Dunaliella* and *Monochrysis*. Here the slopes are not significantly different from zero and mean N/cell $\pm$ 2 S.D. is 0·078 $\pm$ 0·017 and 0·047 $\pm$ 0·011 $\mu\mu$g at/cell respectively.

The five population measures involved in most of these ratios are, therefore, not interchangeable over a range of growth rates. Since we are interested in ascertaining the nutritional state (internal nitrogen per unit population) of the population at various growth rates studied, the choice of an appropriate measure of population is a very real problem.

Values for the nitrogen content of the population as measured with the CHN analyzer did not always agree with values calculated from chemical analysis using the difference between the incoming and growth chamber nutrient concentration (see Table 1). If this discrepancy were due to excretion of organic nitrogen, one would expect the chemistry values to be consistently higher than the CHN values. A linear regression of CHN values on chemistry values for the series of steady states would then have a non-zero intercept and/or a slope different from 1. This was not the case for ammonium-limited *Dunaliella* and *Monochrysis*, and the regression for *Cyclotella* was not significant (correlation coefficient = 0·05). Thus, the observed differences in the estimates of nitrogen content of the three populations do not clearly indicate that excretion of organic nitrogen occurred. The agreement between the two methods is good for *Dunaliella* and *Monochrysis* on nitrate or ammonium, but not satisfactory for *Cyclotella*. Chemistry values for *Coccochloris* are not available for comparison. We have used the more indirect, but equally valid, chemistry estimate of population nitrogen content only for *Cyclotella*.

Using observed nitrogen to carbon atom ratio as a measure of the nutritional state of the growing population, the growth rate can be satisfactorily related to population nutritional state by a Michaelis–Menten type hyperbola of the form

$$\mu = \mu_m (q - q_0)/[K + (q - q_0)]$$

where μ is specific growth rate and q is N/C. The constants μ_m, K, and q_0 are estimated from the data and represent, respectively, maximum specific growth rate, the half-saturation constant and the nutritional state (N/C) at zero growth rate. Figures 1, 2 and 3 give examples of best fit hyperbolas to the data in Table 1 for *Cyclotella* and *Coccochloris* on nitrate and *Dunaliella* on ammonium. The three sets of growth constants, μ_m, K and q_0, together with values of μ_m determined from batch culture experiments are given in Table 3. The fit of the data to their respective hyperbolas is adequate and the values of μ_m estimated in this way compare well with the values determined independently from batch culture work. The relationship given in Fig. 4 between μ and N/C for *Monochrysis* on ammonium seems linear rather than hyperbolic. A linear fit would do as well as a hyperbola for *Coccochloris* also, but this is to be expected where all growth rate studies are below 1/2 μ_m.

For any other measure of population, a hyperbola of the above form will be an adequate description only if a linear relationship, $q = aq' +$ b, exists with

$$q_0 = aq_0' + b, \ a \neq 0.$$

Table 1. Steady-state population characteristics. Data are presented as mean ± 2 standard deviations.

Growth rate (hr⁻¹)	Limiting nutrient conc. (μg at/l.)	Population carbon (μg at/l.)	Population N/C ratio (atom/atom)		Cells/C [(no./μg at) × 10⁻⁶]	Chl-a/C (μg/μg at)	Cell vol./C [(μ³/μg at) × 10⁻⁹]
			CHN	Chemistry			
Dunaliella (NO₃⁻)							
0·0127 ± 0·0002	0·08 ± 0·08	100·0 ± 9·0	0·051	0·052	0·709	0·082	n.a.
0·0216 ± 0·0004	0·05 ± 0·02	79·0 ± 1·6	0·062	0·065	0·540	0·111	0·029
0·0324 ± 0·0004	0·06 ± 0·04	74·4 ± 8·3	0·080	0·069	0·490	0·127	0·026
Dunaliella (NH₄⁺)							
0·0090 ± 0·0006	0·05 ± 0·13	79·4 ± 17·8	0·044	0·046	0·579	n.a.	0·034
0·0175 ± 0·0012	0·14 ± 0·24	109·5 ± 8·6	0·055	0·043	0·586	n.a.	0·036
0·0266 ± 0·0008	0·14 ± 0·05	55·8 ± 15·6	0·062	0·082	0·848	0·110	0·046
0·0460 ± 0·0008	0·05 ± 0·02	33·5 ± 7·6	0·095	0·122	1·26	0·270	0·084
0·0534 ± 0·0028	0·04 ± 0·10	12·4 ± 3·8	0·127	0·129	1·76	0·411	0·108
Monochrysis (NO₃⁻)							
0·0189 ± 0·0011	0·24 ± 0·11*	73·0 ± 8·0	0·056	0·049	1·11	n.a.	n.a.
Monochrysis (NH₄⁺)							
0·0081 ± 0·0006	0·14 ± 0·07	155·0 ± 15·0	0·051	0·057	1·05	0·036	0·029
0·0168 ± 0·0005	0·06 ± 0·18	135·0 ± 2·4	0·055	0·032	1·15	n.a.	0·033
0·0221 ± 0·0006	0·00	90·9 ± 10·8	0·063	0·062	1·34	0·031	0·045
0·0303 ± 0·0006	0·09 ± 0·06	52·1 ± 6·4	0·058	0·044	1·53	0·060	0·051
0·0330 ± 0·0006	0·15 ± 0·22	66·5 ± 6·4	0·065	0·061	1·47	0·069	0·048
0·0412 ± 0·0010	0·33 ± 0·44	36·3 ± 13·0	0·077	0·092	1·59	n.a.	0·062
0·0501 ± 0·0006	0·38 ± 0·38	51·3 ± 3·8	0·082	0·081	1·50	0·154	0·064
0·0615 ± 0·0008	0·02 ± 0·08	32·2 ± 1·6	0·092	n.a.	1·72	n.a.	0·085

Cyclotella (NO$_3^-$)							
0·0087 ± 0·0004	0·14 ± 0·07	66·8 ± 7·8	0·050	0·061	2·60	0·024	0·030
0·0176 ± 0·0004	0·45 ± 0·17*	92·4 ± 13·4	0·080	0·053	3·54	0·036	0·045
0·0314 ± 0·0003	0·04 ± 0·01	44·4	0·100	0·076	3·53	n.a.	0·054
0·0363 ± 0·0010	0·21 ± 0·09*	31·6 ± 7·4	0·081	0·112	2·88	0·046	0·062
0·0402 ± 0·0005	0·13 ± 0·03*	62·6 ± 6·4	0·111	0·058	4·07	0·087	0·084
0·0445 ± 0·0008	0·12 ± 0·04	44·9 ± 1·4	0·108	0·117	5·10	0·106	0·080
0·0478 ± 0·0009	0·08 ± 0·06	26·4 ± 3·1	0·142	0·168	6·55	0·177	0·099
0·0632 ± 0·0003	0·17 ± 0·04	31·1 ± 3·4	0·120	0·159	5·40	0·169	0·076
0·0676 ± 0·0016	0·19 ± 0·04	23·7 ± 5·0	0·125	0·194	5·40	n.a.	0·076
0·0708 ± 0·0016	0·26 ± 0·07	21·3 ± 3·8	0·158	0·231	5·63	0·218	0·114
0·0768 ± 0·0011	0·22 ± 0·02	32·8 ± 2·8	0·130	0·131	4·82	0·157	0·084
Coccochloris (NO$_3^-$)							
0·0072 ± 0·0007	0·15	480·0 ± 142·0	0·075	n.a.	13·8	n.a.	0·017
0·0131 ± 0·0006	0·12 ± 0·04	388·0 ± 42·0	0·084	n.a.	15·8	n.a.	0·030
0·0228 ± 0·0006	0·17 ± 0·11	331·0 ± 38·0	0·097	n.a.	17·4	n.a.	0·027
0·0348 ± 0·0018	0·22 ± 0·02	311·0 ± 32·0	0·109	n.a.	16·8	n.a.	0·029
0·0428 ± 0·0006	0·10 ± 0·03	278·0 ± 25·0	0·127	n.a.	17·7	n.a.	0·032

*The mean of a series of duplicate nitrate measurements.

n.a.: Data are not available.

Table 2. *Linear regressions of selected ratios of population characteristics* (x) *on steady-state growth rate* (x = a + bμ). *The* r *values are correlation coefficients, and the* |t| *values are for tests of a non-zero slope. Significance levels for* n *points are in parentheses. The dimensions of the ratios correspond to the dimensions in Table* 1.

Ratio (x)	a	b	n	r	|t|
Dunaliella (NO₃⁻)					
N/C	0·0316	1·46	3	0·996 (0·05)	10·76 (0·1)
N/cell	−0·0134	4·62	3	0·999 (<0·01)	181·20 (<0·01)
N/chl-*a*	0·587	0·63	3	0·157 (>0·1)	0·16 (>0·8)
cells/C	0·824	−11·00	3	0·943 (>0·1)	2·85 (>0·2)
chl-*a*/C	−0·0569	2·24	3	0·979 (>0·1)	4·76 (>0·1)
chl-*a*/cell	0·0325	7·24	3	0·983 (>0·1)	5·40 (>0·1)
Cyclotella (NO₃⁻)					
N/C	0·0541	1·21	11	0·868 (<0·01)	5·25 (0·01)
N/cell*	0·0163	0·23	11	0·569 (<0·1)	2·08 (<0·1)
N/vol*	1·30	8·33	11	0·361 (>0·1)	1·16 (<0·4)
N/chl-*a**	2·26	−20·50	9	0·675 (<0·05)	2·42 (<0·05)
cells/C	2·56	40·46	11	0·776 (<0·01)	3·69 (<0·01)
vol/C	0·0330	0·87	11	0·801 (<0·01)	4·02 (<0·01)
vol/cell	0·0138	0·056	11	0·365 (>0·1)	1·18 (<0·3)
chl-*a*/cell	0·0032	0·45	9	0·948 (<0·01)	7·85 (<0·01)
chl-*a*/C	−0·0093	2·72	9	0·887 (<0·01)	5·08 (0·01)
chl-*a*/vol	0·419	21·49	9	0·862 (<0·01)	4·50 (0·01)
Coccochloris (NO₃⁻)					
N/C	0·0653	1·38	5	0·992 (<0·01)	13·86 (<0·01)
N/cell	0·0048	0·05	5	0·939 (0·02)	4·75 (<0·02)
N/vol	3·58	5·22	5	0·136 (>0·1)	0·24 (<0·9)
cells/C	14·15	89·70	5	0·836 (<0·1)	2·64 (<0·1)
vol/C	0·0201	0·28	5	0·737 (>0·1)	1·89 (<0·2)
vol/cell	0·0014	0·009	5	0·522 (>0·1)	1·06 (<0·4)
Dunaliella (NH₄⁺)					
N/C	0·0230	1·76	5	0·970 (<0·01)	6·88 (<0·01)
N/cell	0·0848	−0·22	5	0·476 (>0·1)	0·94 (>0·3)
N/chl-*a*	0·827	−9·93	3	0·993 (<0·1)	8·41 (<0·1)
N/vol	1·48	−6·13	5	0·675 (>0·1)	1·58 (<0·3)
cells/C	0·220	25·70	5	0·961 (<0·01)	6·00 (<0·01)
vol/C	0·0098	1·70	5	0·971 (<0·01)	7·02 (<0·01)
vol/cell	0·0568	0·12	5	0·508 (>0·1)	1·02 (<0·3)
chl-*a*/cell	0·0249	3·97	3	0·996 (0·05)	11·28 (<0·1)
chl-*a*/C	−0·183	10·60	3	0·976 (>0·1)	4·52 (<0·2)
chl-*a*/vol	0·964	51·40	3	0·985 (0·1)	5·72 (<0·2)
Monochrysis (NH₄⁺)					
N/C	0·0417	0·76	9	0·968 (<0·01)	10·30 (<0·01)
N/cell	0·0421	0·16	8	0·499 (>0·1)	1·41 (>0·1)
N/chl-*a*	1·96	−27·20	5	0·735 (>0·1)	1·87 (<0·2)
N/vol	1·76	−11·67	8	0·834 (<0·05)	3·70 (0·01)
cells/C	1·03	11·70	8	0·916 (<0·01)	5·59 (<0·01)
vol/C	0·0192	0·99	8	0·980 (<0·01)	11·99 (<0·01)
vol/cell	0·0224	0·40	8	0·967 (<0·01)	9·35 (<0·01)
chl-*a*/cell	0·0008	1·67	5	0·833 (<0·1)	2·61 (<0·1)
chl-*a*/C	−0·013	2·89	5	0·894 (<0·05)	3·46 (<0·05)
chl-*a*/vol	0·501	30·70	5	0·752 (>0·1)	1·98 (<0·2)

*Nitrogen values are based on chemical rather than CHN analysis.

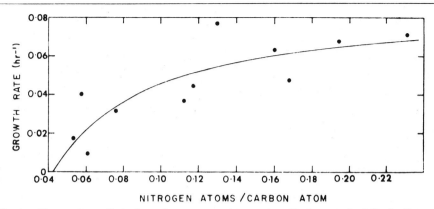

Fig. 1. Eleven nitrate-limited steady-state growth rates (or dilution rates) of *Cyclotella nana* plotted against population nitrogen/carbon ratio. The curve is a least squares fit of the hyperbola $\mu = \mu_m (q - q_0)/(K + q - q_0)$. K, q_0 and μ_m are constants, and μ and q represent, respectively, growth rate and nitrogen/carbon ratio. K is the half-saturation constant; μ_m is the maximum growth rate; and q_0 is the N/C ratio at zero growth rate. q_0 was estimated by eye from the data, and K and μ_m were determined from the least squares fit of the hyperbola to the data. Values for these constants are given in Table 3.

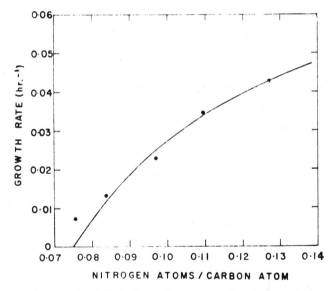

Fig. 2. Five steady-state nitrate-limited growth rates plotted against population nitrogen atoms per carbon atom for *Coccochloris stagnina*. The curve is a least squares fit of the hyperbola, $\mu = \mu_m(q - q_0)/(K + q - q_0)$, to the five points.

Then

$$\mu = \mu_m (q' - q_0')/[(K/a) + (q' - q_0')].$$

The slope, a, of the linear regression of nitrogen per unit population carbon, q, on nitrogen per cell, q', is significantly different from zero for all nitrate-limited experiments: *Coccochloris* (5 points) at the 5 % level, *Dunaliella* (3 points) at the 8 % level and *Cyclotella* (11 points) at the 1 % level, with correlation coefficients of 0·941, 0·955 and 0·804, respectively. Thus, a hyperbolic relation between growth rate and the nutritional state

Table 3. *Constants for hyperbolas relating growth rate to nitrogen and chlorophyll-a content of the population:*

$$\mu = \mu_m(q - q_0)/(K + (q - q_0)),$$

where μ *is growth rate* (hr^{-1}) *and* q *is the nitrogen or chlorophyll-a content of the population (atoms N/atom C or* μ*g chl-a/*μ*g at C).* μ_m *is the maximum specific growth rate* (hr^{-1}), q_0 *is the population nitrogen or chlorophyll-a content at zero growth rate (atoms N/atom C or* μ*g chl-a/*μ*g at C), and* K *is the half-saturation constant (atoms N/atom C or* μ*g chl-a/*μ*g at C).*

| Experiment | Chemostat: population nitrogen | | | Chemostat: population chlorophyll-a | | | Batch culture |
	μ_m	q_0	K	μ_m	q_0	K	(μ_m)
Coccochloris (NO$_3^-$)	0·090	0·075	0·057	n.a.	n.a.	n.a.	0·086
Cyclotella (NO$_3^-$)	0·087	0·043	0·053*	0·088	0·017	0·065	0·086
Dunaliella (NH$_4^+$)	0·076	0·044	0·034	0·063	0·065	0·069	0·070
Monochrysis (NH$_4^+$)	n.h.	n.h.	n.h.	0·062	0·025	0·034	0·051

*Values are based on chemical rather than CHN analysis.

n.h.: Data are not fit by a hyperbola.

n.a.: Data are not available.

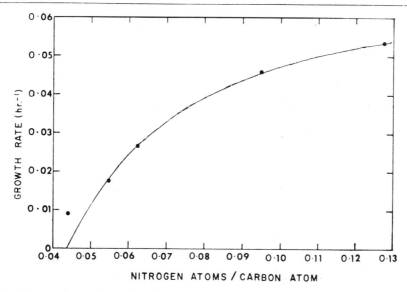

Fig. 3. Five ammonium-limited steady-state growth rates plotted against population nitrogen atoms per carbon atom for *Dunaliella tertiolecta*. The curve is a least squares fit of the hyperbola,
$$\mu = \mu_m(q - q_0)/(K + q - q_0).$$

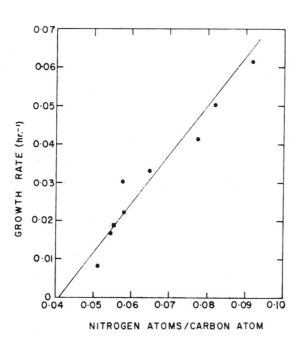

Fig. 4. Eight ammonium-limited (filled circles) and one nitrate-limited (filled square) steady-state growth rates plotted against population nitrogen atoms per carbon atom for *Monochrysis lutheri*. The straight line is a linear regression of ammonium-limited growth rate on nitrogen/carbon ratio, $\mu = 1.24 \, (N/C) - 0.0493$, with a correlation coefficient of 0.968.

of the population is consistent with our results for *Coccochloris*, *Dunaliella* and *Cyclotella* growing on limiting nitrate using either carbon or cell concentration as the population measure. This agrees with the results of earlier experiments with nitrate-limited *Isochrysis galbana* (CAPERON, 1968). The slope of the regression of nitrogen per unit carbon on nitrogen per cell is not significantly different from zero for ammonium-limited *Dunaliella*, and the mean nitrogen per cell $\pm$ 2 S.D. over all growth rates studied for this population is 0.078 ± 0.017 ng at/cell. For ammonium-limited *Monochrysis* the slope of this regression is significantly different from zero at the 7% level with a correlation coefficient of 0.668. The 7% level is not as low as one would like for a regression involving 8 points, and the mean nitrogen per cell $\pm$ 2 S.D. is 0.047 ± 0.011 ng at/cell. Thus, the nitrogen per cell in ammonium-limited cases can be considered invariant over the growth rates examined.

In four experiments the mean environmental nitrate concentrations (Table 1) (denoted by the superscript *) represent 4 to 8 sets of duplicate samples from the growth chamber. In all four cases analysis of variance shows significant (at the 1% level) variation among the sets of duplicates. This variation is, in all cases, without trend. Thus, there is real detectable variation in limiting nutrient concentration at steady state even over short periods of 2 to 6 hours. The means $\pm$ 2 S.D. (Table 1) give a measure of this variability. During one nitrate-limited steady state with *Dunaliella*, the nitrate concentration in the growth chamber was determined from hourly samples over a 52 hour period. The mean nitrate concentration $\pm$ 2 S.D. was $0.117 \pm 0.044 \mu g$ at/l. Analysis of variance shows significant temporal variability among the sets of duplicates ($p < 0.005$). The variability of mean steady-state nutrient concentration among the several steady-state growth rates studied for each population can also be investigated by analysis of variance. For *Cyclotella* (nitrate-limited), *Coccochloris* (nitrate-limited) and *Monochrysis* (ammonium-limited) the differences in mean limiting nutrient concentration at various steady-state growth rates are significant (at the 0.1% level), and these differences show no correlation with growth rate. There, thus, appears to be real detectable variation in limiting nutrient concentration that cannot be related to steady-state growth rate. Analysis of variance of the steady-state nutrient concentrations for *Dunaliella* growing with either nitrate or ammonium ion as limiting nutrient showed no significant (19 and 6%, respectively) variation among mean nutrient concentrations at the steady-state growth rates studied. Here, then, the variability—even over the few hours during which limiting nutrient concentration was observed—was comparable to the variability of the mean values over all growth rates studied.

The relationship between μ and chlorophyll-*a* per unit carbon appears to be hyperbolic where we have sufficient data to make an analysis. Figures 5, 6 and 7 present best fit hyperbolas to the chlor-*a*/C vs. μ data for the nitrate-limited *Cyclotella* experiments, the ammonium-limited *Monochrysis* experiments and the composite nitrate and ammonium *Dunaliella* results. Table 3 includes the parameters of these three hyperbolas. Here again the comparison between the derived μ_m values and batch culture μ_m values is good.

DISCUSSION

There is no evident relationship between nutrient concentration in the growth chamber and steady-state growth rate for any of the populations studied. This supports

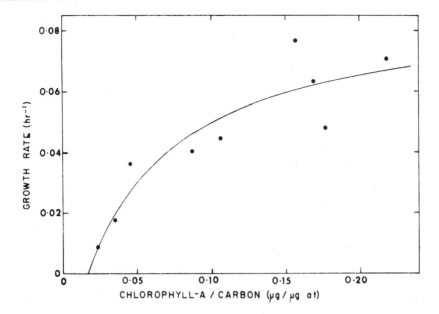

Fig. 5. Nine steady-state growth rates for nitrate-limited *Cyclotella nana* plotted against population chlorophyll-*a* per unit carbon. The curve is a least squares fit of the hyperbola, $\mu = \mu_m (q - q_0)/(K + q - q_0)$, where K, q_0 and μ_m are constants, μ is growth rate and q is chlorophyll-*a*/carbon ratio.

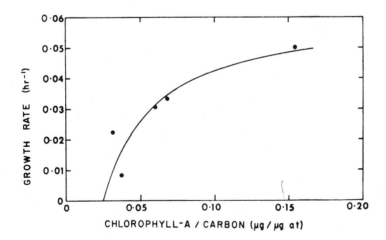

Fig. 6. Five ammonium-limited steady-state growth rates plotted against chlorophyll-*a*/carbon ratio for *Monochrysis lutheri*. The curve is a least squares fit of the hyperbola, $\mu = \mu_m (q - q_0)/(K + q - q_0)$.

163

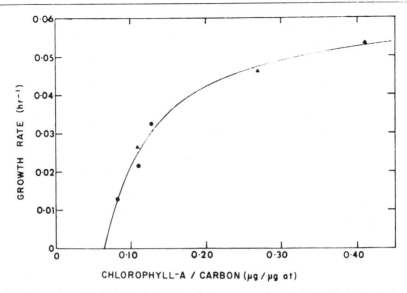

Fig. 7. Three nitrate-limited (filled circles) and three ammonium-limited (filled triangles) steady-state growth rates for *Dunaliella tertiolecta* plotted against population chlorophyll-*a* per unit carbon. The curve is a least squares fit of the hyperbola, $\mu = \mu_m (q - q_o)/(K + q - q_o)$, to all six points.

an earlier result for *Isochrysis galbana* (CAPERON, 1968). The relationship between growth rate and limiting nutrient is provided only by the variable internal nutritional state. This does not explain the failure to detect differences in concentration in the growth medium itself, since the only possible control over population growth is environmental concentration. Abundant evidence from batch culture uptake experiments (e.g. EPPLEY, ROGERS and McCARTHY, 1969) indicates that Michaelis–Menten kinetics are at least descriptively adequate in relating uptake rate to environmental nitrate or ammonium ion concentration. Continuous culture experiments should duplicate these results. Failure to do so could be because the nutrient concentrations involved are below the level of detectability of the analytical techniques used, as FUHS (1969) has shown for phosphate-limited populations. This does not seem to be the case for the nitrate-limited populations we have studied. The steady-state nitrate concentrations are uniformly within the range of detectability and show significant non-growth-related variability. The ammonium ion experiments differ in that the nutrient was often below the range where this method gives adequate reproducibility. Thus, the steady state seems, in fact, to be a steady state only in the long-term average sense.

At steady state the uptake rate is proportional to the growth rate. Thus, for populations where both the uptake kinetics and growth rate kinetics are known, steady-state environmental nutrient can be determined indirectly from this equation as DROOP (1968) has done for vitamin B-12-limited growth of *Monochrysis lutheri*. The differences in these calculated steady-state concentrations can be compared to the resolvability afforded by the employed analytical technique. We shall return to this matter in the second part of this paper.

The observed nutrient variation at steady state may result from the one or two per cent variation in pumping rate and/or slow response of the uptake mechanism to environmental change. It is very possible that our chemostats are as close to a truly steady-state environment as most natural situations, and we suggest that direct inference of nitrate- or ammonium-limited growth rate from anything short of an extended time series of nutrient observations in nature is likely to be of quite low precision. Apparently one can decide whether or not the nitrate concentration is in the limiting range, but little more.

Isochrysis galbana (CAPERON, 1968), *Coccochloris*, *Cyclotella* and *Dunaliella* have all been shown to exhibit a range of steady-state growth rates in continuous culture under low (0·05 to 0·45 μg at/l.) but detectable and quite variable environmental nitrate concentration. The variation is real, i.e. not attributable to analytical techniques, and shows no relationship to growth rate. These organisms represent a rather broad taxonomic spectrum. Even though they were chosen more for experimental convenience than for the fact that they are representative of the phytoplankton, they do represent what can be in the phytoplankton. Our results, thus, imply that environmental nitrate concentrations in excess of 0·5 μg at/l. means that fixed nitrate nitrogen is not limiting steady-state growth in the ocean, or that the phytoplankton must be dominated by species less effective at low nutrient concentration than would be expected if low nutrient were a persistent attribute of the environment. Also significant is the fact that detectable amounts (i.e., greater than 0·05 μg at/l. of nitrate) are found at the lowest observed steady-state growth rates. Thus, the frequently observed undetectable amounts of nitrate nitrogen in the euphotic zone imply a very low supply from deep water and growth rates limited by other forms of fixed nitrogen or lower than any of those observed in continuous culture. Environmental ammonium concentrations greater than 0·40 μg at/l. imply that the growth rate of neither *Dunaliella* nor *Monochrysis* is limited by fixed nitrogen, since ammonium is preferentially used over nitrate.

The internal nitrogen per cell for nitrate-limited populations shows systematic variation with growth rate and is linearly related to internal nitrogen per unit population carbon. This implies that the hyperbolic relationship between growth rate and nutritional state of the population is appropriate with either cell numbers or carbon as the measure of population size. It is likely that much of this difference in nitrogen per cell is due to an internal reservoir of nitrate nitrogen. Evidence for this is provided by EPPLEY and COATSWORTH (1968) who found 40% of the nitrate recoverable after rapid uptake in light. Another line of evidence will be presented in the second part of this paper. At steady state, nutrient uptake rate and rate of incorporation into protoplasm (growth) must be equal, but in a non-steady state environment, the evident decoupling of the two steps makes possible the maintenance of a higher mean growth rate than would be possible without a rapid uptake ability and an internal storage mechanism. It represents clear ecological advantage to the phytoplankton.

A theoretical rationale for the applicability of the hyperbola relating growth rate to nutritional state of the population has been provided by CAPERON (1967; 1968). This treatment requires an internal reservoir of limiting nutrient analogous to that hypothesized by DROOP (1968) and WILLIAMS (1971). The additional studies on nitrate-limited growth presented here are consistent with this earlier work. The ammonium-limited studies are not clearly interpretable in the same way. The internal nitrogen per cell for ammonium-limited populations does not show significant variation with growth rate, and the changing nutritional state of the population is reflected only in the nitrogen-to-carbon ratio. The carbon per cell shows a linear decrease over the range of growth rates examined. It is possible that incorporated nitrogen is also decreasing and that an increasing concentration in an internal reservoir is compensating in an amount sufficient to give the observed, rather uniform nitrogen per cell over the range of growth rates studied. This seems an unnecessarily complicated attempt to fit ammonium-limited growth into an interpretative model simply because it works for other nutrient species. When ammonium is the limiting nutrient, the relationships between nitrogen-to-carbon ratio and growth rate—linear in the *Monochrysis* experiments and hyperbolic in the *Dunaliella* experiments—must be viewed in the same context as any of the other ratios indicating the physiological state of the population at various nutrient-limited growth rates.

The nitrogen per cell at the lowest nitrate-limited growth rates studied for *Monochrysis* ($\mu = 0.0189$ hr^{-1}) and *Dunaliella* ($\mu = 0.0127$ hr^{-1}) was 0.072 and 0.050 ng at/cell, respectively. These two values compare with the mean nitrogen per cell over all ammonium-limited growth rates studied for these two organisms of 0.078 and 0.047 ng at/cell, respectively. We have interpreted nitrogen content per cell at low nitrate-limited growth rates as representing growth with a nearly empty reservoir (N/C approaching q_0). Thus, ammonium-limited growth, even at the highest steady-state growth rates, takes place with a cellular nitrogen content that is minimal when compared to nitrate-limited growth. At the intermediate growth rate of 0.0324 hr^{-1} for nitrate-limited *Dunaliella*, the cell nitrogen has already reached 0.163 ng at/cell —more than twice the value of 0.0725 ng at/cell for ammonium-limited growth at 0.0534 hr^{-1} for this organism. These comparisons strengthen both the argument for an internal nitrate reservoir and against one for ammonium.

When ammonium is the limiting nutrient, growth in terms of cell numbers keeps pace with nutrient assimilation. This probably reflects the fact that the ammonium ion is already in the appropriate oxidation state. Cell division takes place with successively smaller amounts of non-essential carbohydrates and lipids as the supply rate of ammonium ion increases. Photic energy can be channeled into growth (cell division type growth) or into storage products (lipids and carbohydrates). As growth rate approaches maximum specific growth rate, the cell composition likely represents the minimum essential carbohydrate component. Thus, ammonium-limited phytoplankton growth exhibits a distinctly different mechanism from that for nitrate, vitamin B-12 (DROOP, 1968), phosphate (FUHS, 1969) or iron (DAVIES, 1970).

Interestingly, maximum specific growth rates from batch culture experiments in our laboratory are the same on ammonium or nitrate ion substrates. It appears that the rate at which a new cell can be built is the same with a full reservoir(s) derived from nitrate nitrogen or a saturating external ammonium nitrogen supply. The internal nitrate (or other intermediate oxidation state products) reservoir(s) duplicates the supply afforded by saturation concentrations of environmental ammonium ion, and when adequate light energy is available, a population is able to make as effective use (in the growth rate sense) of nitrate as ammonium ions.

Growth rate, μ has units of time^{-1}, and is equal to $(1/n)(dn/dt)$ where n is the population measure. μ will be the same for all measures of population only when their ratios are constant over all growth rates. Our work has shown that these ratios are not constant. Only population nitrogen and cell numbers under ammonium-limited conditions give the same μ. In all other cases, μ is clearly dependent on the population measure. Carbon-14 measurements using either carbon or chlorophyll for population cannot be equated to growth rate predictions on the basis of kinetics of ammonium ion uptake (population measure is nitrogen) or the growth kinetics developed from cell counts (THOMAS and OWEN, 1971). This comparison of ^{14}C measurements with predictions from nutrient-limiting kinetics is an excellent approach and an essential test of nutrient models, but our data indicate that the comparison must take into account the variability of q.

Throughout we have used a modified form of the Michaelis–Menten hyperbola. For data that require a curve not passing through the origin the simpler two-parameter rectangular hyperbola, $\mu = \mu_m (q - k)/q$, used by DROOP (1968) would be preferable provided it fits the data equally well. In this expression k is equivalent to our q_o and the value of q at $\mu = \mu_m/2$ is $2k$, i.e. $K = q_o$. The K values in Table 3 are not greatly different from q_o except for the chlor-a/C vs. μ for $Cyclotella$; however, we do not feel that our data are sufficient to make a clear choice between the two hyperbolic forms in the remaining 5 cases.

When we depart from the kinetics of nutrition as we have in relating growth rate to physiological states, there is no *a priori* reason to expect a hyperbola. A hyperbolic curve under these conditions provides a general descriptive tool, and in all cases where a hyperbola is appropriate, there is a straight line limiting case (CAPERON, 1967). The linear relationship between N/C and μ for $Monochrysis$ on ammonium can be viewed in this context. The fitted curves in Figs. 5, 6 and 7 are viewed similarly. Thus, chlor-a/C is as acceptable an indicator of nutrient-limited growth rate as N/C or any of the other physiological variables. Only for nitrate-limited growth does internal nitrogen have a clear rational interpretation that makes it the measure of choice.

Many of the observed population parameters exhibit a predictively useful relationship to growth rate for the particular population studied, but no precise general relationship is available for both ammonium and nitrate-limited growth for all populations studied. In rare instances where the particulate organic material in sea water is heavily dominated by phytoplankton, the N/C ratio could be a meaningful indicator of steady-state growth rate. The differences in this relationship for the $Coccochloris$, $Cyclotella$ and $Dunaliella$ populations growing on nitrate are surprisingly small for such diverse organisms. Even the N/C vs. growth rate relationships for the ammonium experiments with $Dunaliella$ and $Monochrysis$ are much less different from nitrate relationships than the evidently different nutritional mechanisms would suggest. It appears that N/C ratio is as good an indicator of fixed nitrogen-limited growth rate as chlorophyll is of standing crop.

REFERENCES

CAPERON J. (1965) The dynamics of nitrate-limited growth of *Isochrysis galbana* populations. Ph. D. Thesis. Univ. of Calif., San Diego.
CAPERON J. (1967) Population growth in micro-organisms limited by food supply. *Ecology*, **48**, 715–722.
CAPERON J. (1968) Population growth response of *Isochrysis galbana* to nitrate variation at limiting concentrations. *Ecology*, **49**, 866–872.
CAPERON J. (1969) Time lag in population growth response of *Isochrysis galbana* to a variable nitrate environment. *Ecology*, **50**, 188–192.

DAVIES A. G. (1970) Iron, chelation and the growth of marine phytoplankton. 1. Growth kinetics and chlorophyll production in cultures of the euryhaline flagellate *Dunaliella tertiolecta* under iron-limiting conditions. *J. mar. biol. Ass. U.K.*, **50**, 65–86.

DROOP M. R. (1961) Vitamin B-12 and marine ecology: the response of *Monochrysis lutheri. J. mar. biol. Ass. U.K.*, **41**, 69–76.

DROOP M. R. (1968) Vitamin B-12 and marine ecology. IV. The kinetics of uptake, growth and inhibition in *Monochrysis lutheri. J. mar. biol. Ass. U.K.*, **48**, 689–733.

DUGDALE R. C. (1967) Nutrient limitation in the sea: dynamics, identification, and significance. *Limnol. Oceanogr.*, **12**, 685–695.

EPPLEY R. W. and J. L. COATSWORTH (1968) Uptake of nitrate and nitrite by *Ditylum brightwelli*—kinetics and mechanisms. *J. Phycol.*, **4**, 151–156.

EPPLEY R. W., J. N. ROGERS and J. J. MCCARTHY (1969) Half-saturation constants for uptake of nitrate and ammonium by marine phytoplankton. *Limnol. Oceanogr.*, **14**, 912–920.

EPPLEY R. W. and W. H. THOMAS (1969) Comparison of half-saturation constants for growth and nitrate uptake of marine phytoplankton. *J. Phycol.*, **5**, 375–379.

FUHS G. W. (1969) Phosphorus content and rate of growth in the diatoms *Cyclotella nana* and *Thalassiosira fluviatilis. J. Phycol.* **5**, 312–321.

GORDON D. C., JR. (1969) Examination of methods of particulate organic carbon analysis. *Deep-Sea Res.*, **16**, 661–665.

LORENZEN C. J. (1966) A method for the continuous measurement of *in vivo* chlorophyll concentration. *Deep-Sea Res.*, **13**, 223–227.

MACISAAC J. J. and R. C. DUGDALE (1969) The kinetics of nitrate and ammonium uptake by natural populations of marine phytoplankton. *Deep-Sea Res.*, **16**, 445–457.

MONOD J. (1942) *La Croissance des Cultures Bacteriennes*. Herman. Paris, 210 pp.

RILEY G. A. (1963) Theory of food chain relations in the ocean. In: *The sea*, M. N. HILL editor, Interscience, New York, **2**, 438–463.

RYTHER J. H. and W. M. DUNSTAN (1971) Nitrogen, phosphorus, and eutrophication in the coastal marine environment. *Science*, **171**, 1008–1013.

SOLÓRZANO L. (1969) Determination of ammonia in natural waters by the phenolhypochlorite method. *Limnol. Oceanogr.*, **14**, 799–801.

STEELE J. H. (1959) The quantitative ecology of marine phytoplankton. *Biol. Revs. Cambridge Phil. Soc.*, **34**, 129–159.

STRICKLAND J. D. H. and T. R. PARSONS (1968) A practical handbook of seawater analysis. *Bull. Fish. Res. Bd Can.*, **167**, 311 pp.

SVERDRUP H. U. (1955) The place of physical oceanography in oceanographic research. *J. mar. Res.*, **14**, 287–294.

SYRETT P. J. (1962) Nitrogen assimilation. In: *Physiology and biochemistry of algae*, R. LEWIN, editor, Academic Press, New York, 171–188.

THOMAS W. H. (1970) On nitrogen deficiency in tropical Pacific oceanic phytoplankton: photosynthetic parameters in poor and rich water. *Limnol. Oceanogr.*, **15**, 380–385.

THOMAS W. H. and R. W. OWEN (1971) Estimating phytoplankton production from ammonium and chlorophyll concentrations in nutrient-poor water for the eastern tropical Pacific Ocean. *Fish. Bull.*, **69**, 87–92.

WILLIAMS F. M. (1965) Population growth and regulation in continuously cultured algae. Ph.D. Thesis, Yale University.

WILLIAMS F. M. (1971) Dynamics of microbial populations. In: *Systems analysis and simulation in ecology*, B. C. PATTEN, editor, Academic Press, New York, **1**, 197–267.

WOOD E. D. F., F. A. J. ARMSTRONG and F. A. RICHARDS (1967) Determination of nitrate in sea water by cadmium–copper reduction to nitrite. *J. mar. biol. Ass. U.K.*, **47**, 23–31.

III
Compartmental Analysis

Editor's Comments on Papers 14 Through 20

With the rise of tracer methods, first in physiology and biochemistry, and a little later in ecology, mathematical techniques were developed to calculate the rate of movement of materials from one part, or compartment, of a system to another. This became known as "compartmental analysis." The same thing was being done in the fields of meteorology and geochemistry, where radioactivity from atomic weapons tests was providing opportunities for tracer studies. The mathematical approaches turned out to be almost identical, although different notation was developed in each field. The important early papers in compartmental analysis are cited in the papers in this part. More complete and up-to-date treatment is provided by Steele (1972).

One of the problems to be faced in tracer experiments is the distinction between a one-way process of uptake or loss and a two-way exchange. A number of early tracer studies were erroneous, because only half of a two-way process was detected and measured. What was reported to be uptake was really half of an exchange. Compartmental analysis, and the use of specific activity (radioactivity per unit weight of the tagged element) as the measure of the amount of label in a compartment, is helpful in preventing this kind of error. The problem does not arise in the papers reproduced here.

After the pioneer studies in Linsley Pond by Hutchinson and Bowen, a number of more detailed compartmental studies were made of lakes and other aquatic systems. An important and early contribution was made by F. R. Hayes and his associates at Dalhousie University. The paper reproduced here is the first of a series on the kinetics of phosphorus in lakes. Not only did this work veryify the rapidity of some of the cycles of phosphorus in natural waters, but it showed the important roles played by bottom sediments and benthic organisms. The importance of the bottom in lake chemistry was not a new idea, for it had been examined on a purely chemical level in the 1930s by Mortimer (1941, 1942) and several European limnologists. The tracer approach showed that the exchange of phosphorus between water and sediments was not limited to the seasonal regime that had been revealed by chemical measurements of standing stock. Again, the seasonal regime was indicative of shifting equilibria. In shallow water or well-mixed lakes the exchange between water and sediments is continuous and rapid. The thermocline does inhibit exchange in stratified lakes, but not as completely as limnologists had thought.

The compartmental approach to cycles of essential elements probably did much to accelerate the trend toward a systems approach. One of the end results of a good compartmental study was a compartmental model of the system, showing the standing stock in each compartment and the fluxes between compartments. During the 1950s and 1960s such compartmental models were the end products of a number of studies. Yet that kind of information was still fragmentary or even lacking for a number of ecosystem types at the beginning of the International Biological Program (IBP). The IBP is filling that gap, although many of the results are not yet available in the open literature. In this part we look at compartmental analysis of a number of ecosystems, sometimes done with radioactive tracers and sometimes by chemical means alone, but always with recognition of the importance of both living and nonliving components of the system.

Bormann, Likens, and their associates have been doing a thorough analysis of the Hubbard Brook watershed in New Hampshire. The use of a drainage unit, such as a watershed, makes it possible to do an input–output study by purely chemical means. Bormann and Likens have now invested nearly a decade of effort in a detailed study of one watershed, looking not only at the natural situation and its year-to-year variation, but the effects of a variety of man-induced changes, including clear-cutting of forests, herbicides, and fallout of pollutants from the atmosphere. The paper chosen for this volume is not their first but one of their better summations of one part of the work dealing with the effects of removal of vegetation on the loss of both water and essential elements from the watershed.

The work of Bormann and Likens is somewhat controversial at this point. Some of their early studies were criticized on the grounds that the results tended to cast doubts on forestry practices, although the experimental methods did not duplicate those practices in every detail. Moreover, it is becoming apparent that what is true for the Hubbard Brook watershed in New Hampshire is not necessarily true of all forests or even of all eastern deciduous forest. A number of other studies of the watershed type are at variance with the Hubbard Brook study. One of the most intensive of these is the Coweeta watershed study in North Carolina, which is part of the Eastern Deciduous Forest Biome study of the IBP. The paper by Johnson and Swank is representative of the variations on the theme of Hubbard Brook which are being developed on other sites. Whether or not these experiments with whole landscapes have revealed how the systems should be managed by man, they have brought out some important points. It is clear that ecosystems must be studied on an ecosystem scale, and that natural units, such as drainage basins, are the units of choice. Experiments with units of that size are possible and need not be prohibitively expensive. When enough results are available from a sufficiently wide variety of study sites, we shall indeed be able to make recommendations for the long-term management and use of forest ecosystems. We shall be concerned both with what we take out and with what we put in, such as acid rain from the combustion of coal in power plants.

171

The tundra studies of Schultz and his associates. which also began well before IBP, have been influential in a different way. They have provided important comparative material from which we may hope to extract principles of ecosystem function. The tundra is a system with unusually severe limits and stresses. Set on permafrost, tundra has little reserve of essential elements in the soil. Low temperature limits and shapes metabolic processes, and it probably limits the roles of microorganisms. The result, as Schultz has shown, is a system that oscillates in a quite regular way. One of the manifestations of this oscillation, known as the "lemming cycle," is well known but often misrepresented in popularized versions, which depict the voles committing mass suicide in the sea. There are many attempts at scientific explanations for the cycle in the literature (see, for example, Watt, 1968 and Deevey, 1960). Schultz shows that, whatever interaction there may be on the population level, one overriding control on the oscillation is the shift of a suite of essential elements from one part of the system to another. This is an enticing system to simulate, and some modeling has been done as part of the IBP Tundra Biome study, but it is not yet in the literature.

Rain forests are another system of great value for comparative studies. In many ways they provide a contrast to the tundra. Where the tundra is simple in structure, rain forest is bewilderingly diversified. Not only are there a great many species of trees, with vegetation arranged in multiple layers, but the food webs and pathways of both organic and inorganic materials are quite complex. The tundra has an ice-sheathed, impervious bottom under it which retains in the system the small stocks of essential elements that are present. The rain forest usually sits on an open sink. Essential elements must be kept in the biomass of the forest, or they will be lost in the downward rush of water. As the rainwater moves down through the forest, it must pass through a series of separation networks where nutrients can be retained. Considering their extent and importance, rain forests have received little attention from ecosystem-oriented ecologists. This is not mere oversight but a matter of practicality. Modern ecology requires modern laboratories and instruments. These things are not ordinarily available in proximity to rain forests.

One place where good facilities do exist close to a rain forest is Puerto Rico, where the Puerto Rico Nuclear Center is reasonably close to the Luqillo National Forest. Here in the 1960s a comprehensive rain forest study was undertaken under the leadership of H. T. Odum. The initial report of the results (Odum, 1970) is too extensive to be reproduced in this volume, but a succinct coverage of the cycling of the essential elements is provided by the more recent review of Jordan and Swank, who were involved in the research in Puerto Rico.

The concentration of most of the good laboratories in the temperate zones has limited the extent of modern ecological research, not only on rain forests but on all polar and tropical ecosystems. There are a number of small laboratories adjacent to coral reefs, but most of them are very primitive. One of the best has been the Eniwetok Marine Biological Laboratory, operated by the University of Hawaii for the U.S. Atomic Energy Commission. Established originally to provide baseline data and environmental monitoring for the nuclear weapons tests in the 1950s, the laboratory developed an independent program of basic research on coral reefs. Another means of getting a laboratory to a remote environment was conceived and developed by Per Scholander of the Scripps Institution of Oceanography. Scholander's idea was to put the laboratory on a small ship. The ship, anchored as near as possible to the study site, would provide the laboratories, living quarters, and power plant. Small boats would take the scientists to and from the ship. However, the ship that was actually built, with funds from the National Science Foundation, was designed exclusively with physiology, not ecology, in mind. Only recently have ecologists been allowed to use the R/V Alpha Helix. One of the first ecological expeditions was Project Symbios to Eniwetok Atoll. This was the largest and best equipped coral reef expedition, combining the facilities of Alpha Helix with those of the Eniwetok Laboratory, together with the communications and technical services provided by the Department of Defense and its contractor. Twenty-five scientists, most of whom were already familiar with the Eniwetok reefs, participated in Project Symbios.

Coral reefs have at least superficial similarities to tropical rain forests. They are lush, tropical ecosystems of high species diversity. The pathways of elements through them are highly complex, and a large amount of some essential elements is probably retained in the biomass, since both ecosystems exist in a torrent of moving water. Both ecosystems give the impression of great stability, but this is true in only a superficial way. When examined in detail over the years, both rain forests and coral reefs prove to be subject to catastrophic damage, largely from storms but also from animal eruptions, such as that of the crown-of-thorns starfish, *Acanthaster plancki*, on Indo-Pacific coral reefs in the 1960s and 1970s. A modern view of the structure and function of these two important and extensive ecosystems is just beginning to emerge.

Two benchmarks in the study of coral reefs, both too extensive to be included, are the report of the Great Barrier Reef Expedition (Yonge, 1930, 1940) and the monograph on Japtan Reef at Eniwetok Atoll by the Odums (Odum and Odum, 1955). These two studies, about twenty years apart, present conflicting views of the coral reef as a system. Yonge views corals as carnivorous cnidarians feeding on zooplankton. He views the whole coral reef as a community supported by plankton carried in by flowing water. The Odums view individual corals as plant-like, symbiotic associations of corals and dinoflagellates. They view the whole coral reef as a community supported largely by photosynthesis inside the corals, but it is not clear how the photosynthetic products were passed down the food chain. Compared to algae, corals are relatively slow-growing organisms. Although they are eaten by a number of specialized fishes, the rate of consumption of corals is not high, compared to the overall rate of metabolism of the reef community. The results of the Symbios Expedition do not resolve this conflict. Instead, a third view of coral reefs as a community is emerging from it. In this view corals are less dominant as primary producers and less dominant in the flux of nitrogen and phosphorus as well. Free-living algae, although inconspicuous, are important as producers on coral reefs, supporting grazing fish populations directly and most of the invertebrate populations indirectly in an elaborate food web.

In the case of coral reefs it appears that we are still learning the most fundamental facts about the structure of the food web. Good compartmental analysis of the reef as a community will be possible only when the major pathways in the food web are verified.

The Symbios Expedition is interesting as a case study because it was planned on the premise that we indeed did not know the food web. Therefore, it was necessary to begin with input–output measurements of the reef as a whole. This proved to be an effective strategy, not only for measuring the overall rates of photosynthesis, respiration, and flux of nitrogen and phosphorus but for getting at some of the missing links in the food web as well.

Studies of essential elements took their first impetus from the concern of Justus Liebig about the fertilization of field crops and forests. Agronomic studies have continued in parallel with ecological ones and with some occasional interaction between the two fields. Agricultural research has added much to our knowledge of the cycles of elements in terrestrial soils. It is beyond the scope of this volume to mention all the benchmarks in that area. However, the review by Alexander presents a pre-IBP statement on the state of the art and shows the extent of the interactions between ecology, biochemistry, and agriculture. It puts some emphasis on the importance of microorganisms in the cycles of elements. What is also significant is the emphasis on biochemistry in an ecologically oriented review. We have seen a revolution in biochemistry over the past several decades. Many of the discoveries in biochemistry are potentially applicable to the solution of problems at the ecosystem level. Both the biochemical methodology and the understanding of metabolic processes at the cellular and molecular levels are of great potential value for solving problems at the community level. Most of that potential has gone unexploited. Where it has been utilized best is in microbial ecology, and Alexander shows how ecology and biochemistry fit together in useful ways. We can expect to see this approach applied to ecology in the study of other ecosystems and at levels other than the microbial one in the near future.

14

Reprinted from *J. Ecol.*, **40**(1), 202–216 (1952)

ON THE KINETICS OF PHOSPHORUS EXCHANGE IN LAKES

By F. R. HAYES, J. A. McCARTER, M. L. CAMERON
AND D. A. LIVINGSTONE

Dalhousie University, Halifax, Nova Scotia, Canada

A number of workers have added commercial fertilizer to lakes and followed the results by analyses of nutrients in the water. There is general agreement about the fate of the nutrients. They disappear from the water, the curve of their decline being rapid at first, and later tending to flatten out, so that within a period of a few days or weeks scarcely any extra phosphorus or nitrogen can be detected. When large amounts of fertilizer are added, it has been usual to assume that their rapid disappearance is evidence of extensive utilization by animals and plants in the lake; that is to say, it is commonly taken as proof of the nutrient value of the extra material.

Further information can be gained by the use of radioactive phosphorus, ^{32}P. The first such experiment was reported by Hutchinson & Bowen in 1947, who added 10 mc. to a eutrophic, stratified, 14-acre lake in Connecticut. Their results showed mixing in the epilimnion and they inferred that there was penetration into the hypolimnion. They also record a 1000-fold concentration of material in the tissues of such littoral plants as *Potamogeton* within a period of 1 week. An additional experiment by the same authors (1950) indicated that complete equilibration between water and solids occurred within a week. Following this there was a steady gain of radio-phosphorus by the hypolimnion, which the authors attribute partly to sedimentation of dead plankton and faeces, and partly to removal from the mud.

In the work to be reported in this paper, tracer phosphorus was added to a lake and the customary loss from the water was observed. In such experiments there is no appreciable initial increase in the phosphorus present, nor does any later decline occur. One is simply following the behaviour of a few marked atoms. Hence the proposition that some enhancement of plant or animal growth takes place is untenable.

In the light of the tracer results, we were led to some doubt as to whether the disappearance, even of massive additions of fertilizer to lakes, should be taken as evidence of growth stimulation. Detailed consideration of several such experiments, together with a proposed general interpretation, is given below.

THE LAKE

Bluff Lake (Fig. 1), which was used for the experiment, is situated about 7 miles west of Halifax, and a few hundred yards from the Old Sambro Road. It is about 10 acres in area and 22 ft. deep. It is a seepage lake with no visible inlet or outlet. The small graph incorporated in Fig. 1 shows the midsummer values for oxygen and temperature at various depths, and indicates that the water is completely mixed from top to bottom. Bluff Lake is classified as geologically primitive, being surrounded by a rocky shore with practically no sphagnum growth. The bottom contains some gravel patches and large areas of mud. The only plants in the shallow waters are a small number of bottom-rooted forms, such as *Lobelia* and a few sedges. The bottom mud, even in the deepest parts, has numerous specimens of the bladderwort, *Utricularia*, while sponges (*Spongilla* and *Heteromeyenia*), in both green and colourless phase, are locally abundant on the rocks. The water is colourless and extraordinarily clear, a Secchi disk being clearly visible to the

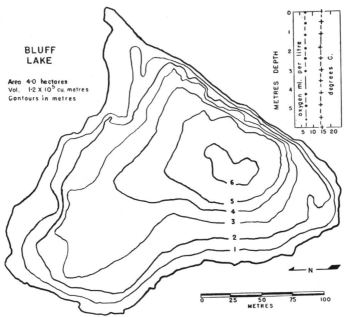

Fig. 1. Contour map of the lake used for experiments. The small inserted graph for midsummer values of oxygen and temperature indicates that the lake is without stratification.

bottom. The only fish observed in the lake were trout (*Salvelinus fontinalis*) and *Fundulus diaphanus* (not to mention the ubiquitous eel). The small variety of fish is not a special characteristic of this lake, for there are only about three dozen species in the fresh waters of the province, and only about five in the granite area in which Bluff Lake occurs.

METHODS

On 13 July 1949, 1000 mc. of radioactive phosphorus, ^{32}P, were added to the surface of the lake. This material, dispersed in 40 g. of KH_2PO_4, was dissolved in containers of 5-gallon capacity. The lake was divided into sections, and each large container of active phosphate was placed on the platform of a rubber boat, and pushed back and forth across a section by another boat, from which was operated a syringe bulb, that pumped air into the jar and started a siphon, by means of which active liquid was delivered into the water. The actual deposition occupied eight people for about 5 hr.

At the time of the experiment, the total phosphorus amounted to 31 parts per billion, or 3.7×10^3 g. in the whole lake. The added 40 g. of KH_2PO_4 would contain 9·1 g. of phosphorus, which would increase the natural store by 0·25 %.

Samples of water, of mud and of aquatic plants and animals were taken at intervals. The methods of collection and treatment have already been described (Coffin, Hayes, Jodrey & Whiteway, 1949). The estimation of radio-phosphorus, correction for decay, etc., are discussed in the same paper.

Samples of aquatic organisms were removed from the lake from time to time and the radio-phosphorus in them estimated. The results confirmed those previously reported (Coffin *et al.* 1949), that is, there was an appreciable uptake of material within a few hours, followed by a steady rise to the same orders of magnitude as in our first experiment. The organisms studied were as follows: *Eriocaulon*, leaves and roots, the leaves showing uptake before the roots. It would thus appear that material does not have to go through the roots in order to reach the leaves. *Utricularia* showed an initially rapid uptake followed by a tendency to level off. In *Pontederia* analyses were made on stems, leaves and roots. In this plant, as in *Eriocaulon*, it is evidently unnecessary for phosphorus to enter the roots before it can reach the stem. The alga, *Batrachospermum*, and the sponge *Heteromeyenia ryderi* displayed a rapid uptake from the first hours, as had been suspected from previous results in which, however, early observations were lacking. Zooplankton, which was over 99 % *Diaptomus minutus*, exhibited the same behaviour to the phosphorus as in the earlier experiment. A more detailed discussion is deferred, pending amplification of certain features by laboratory aquarium tests.

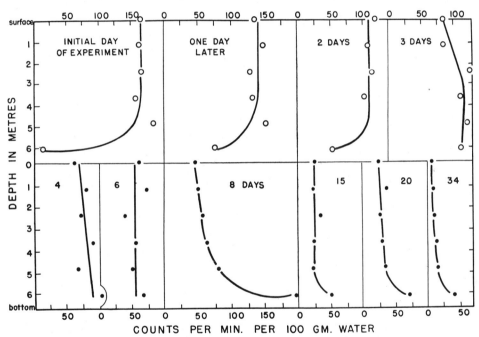

Fig. 2. Counts per minute of radio-phosphorus at different depths and at increasing intervals after the time of addition. Samples were collected at three stations on the lake. Most points are based on six determinations, being duplicates from each of the stations. The points in deeper water were based on four or two samples.

PENETRATION INTO THE DEPTHS

Three sampling stations were selected on the lake, and duplicate water samples taken periodically at 4 ft. depth intervals. The deepest water samples were all from one station, within the 6 m. contour line. The results are shown in Fig. 2, from which it appears that by the time of the first observation, mixing was practically complete in all but a small pit below the 5 m. contour. In this deep spot it took about 3 days for equilibration to be reached, which suggests a slight stratification. What is somewhat surprising is that from 8 days on, the bottom water had more radio-phosphorus in it than the lake generally. In view of the fact that mud exhibited a strong uptake of radioactive phosphorus, this might have been due to a slight admixture of mud with the lowest water samples. However, the deepest water taken was 2 ft. above bottom, and every care was exercised to avoid contamination. Moreover, all samples were filtered before analysis.

If the delayed loss of radio-phosphorus from the deeper layers is genuine (as appears likely), a tentative explanation is at hand in the light of the exchange theory outlined below. Suppose (1) that the equilibration rate between mud and water is lower than between living material and water, and (2) that there are fewer plants, etc. at 6 m. depth than in the shallows. Under such circumstances, and given slight stratification, the radio-phosphorus would not disappear so quickly from the deep pit. There is some basis for the first of the two assumptions in laboratory experiments which are now in progress. The second assumption would be valid for almost any lake.

Loss from lake water

Fig. 3 shows the course of disappearance of the added phosphorus from the lake water. The counts here are for the entire lake. It is immediately clear from Fig. 3A that loss of radio-phosphorus from the lake is at first rapid but later tends to flatten out and reach a plateau after a month. The plateau value is some 10 % of the original count. In Fig. 3B, in which a semi-logarithmic scale is used, the upper or observed line is still curved. This shows that the loss is not merely logarithmic, i.e. so much per cent per day. After considering various possible explanations for the data shown in Fig. 3, we are inclined to suggest that there is a continuous exchange of phosphorus going on between the water and solids of the lake. (In this paper the term 'solids' is meant to include the plants and animals in the lake, as well as whatever thickness of mud enters into the equilibration.)

Thus, if a number of marked atoms are placed in the water, they will tend to leave the water exponentially and enter the solids; such a theoretical curve of disappearance is shown in the lowest line of Fig. 3. At the same time there is a return of material from the solids, which also proceeds exponentially. However, in the first stages after addition of material, little can return from the solids, because little has yet entered them. This initial phase is shown in the first 5 days or so of the present experiment in which the declining water curve is not supported by any perceptible return from the solids. During this interval the curve of decline is indistinguishable from a logarithmic curve, a fact which has been made use of in some of the calculations to follow. As time goes on, and there is more material present in the solids, the return from them becomes appreciable and tends to hold up the phosphorus content of the water, as shown in the upper line of B in Fig. 3. Looking at Fig. 3B, there is seen, as the central space between the two theoretical curves, the amount of phosphorus delivered back from the solids to the water.

The argument in the foregoing paragraph can be expressed by the differential equation

$$\frac{dN}{dt} = -N(\lambda + \mu) + \mu N_0,$$

where λ and μ are respectively the percentage per day losses from water and solids, N is the amount present after time t, and N_0 is the concentration in the water at time t_0. By the use of the equation a half-life is estimated for water and for solids. From a half-life one may obtain the corresponding turnover time, t_t, which is the time required for the

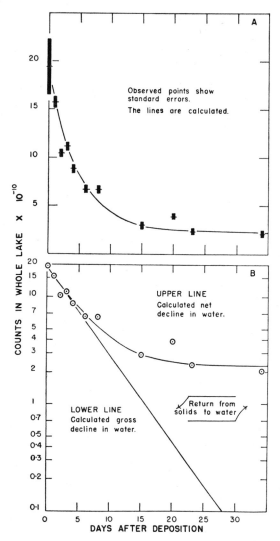

Fig. 3. Fate of the phosphorus added to Bluff Lake. The points in the curves were obtained from the data plotted in Fig. 2. Observations recorded in Fig. 2 were made at depth intervals of 1·22 m. (4 ft.). The lake was conceived as a number of layers, each 1·22 m. thick, and for each layer the radio-phosphorus concentration was taken as the average of the values at its upper and lower boundaries. From a contour map of the lake (Fig. 1) the volume in each layer was established. The total counts in each layer for a given day are:

$$\frac{\text{Average of two points in Fig. 2} \times \text{total ml. water in layer of lake}}{100}.$$

Summing up the values in all layers, one obtains the total counts for the lake as set down in Fig. 3. The counts in the deeper layers were of course small, because the contours were closing in.

The upper part of Fig. 3 is plotted arithmetically, the lower part on an arith-log scale. In the upper part the standard error of each point is shown. As noted, the readings on the ordinate should be multiplied by 10^{-10}.

appearance or disappearance of an amount of phosphorus equal to the amount in the lake. The equation, based on Zilversmit, Entenman & Fishler (1943) and Zilversmit, Entenman, Fishler & Chaikoff (1943) is

$$t_t = \frac{t_{0.5}}{\log 2} \log e = 1.44 t_{0.5}.$$

Alternatively, t_t for water is $1/\lambda$ and t_t for solids is $1/\mu$.

The turnover time for the phosphorus in Bluff Lake water works out at 5·4 days (half-life 3·73 days) and for the participating phosphorus in the solids the turnover time is 39 days (half-life 27 days). Of the phosphorus which is turning over in the system less than one-sixth of the total is in the water. (Details of the calculations used to derive quoted values will be furnished by the authors upon request.)

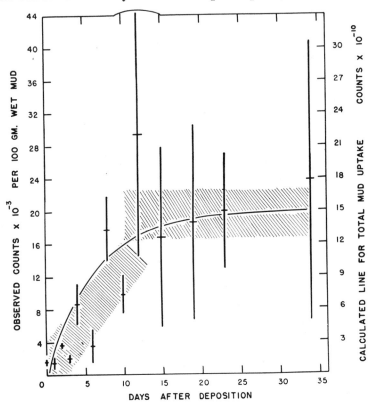

Fig. 4. Changes in the concentration of ^{32}P in mud. About four mud samples were averaged to establish each point. The points (or rather vertical lines) are based on direct observations, and the standard error is shown in each. The shaded portion, whose height is 2σ, shows the zone of standard error, based on the best slope of the observed points for the first 10 days (the slope being significant), and the average value of the points from 10 days on (assuming no significant slope). The curve as drawn is derived from Fig. 3 and represents the radio-phosphorus which has been lost from the water, on the assumption of an exchange between water (half-life 3·73 days) and solids (half-life 27 days). The calculated line is evidently not inconsistent with the direct observations.

GAIN BY MUD

By means of the ooze sucker, as described by Welch (1948, p. 181), samples of surface mud were regularly collected from three lake stations, and counts of radio-phosphorus made on them. Results are given in Fig. 4, from which it is seen that the scatter of the samples is quite great. Errors are no doubt due in large part to variations in the thickness of the mud picked up by the ooze sucker. However, it can be shown that there is a significant slope during the first 10 days, and this slope has been calculated and its standard error indicated by cross-hatching. After the first 10 days, no significant slope exists and the points have been averaged and their standard error shown by cross-hatching. The line as drawn comes from the calculated values as mentioned in the last section, giving solids a half-life of 27 days. It is seen that the theoretical line is not inconsistent with observed values as far as its general shape is concerned.

To superimpose the line of Fig. 4 on the points, 15×10^{10} on the right ordinate has been made to fall over 20×10^3 on the left ordinate. This provides no reason at all for supposing that the mud is wholly or largely responsible for the exchange mechanism. Indeed it is known that the aquatic organisms are very active in the process. All that Fig. 4 suggests is that mud is participating to some degree with (so far as the evidence goes) the half-lives which were calculated from the water values.

It is of interest to set an upper limit to the thickness of mud which might be participating in the exchange. For this one imagines that mud constitutes all the solids involved. Using data from Figs. 1–4, a calculation was made which gave a layer of mud 1·9 cm. thick. This is what might be looked for if the complete exchange phenomenon were credited to mud alone. However, laboratory tests on undisturbed mud cores taken with the Jenkin sampler (Mortimer, 1941, 1942) indicate that the actual exchanging thickness is of the order of 1 mm., which would suggest that some nine-tenths of the phosphorus exchange is with the living organisms in the lake.

DISAPPEARANCE OF ADDED PHOSPHORUS FROM OTHER LAKES

It will perhaps be useful to examine the exchange mechanism between water and solids in some lakes other than Bluff Lake. Several have been selected from the literature, and the relevant data concerning them are set down in Table 1.

The Punchbowl is a small acid bog lake, west of Halifax, to which radio-phosphorus was added in 1948 (Coffin *et al.* 1949). The lake is highly stratified and surrounded by a sphagnum margin, and little of the added phosphorus penetrated to the depths. In this case, 'solids' are dominated by sphagnum, and include other plants and plankton, but do not include much mud. Instead of reaching the bottom, phosphorus added to the Punchbowl is stopped at a thermocline. None the less, the curve of disappearance (Fig. 5, upper part) is of the same form as in Bluff Lake, and permits the calculation of turnover times of 7·6 and 37 days for water and solids respectively. It is further calculated that for every gram of phosphorus in lake water, there are 3·7 additional grams of exchanging phosphorus in the solids.

The lower portion of Fig. 5 is from Smith (1945). Crecy Lake in New Brunswick appears to be comparable to Bluff Lake in general characteristics; in fact the description which has been given of the fauna and flora of Bluff Lake would fit Crecy very well.

To Crecy Lake was added one ton of solid ammonium phosphate and 500 lb. of potassium chloride. If adequately mixed, an initial concentration of 390 mg./cu.m. (p.p.b.) of phosphorus would have been produced. Probably this was not achieved and some solid material went to the bottom. Orr (1947) and Pratt (1949) both failed to get theoretical quantities of phosphorus into solution when solid fertilizers were added.

We have ascertained the best line to fit the Crecy points, and from it the turnover times for water and solids emerge as 17 and 176 days, with the solids containing nearly eight times as much exchanging phosphorus as the water. The general level of phosphorus in the water before the experiment was about 15 p.p.b.; the new equilibrium level as shown in Fig. 5 was some 25 p.p.b., an increase of 67 %. Presumably the participating phosphorus in the solids was also increased by 67 %. If all the exchange were credited to mud, the thickness of the participating layer would work out at 1·24 cm.

Table 1. *Data on lakes discussed in the text*

(The values in col. 4 are rough approximates, especially where solid fertilizer was added. They are supposed to represent the material which went into solution.)

Lake area (ha.) (1)	Max. depth (m.) (2)	Strati-fication (3)	Increased P as % of initial value (4)	Turnover time for P in water (days) (5)	Turnover time for P in solids (days) (6)	Ratio of total P turning over to P in water (7)	Remarks (8)	Reference (9)
Bluff (4·0)	7	None	0·25	5·4	39	6·41	Dissolved radio-phosphorus	This paper
Punchbowl (0·3)	6·2	Marked	3·0	7·6	37	4·7	Dissolved radio-phosphorus, remained in epilimnion	Coffin *et al.* (1949)
Crecy (20·4)	3·8	None	1600	17	176	8·7	Dry ammonium phosphate and potassium chloride	Smith (1945)
Ullswater (900)	63	Marked	1000	29	—	—	Normal spring maximum. Nothing added	Pearsall (1930)
Eight English lakes (100–900)	19–79	Marked	400–1500	26–40	—	—	Normal spring maximum. Nothing added	Pearsall (1930)
Loch Craiglin (sea loch) (7·3)	6	None	500–2000	3·2	—	—	Dry superphosphate and sodium nitrate. Surface values used	Orr (1947)
Cohasset Pond (marine)	0·7	None	2000	2·4	—	—	Dry superphosphate and sodium nitrate	Pratt (1949)

A phosphorus increase of 67 % in water could usually be detected despite summer fluctuations. Mud samples, however, are in general taken by a dredge, which would scoop up a layer 10 or 15 cm. thick. In such a sample the average increase in phosphorus would be reduced to perhaps 5 % and lost in the experimental error. The Crecy Lake results might be summed up as follows:

Theoretical increase in water phosphorus by solid fertilizer added	2500 %
Quantity which failed to go into solution, perhaps	900 %
Actual initial increase in water phosphorus (± 100 %)	1600 %
Residual increase after equilibration	67 %
Increase detectable in mud, probably	Nil
Increase detectable in bodies of plants and animals, probably	Nil

These values show how a fertilization experiment can easily lead to the inference that, since little of the added phosphorus is left in the water, and none detectable in the mud, there must have been a great stimulus to growth and consequent increase in the aquatic fauna and flora. What is actually available to contribute to eutrophication is an extra 67 %, not 2500 %.

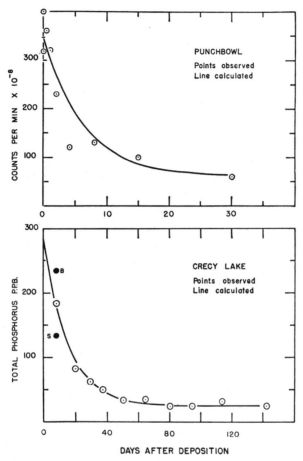

Fig. 5. Original observations and calculated curves for the disappearance of phosphorus from lakes. In the Punchbowl (Coffin *et al.* 1949), the enhancement of the water was 3 % and the calculated turnover times for water and solids are 7·6 and 37 days. The mud in this case probably did not enter appreciably into the exchange, as the lake was highly stratified and the phosphorus remained largely in the epilimnion.

In Crecy Lake (Smith, 1945), the increase by fertilization amounted to some 1600 % and the turnover times for water and solids are taken as 17 and 176 days. The lake is unstratified. Solid circles at 8 days are surface and bottom values which, on this occasion only, showed considerable divergence.

DISAPPEARANCE OF INORGANIC PHOSPHORUS FROM LAKES

There are some instances in which the fate of inorganic phosphorus (as distinct from total phosphorus above) has been followed in lakes. In the type of lake that supports trout, the standing summer level of inorganic phosphorus is usually too low to be measured by current methods, being less than 5 p.p.b., below which the accuracy of detection diminishes.

Pearsall (1930) followed the normal annual cycle in nine English lakes, one of which, Ullswater, is shown in Fig. 6. In all the lakes there was a sharp spring maximum after the ice cover broke up, followed by a decline during the summer. As to the cause of the maximum, it would seem that the mixing of the water after winter stagnation might have dispersed through the lake the phosphorus from material which had died during the winter. The points which follow the spring maxima in all the lakes can be fitted quite well by the formula of equilibration. Unfortunately, however, the formula cannot

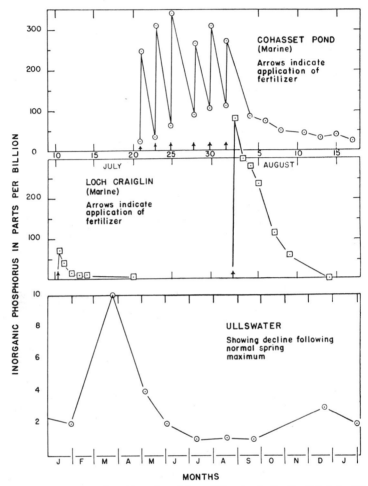

Fig. 6. Disappearance of inorganic phosphorus from lakes. Ullswater in the English Lake District shows the normal cycle, including decline after the spring maximum (Pearsall, 1930). The turnover time for water is calculated at 29 days.

Loch Craiglin in Scotland (Orr, 1947) and Cohasset Pond, Massachusetts (Pratt, 1949) are both marine and were fertilized repeatedly. For the water in Craiglin the turnover time is estimated as 3.2 ± 0.7 days and for Cohasset 2.4 ± 0.5.

be used to determine numerical values for turnover times, because one of the parameters is lacking, namely, the actual date of the highest level of phosphorus reached during the spring maximum. An approximate turnover time for water can, however, be found by assuming (as mentioned and illustrated with reference to Fig. 2) that the return from solids is, in the early stages of decline, negligible. In this phase the water decline is an exponential curve, so that

$$t_t = \frac{t \log e}{\log A_0 - \log A},$$

where t_t is the turnover time sought, A_0 is the phosphorus in the water at any time zero, and A is the concentration t days later.

The turnover time for Ullswater is calculated to be 29 days. This is the largest value so far encountered, and is representative of those obtained for the remaining eight, as Table 1, col. 5, shows. A long turnover time might be related to some of the following factors:

(*a*) these are larger lakes than those discussed above;

(*b*) the actual spring maximum may have been earlier and higher than the one detected;

(*c*) it was too early in the season for aquatic vegetation to flourish;

(*d*) a difference in behaviour may exist between organic and inorganic phosphorus.

The remaining two lakes, Craiglin and Cohasset, are both marine, but have only a limited exchange of water with the sea at large. Both were subjected to repeated applications of commercial fertilizers, and Fig. 6 shows a portion of each experiment. The loss of added material from the water was so rapid that the return mechanism from the solids can almost certainly be neglected in calculations of water turnover times. In fact, the tails of the curves to the right are found to decline exponentially, so that no return mechanism can be observed. This suggests that the reservoir of exchanging phosphorus in the solids is much larger than in the fresh-water lakes considered. Probably the live material is responsible, being greater in the sea than in fresh water. It might be, of course, that if the measurements were of total phosphorus instead of inorganic phosphorus, a return from solids would be shown. If this were so it would indicate a rapid synthesis of organic phosphorus, which seems quite likely.

Loch Craiglin (Orr, 1947) was fertilized in July with dissolved phosphate. The turnover time for water works out at $1\cdot61 \pm 0\cdot49$ days. In the August experiment, where a larger quantity of solid fertilizer was added, the turnover time was $4\cdot68 \pm 0\cdot58$ days. It is unlikely that the difference between the two values has any significance, and they may be combined to yield a general result of $3\cdot2 \pm 0\cdot7$ days as the turnover time. This is almost twice the disappearance rate from Bluff Lake, and there is no return.

Cohasset Pond (Pratt, 1949) was fertilized a number of times in rapid succession, with a sharp drop following each application. The turnover time has been calculated from each of these drops, and also from those in a similar experiment conducted in May (Pratt's experiments, nos. 4 and 5). No significant trend is discernible during the progress of either series, and the turnover time from fourteen tests works out at $2\cdot40 \pm 0\cdot46$ days. This is so close to the combined Loch Craiglin results, that they appear to be the same, so that they may be combined to give a general turnover time for water in these marine ponds as $2\cdot7 \pm 0\cdot9$ days.

Factors affecting the rate of equilibration

If the equilibrating medium were mainly mud, the dimensions of the lake would probably be decisive in determining the rate of turnover of phosphorus in the water (provided there were no stratification). Turnover rate might be proportional to the ratio of water volume to mud surface. Some calculations of the ratio have been made, based on the formula for a cone, which are shown in Fig. 7. It is seen that when the depth is less than 3 m. the surface is always relatively large (shallow lakes would have a rapid phosphorus turnover). Below 5 m. depth the ratio is inversely proportional to the area of the lake (large lakes would have a slow phosphorus turnover).

If the equilibrating solids were mainly live organisms, the number of these might be decisive. Equilibration would be more rapid in a eutrophic lake, and more rapid in summer. Also, the effect of summer stratification would not greatly affect the turnover time in the epilimnion.

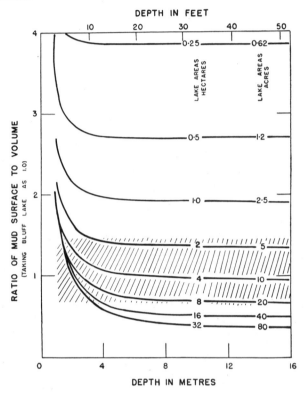

Fig. 7. Curves to show the way in which mud surface might be expected to vary in relation to lake volume with changing dimensions. All curves are constructed from the formula for a cone, and are made relative by supposing that in Bluff Lake the ratio of mud surface to water volume is 1·0. The shaded portion takes in 50 % above and below the Bluff Lake value. It is seen that lakes from 4 to 20 acres (except when very shallow) fall within the shaded area.

Mortimer (1941, 1942) has shown that various nutrients accumulate in the hypolimnion during summer stagnation, and disappear rapidly when the stratification of a lake is broken up in the autumn. The release of the nutrients coincides with solubility increases which accompany reduction at the mud surface. The autumn disappearance has usually been attributed to the opposite cause, namely, precipitation following readmission of oxygen to the system (Einsele, 1938). It may be, however, that as soon as the nutrients reach shallow waters they are equilibrated with living organisms. Volatile substances such as ammonia might at the same time be lost to the atmosphere.

Under conditions of equilibrium, as many atoms of phosphorus must be returning to the water per day as are leaving it. If the participating solids were doubled (and the same number of atoms still returned to the water), it would in effect double the turnover time for the solids. Now it follows from the formula of equilibration, that the longer the half-life for the solids, the lower will be the amount of phosphorus in the water at equilibrium. Thus it might be expected that in a eutrophic lake, the summer concentration of phosphorus would be lower than in an oligotrophic lake. This appears to be the case in English lakes studied by Pearsall (1930), as shown in Table 2.

Table 2

Lake	Percentage of drainage system cultivable	Inorganic phosphorus in surface water, p.p.b. June, July and August average
Wastwater	5·2	2·0
Ennerdale	5·4	2·6
Crummock	8·0	2·6
Derwentwater	10·0	1·8
Ullswater	16·6	1·4
Lowes Water	24·0	1·1
Bassenthwaite	29·4	0·9
Esthwaite	45·4	1·1

(In the above table and in Table 1, the findings on Lake Windermere are omitted. This is because Pearsall stated that the lake was 'certainly much contaminated by sewage effluents' and did not himself use the spring maximum value for phosphorus, which he considered unreasonable.)

EUTROPHICATION

After fertilizer is added to a lake, there is a period of weeks during which the phosphorus content of the water remains high. During this interval, blooms of zooplankton and algae are generally observed (Smith, 1945; Langford, 1948). There may also be blooms during the slight excess that is maintained in the water after equilibrium is established.

Probably during the winter following fertilization, there is an equilibration of a slower kind which carries some of the added phosphorus into deeper layers of the mud. It is sometimes said that there is enough left in the water to produce a stimulus to plankton growth the following season. However, in view of the occurrence of a normal spring maximum of phosphorus (Pearsall, 1930) and the extreme variation in quantity of plankton from year to year (Langlois, 1946), we must doubt whether there is any valid evidence on this point.

It appears unlikely that any single or periodic application of fertilizer could have more than a very brief effect, however great in quantity it was. Thus, in Loch Craiglin, a 3-year programme of adding fertilizers threw the lake out of equilibrium and actually reduced the growth of fish below the normal level (Gross, 1949). Also 'the zooplankton which during the first year of fertilizations was richer in Loch Craiglin than in the outside loch diminished in the second year and remained poor thereafter' (Marshall & Orr, 1948). The bottom fauna, however, was still at a higher level than normal, 2½ years after the programme of fertilization terminated (Raymont, 1949). This might be because a good deal of the fertilizer added reached the bottom in solid form.

Permanent examples of eutrophication, as described by Hasler (1947), have occurred when there is a constant addition of sewage to a lake, which presumably keeps the nutrients in the water continuously above the level of equilibration with the solids. Such lakes often promote the growth of coarse fish and eliminate trout. It is reasonable to suppose that if the addition of sewage were stopped, the lakes would soon exhibit a lower level of phosphorus in the water than before the treatment began, and become oligotrophic. In other words, eutrophication may consist of the drainage into a lake of more nutrients than can be equilibrated with the solid phase.

SUMMARY

One thousand millicuries of radioactive phosphorus, ^{32}P, were added to the surface of an unstratified, primitive lake of area 4 ha. (10 acres) and depth 7 m., situated near Halifax, Nova Scotia. There was a rapid loss from the water, and uptake by aquatic life and mud.

As the added phosphorus increased the normal water content by only 0·25 %, a mechanism of removal does not explain the results. Instead, an active exchange is postulated, between the water phosphorus and participating phosphorus in the lake solids, which together make up a single system.

An equation is given to describe the equilibrium reached by an exponential decline of water phosphorus, balanced by a return from the solids, also exponential. The turnover time for the water phosphorus is calculated as 5·4 days; for the solids 39 days.

The equation is applied to several other lakes where radio-phosphorus or fertilizer has been added and followed by water analyses, and also to the decline in phosphorus, following the spring maximum. The results of adding fertilizer to two marine ponds are also examined. The turnover times for water vary from 2·4 to 30 days or more, and for the solids from 37 days to at least 176 days.

The relation of the results to the general problem of eutrophication is discussed.

This work was supported by grants from the National Research Council, the Nova Scotia Research Foundation and the Nova Scotia Department of Trade and Industry, to all of whom our best thanks are due.

The lake itself was selected as a suitable one for the present experiment by Prof. H. P. Bell, whose botanical advice has been freely available at all times. We are indebted to Mr James Lewin for a survey of the lake and for much assistance with the field work. Many of the collections were made by Miss Shirley Mason, and a large proportion of the laboratory analyses were carried out by Miss Louise Jodrey.

REFERENCES

Coffin, C. C., Hayes, F. R., Jodrey, L. H. & Whiteway, S. G. (1949). Exchange of materials in a lake as studied by the addition of radioactive phosphorus. *Canad. J. Res.* D, **27**, 207–22.

Einsele, W. (1938). Über chemische und kolloidchemische Vorgänge in Eisen-Phosphat-Systemen unter limnochemischen und limnogeologischen Gesichtspunkten. *Archiv. Hydrobiol.* **33**, 361–87.

Gross, F. (1949). Further observations on fish growth in a fertilized sea loch (Loch Craiglin). *J. Mar. Biol. Ass. U.K.* **28**, 1–8.

Hasler A. D. (1947). Eutrophication of lakes by domestic drainage. *Ecology*, **28**, 383–95.

Hutchinson, G. E. & Bowen, V. T. (1947). A direct demonstration of the phosphorus cycle in a small lake. *Proc. Nat. Acad. Sci., Wash.*, **33**, 148–53.

Hutchinson, G. E. & Bowen, V. T. (1950). Limnological studies in Connecticut. IX. A quantitative radio-chemical study of the phosphorus cycle in Linsley Pond. *Ecology*, **31**, 194–203.

Langford, R. R. (1948). Fertilization of lakes in Algonquin Park, Ontario. *Trans. Amer. Fish. Soc.* **78**, 133–44.

Langlois T. H. (1946). The herring fishery of Lake Erie. *Inland Seas*, **2**, 101–4.

Marshall, S. M. & Orr, A. P. (1948). Further experiments on the fertilization of a sea loch (Loch Craiglin). The effect of different plant nutrients on the phytoplankton. *J. Mar. Biol. Ass. U.K.* **27**, 360–79.

Mortimer, C. H. (1941, 1942). The exchange of dissolved substances between mud and water in lakes. *J. Ecol.* **29**, 280–328; **30**, 147–201.

Orr, A. P. (1947). An experiment in marine fish cultivation. II. Some physical and chemical conditions in a fertilized sea-loch (Loch Craiglin, Argyll). *Proc. Roy. Soc. Edin.* B, **68**, 3-20.

Pearsall, W. H. (1930). Phytoplankton in the English Lakes. I. The proportions in the waters of some dissolved substances of biological importance. *J. Ecol.* **18**, 306–20.

Pratt, D. M. (1949). Experiments in the fertilization of a salt water pond. *Sears Found. J. Mar. Res.* **8**, 36–59. (Reprinted as Contribution no. 466 in Collected Reprints, Woods Hole Oceanograph. Inst., 1949.)

Raymont, J. E. G. (1949). Further observations on changes in the bottom fauna of a fertilized sea loch. *J. Mar. Biol. Ass. U.K.* **28**, 9–19.

Smith, M. W. (1945). Preliminary observations upon the fertilization of Crecy Lake, New Brunswick. *Trans. Amer. Fish. Soc.* **75**, 165–74.

Welch, P. S. (1948). *Limnological Methods.* Philadelphia: Blakiston.

Zilversmit, D. B., Entenman, C. & Fishler, M. C. (1943). On the calculation of 'turnover time' and 'turnover rate' from experiments involving the use of labeling agents. *J. Gen. Physiol.* **26**, 325–31.

Zilversmit, D. B., Entenman, C., Fishler, M. C. & Chaikoff, I. L. (1943). The turnover rate of phospholipids in the plasma of the dog as measured with radioactive phosphorus. *J. Gen. Physiol.* **26**, 333–40.

15

Reprinted from *Ecol. Monographs*, **40**, 23–47 (1970) with permission of the publisher, Duke University Press, Durham, North Carolina

EFFECTS OF FOREST CUTTING AND HERBICIDE TREATMENT ON NUTRIENT BUDGETS IN THE HUBBARD BROOK WATERSHED-ECOSYSTEM[1]

Gene E. Likens[2]

Department of Biological Sciences
Dartmouth College
Hanover, New Hampshire

F. Herbert Bormann

School of Forestry
Yale University
New Haven, Connecticut

Noye M. Johnson

Department of Earth Sciences
Dartmouth College
Hanover, New Hampshire

D. W. Fisher

U. S. Geological Survey
Washington, D. C.

Robert S. Pierce

Northeastern Forest Experiment Station
Forest Service, U. S. Department of Agriculture
Durham, New Hampshire

(Accepted for publication July 31, 1969)

Abstract. All vegetation on Watershed 2 of the Hubbard Brook Experimental Forest was cut during November and December of 1965, and vegetation regrowth was inhibited for two years by periodic application of herbicides. Annual stream-flow was increased 33 cm or 39% the first year and 27 cm or 28% the second year above the values expected if the watershed were not deforested.

Large increases in streamwater concentration were observed for all major ions, except NH_4^+, $SO_4^=$ and HCO_3^-, approximately five months after the deforestation. Nitrate concentrations were 41-fold higher than the undisturbed condition the first year and 56-fold higher the second. The nitrate concentration in stream water has exceeded, almost continuously, the health levels recommended for drinking water. Sulfate was the only major ion in stream water that decreased in concentration after deforestation. An inverse relationship between sulfate and nitrate concentrations in stream water was observed in both undisturbed and deforested situations. Average streamwater concentrations increased by 417% for Ca^{++}, 408% for Mg^{++}, 1558% for K^+ and 177% for Na^+ during the two years subsequent to deforestation. Budgetary net losses from Watershed 2 in kg/ha-yr were about 142 for NO_3-N, 90 for Ca^{++}, 36 for K^+, 32 for SiO_2-Si, 24 for Al^{+++}, 18 for Mg^{++}, 17 for Na^+, 4 for Cl^-, and 0 for SO_4-S during 1967–68; whereas for an adjacent, undisturbed watershed (W6) net losses were 9.2 for Ca^{++}, 1.6 for K^+, 17 for SiO_2-Si, 3.1 for Al^{+++}, 2.6 for Mg^{++}, 7.0 for Na^+, 0.1 for Cl^-, and 3.3 for SO_4-S. Input of nitrate-nitrogen in precipitation normally exceeds the output in drainage water in the undisturbed ecosystems, and ammonium-nitrogen likewise accumulates in both the undisturbed and deforested ecosystems. Total gross export of dissolved solids, exclusive of organic matter, was about 75 metric tons/km² in 1966–67, and 97 metric tons/km² in 1967–68, or about 6 to 8 times greater than would be expected for an undisturbed watershed.

The greatly increased export of dissolved nutrients from the deforested ecosystem was due to an alteration of the nitrogen cycle within the ecosystem.

The drainage streams tributary to Hubbard Brook are normally acid, and as a result of deforestation the hydrogen ion content increased by 5–fold (from pH 5.1 to 4.3).

Streamwater temperatures after deforestation were higher than the undisturbed condition during both summer and winter. Also in contrast to the relatively constant temperature in the undisturbed streams, streamwater temperature after deforestation fluctuated 3–4°C during the day in summer.

Electrical conductivity increased about 6–fold in the stream water after deforestation and was much more variable.

Increased streamwater turbidity as a result of the deforestation was negligible, however the particulate matter output was increased about 4–fold. Whereas the particulate matter is normally 50% inorganic materials, after deforestation preliminary estimates indicate that the proportion of inorganic materials increased to 76% of the total particulates.

Supersaturation of dissolved oxygen in stream water from the experimental watersheds is common in all seasons except summer when stream discharge is low. The percent saturation is dependent upon flow rate in the streams.

Sulfate, hydrogen ion and nitrate are major constituents in the precipitation. It is suggested that the increase in average nitrate concentration in precipitation compared to data from 1955–56, as well as the consistent annual increase observed from 1964 to 1968, may be some measure of a general increase in air pollution.

TABLE OF CONTENTS

INTRODUCTION

Management of forest resources is a worldwide consideration. Approximately one-third of the surface of the earth is forested and much of this is managed or deforested by one means or another.

Forests may be temporarily or permanently reduced by wind, insects, fire, and disease or by human activities such as harvesting or management utilizing physical or chemical techniques. Management goals range from simple harvest of wood and wood products, to increased water yields, to military stratagems involving defoliation of extensive forested areas.

Despite the importance of the forest resource, there is very little quantitative information at the ecosystem level of understanding on the biogeochemical interactions and implications resulting from large-scale changes in habitat or vegeta-

¹ This is Contribution No. 14 of the Hubbard Brook Ecosystem Study. Financial support for this work was provided by NSF Grants GB 1144, GB 4169, GB 6757, and GB 6742. The senior author acknowledges the use of excellent facilities and resources at the Brookhaven National Laboratory during the preparation of part of this manuscript. Also, we thank J. S. Eaton for special technical assistance, and W. A. Reiners and R. C. Reynolds for critical comments and suggestions. Published as a contribution to the U. S. Program of the International Hydrological Decade, and the International Biological Program.
² Present address: Division of Biological Sciences, Cornell University, Ithaca, New York 14850.

tion. This gap in our understanding results because it is particularly difficult to get quantitative ecological information that allows predictions about the entire ecosystem. The goal of the Hubbard Brook Ecosystem study is to understand the energy and biogeochemical relationships of northern hardwood forest watershed-ecosystems as completely as possible in order to propose sound land management procedures.

The small watershed approach to the study of hydrologic-nutrient cycle interaction used in our investigations of the Hubbard Brook Experimental Forest (Bormann and Likens 1967) provides an opportunity to deal with complex problems of the ecosystem on an experimental basis. The Hubbard Brook Experimental Forest, maintained and operated by the U.S. Forest Service, is especially well-suited to this approach since ecosystems can be defined as discrete watersheds with similar northern hardwood forest vegetation and a homogeneous bedrock, which forms an impermeable base (Bormann and Likens, 1967; Likens, et al., 1967). Thus, the six small watersheds we have used at Hubbard Brook provide a replicated experimental design for manipulations at the ecosystem level of organization.

All vegetation on Watershed 2 (W2) was cut during the late fall and winter of 1965, and subsequently treated with herbicides in an experiment designed to determine the effect on 1) the quantity of stream water flowing out of the watershed, and 2) fundamental chemical relationships within the forest ecosystem, including nutrient relationships and eutrophication of stream water. In effect this experiment was designed to test the homeostatic capacity of the ecosystem to adjust to cutting of the vegetation and herbicide treatment. This paper will discuss the results of this experimental manipulation in comparison to adjacent, undisturbed watershed-ecosystems.

THE HUBBARD BROOK ECOSYSTEM

The hydrology, climate, geology, and topography of the Hubbard Brook Experimental Forest have been reported in detail elsewhere (Likens, et al., 1967).

The climate of this region is dominantly continental. Annual precipitation is about 123 cm (Table 1), of which about one-third to one-fourth is snow. Although precipitation is evenly dis-

TABLE 1. Average annual water budgets for Watersheds 1 through 6 of the Hubbard Brook Experimental Forest. Watershed 2 has been excluded from the averages for 1965–68; 1967–68 is based on Watersheds 1, 3, and 6 only

Water Year (1 June-31 May)	Precipitation (P) (cm)	Runoff (R) (cm)	P-R (Evaporation and Transpiration) (cm)
1963–64	117.1	67.7	49.4
1964–65	94.9	48.8	46.1
1965–66	124.5	72.7	51.8
1966–67	132.5	80.6	51.9
1967–68	141.8	89.4	52.4
1963–68	122.2	71.8	50.4
1955–68	122.8	71.9	50.9

tributed throughout the year, stream flow is not. Summer and early autumn stream flow is usually low; whereas the peak flows occur in April and November. Loss of water due to deep seepage appears to be minimal in the Hubbard Brook area (Likens, et al., 1967). The bedrock of the area is a medium to coarse-grained sillimanite-zone gneiss of the Littleton Formation and consists of quartz, plagioclase and biotite with lesser amounts of sillimanite. The mantle of till is relatively shallow and has a similar mineral and chemical composition to the bedrock. The soils are podzolic with a pH less than 7. Despite extremely cold winter air temperatures, soil frost seldom forms since insulation is provided by several centimeters of humus and a continuous winter snow cover (Hart et al., 1962).

METHODS AND PROCEDURES

Precipitation is measured in the experimental watershed with a network of precipitation gauges,

approximately 1 for every 12.9 hectares of watershed. Streamflow is measured continuously at stream-gauging stations, which include a V-notch weir or a combination of V-notch weir and San Dimas flume anchored to the bedrock at the base of each watershed.

Weekly samples of precipitation and stream water were obtained from the experimental areas for chemical analysis. Rain and snow were collected in two types of plastic containers, 1) those continuously uncovered or 2) those uncovered only during periods of rain or snow. One-liter samples of stream water were collected in clean polyethylene bottles approximately 10 m above the weir in both the deforested and undisturbed watersheds. Chemical concentrations characterizing a period of time are reported as weighted averages, computed from the total amount of precipitation or streamflow and the total calculated chemical content during the period. Details concerning the methods used in collecting samples of precipitation and stream water, analytical procedures, and measurement of various physical characteristics have been given by Bormann and Likens (1967), Likens, et al. (1967), and Fisher, et al. (1968).

During November and December of 1965 all trees, saplings and shrubs of W2 (15.6 ha) were cut, dropped in place, and limbed so that no slash was more than 1.5 m above the ground. No roads were made on the watershed and great care was taken to minimize erosion. No timber or other vegetation was removed from the watershed. Regrowth of vegetation was inhibited by aerial application of the herbicide, Bromacil ($C_9H_{13}Br N_2O_2$), at 28 kg/ha on 23 June 1966. Approximately 80% of the mixture applied was Bromacil and 20% was largely inert carrier (H. J. Thorne, personal communication). Also, during the summer of 1967, approximately 87 liters of an ester of 2, 4, 5-trichlorophenoxyacetic acid (2, 4, 5-T) was individually applied to scattered regrowths of stump sprouts.

The results reported cover the period immediately following the cutting of the vegetation on W2, 1 January 1966 through 1 June 1968.

HYDROLOGIC PARAMETERS

The annual hydrologic regime at Hubbard Brook has varied greatly since we began our study in 1963 (Table 1). The 1964-65 water-year was exceptionally dry, and 1967–68 was very wet. These fortuitous extremes have provided a wide range of hydrologic conditions for our study of the hydrologic-nutrient cycle interactions.

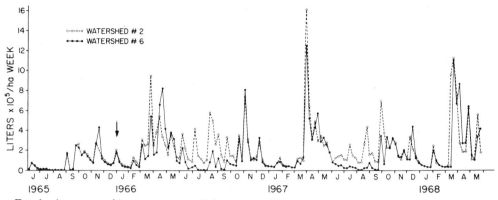

Fig. 1. Average weekly stream water discharge from Watersheds 2 and 6 during 1965–68. The vegetation on Watershed 2 was cut during November and December 1965. The arrow indicates the completion of the cutting.

Chemical input into a watershed from meteorological sources is based in part on the volume of precipitation, thus it is important to determine whether precipitation is distributed randomly throughout the watersheds. Our studies in 1963–64 and 1964–65 indicated no significant difference between rain gauges at different elevations (Likens *et al.*, 1967). Subsequent data indicate that there is generally very little difference in the precipitation pattern with elevation, but one or two storms of high intensity may significantly alter the spatial distribution for total annual precipitation within the area. Since the overall precipitation pattern for 1966–67 and 1967–68 was more variable than previous years, we have calculated the precipitation input for each watershed, for these two years, on the basis of Thiessen averages established for the area (see Thiessen, 1923). In spite of the variation in precipitation and runoff, the amount of water lost by evaporation and transpiration from undisturbed watersheds remained about the same each year during 1963 to 1968 (Table 1).

Cutting the vegetation of W2 produced a significant effect on the distribution of water loss from the watershed. These changes are reported in detail elsewhere (Hornbeck, *et al.*).

The annual runoff in 1967–68 and 1968–69 from the deforested ecosystem increased by 39% and 28%,[3] respectively, over the values expected had the watershed not been cut (Table 2). The greatest difference occurred during June through September, when runoff values were 414% (1966–67) and 380% (1967–68) greater than expected.

[3] The slight discrepancy from data presented by Hornbeck, *et al.*, is due to rounding errors incurred in the conversion of English to metric units.

The increased streamflow during summer is directly attributable to the removal of transpiring surface. In addition to the increased yield of

TABLE 2. Runoff and precipitation for Watershed 2 (cut-over) of the Hubbard Brook Experimental Forest, expressed in cm of water. The watershed was clear-cut in November and December of 1965. Predicted values for runoff were calculated from regression analyses established during a pretreatment calibration period. (After Hornbeck, *et al.*, 1970)

	1965-66	1966-67	1967-68
Precipitation	124.6	132.1	138.7
Predicted runoff	(77.7)	(86.5)	(96.7)
Measured runoff	79.9	119.9	123.4
Evaporation and Transpiration	44.7	12.2	15.4

drainage water from the watershed, the snowmelt was advanced a few days and was more rapid, particularly during 1967–68 (Fig. 1; Federer, 1969; Hornbeck and Pierce, 1969).

In various experiments throughout the world, with 100% reduction in forest cover by clear-cutting or chemical treatment, stream flow increases averaged about 20 cm the first year after treatment (Hibbert, 1967). Detailed studies at the Coweeta Hydrologic Laboratory, Southeastern Forest Experiment Station in North Carolina showed maximum increases in water yield of about 41 cm during the first year following the complete removal of the hardwood forest vegetation on small watersheds (Hoover, 1944).

PRECIPITATION CHEMISTRY

Sulfate and hydrogen ions are the most abundant constituents (in terms of chemical equivalents) in precipitation falling on the watersheds

Table 3. Average weighted ion content of bulk precipitation collected within the Hubbard Brook Experimental Forest from 1 June to 31 May expressed in mg/liter.

	1965-66	1966-67	1967-68
calcium	0.22	0.21	0.20
magnesium	0.04	0.03	0.05
potassium	0.05	0.05	0.05
sodium	0.16	0.10	0.12
aluminum	***	0.1	***
ammonium	0.21	0.18	0.22
hydrogen ion	0.07	0.08	0.07
nitrate	1.41	1.49	1.56
sulfate	3.3	3.1	3.3
chloride	0.21+	0.50##	0.35
bicarbonate	#	#	#
dissolved silica	***	***	***

***Not determined; probably less than 0.1
#Virtually absent
##Based on samples for 9 months. Samples collected from 1 September through 30 November were discarded because plastic screens were used on the collection apparatus (Juang and Johnson 1967; Fisher, et al. 1968).
+Calculated for the period September through August (Juang and Johnson, 1967).

at Hubbard Brook. The pH of rain and snow samples is frequently less than 4.0. Nitrate is next in abundance, and significant amounts of ammonium, chloride, sodium and calcium are usually present. Lesser amounts of magnesium, potassium and aluminum are also found (Table 3).

Since precipitation samples are composited over weekly intervals, it is difficult to identify the origin of chemical impurities in individual air masses or storms.

Our chemical input data are based on bulk precipitation, i.e., a mixture of rain or snow and dry fallout (Whitehead and Feth, 1964). We have been concerned with the contributions from dry fallout since the beginning of the study. Juang and Johnson (1967) suggested that dry fallout was a source of chloride in the Hubbard Brook ecosystem. However, based upon comparisons between precipitation collectors that were continuous open (Likens, et al., 1967) and those that opened only during periods of rain or snow (Wong Laboratories, Mark IV), it is clear that the bulk of chemical input to the ecosystem comes in rain and snow. Our attempts to quantify the much smaller contributions from dry fallout have been thus far inconclusive.

Average values for the ion content of precipitation, interpolated from isopleth maps for central New Hampshire (Junge, 1958; Junge and Werby, 1958), do not agree closely with the average weighted concentration of ions measured in precipitation at Hubbard Brook (Table 3). Considering the yearly variation in precipitation chemistry (Likens, et al., 1967; Fisher, et al., 1968; Table 3), and that our samples were taken in 1965–68, it is not surprising that there is not

better agreement between our values and those obtained in 1955–56 by Junge and Werby. Our cation values are nearly 50% lower than those reported by Junge (1958) and Junge and Werby (1958); whereas our anion values, with the exception of chloride, are somewhat higher.

Soil and road dust are among the principal sources of base metal ions in local precipitation (e.g., Gambel and Fisher, 1966). In the Hubbard Brook area, the almost total forest cover strongly minimizes the generation of soil dust, thereby reducing the concentration of these ions in the local precipitation.

The weighted concentration of nitrate in precipitation at Hubbard Brook is considerably higher than the concentrations reported by Junge (1958) for New England, and in fact, for most regions of the U. S. Since air pollution can affect the concentration of nitrate in precipitation (e.g., Junge, 1963), our data might reflect some measure of increased air pollution between 1955–56 and 1965–66, and its resultant effect on nutrient budgets in rural as well as urban areas. In this regard it is interesting that our average values for nitrate in precipitation (rain and snow plus dry fallout) has increased each year since 1964; whereas most of the other chemical constituents remained constant or fluctuated slightly (Table 3).

CHEMICAL INPUT THROUGH HERBICIDE APPLICATION

Our nutrient budgets are based on the difference between chemical input in precipitation and output in stream water from the watershed-ecosystems (Bormann and Likens, 1967). Thus, it is important to account for any extraneous chemical inputs to the ecosystem, such as an application of herbicide. Approximately 3650 liters of Bromacil solution were sprayed on the cutover watershed in June 1966. Based on a chemical analysis of the herbicide solution, and assuming complete decomposition, 0.04 Ca^{++}/ha, 0.09 kg Mg^{++}/ha, 0.01 kg K^+/ha, 0.11 kg Na^+/ha, and 10.6 kg NO_3^-/ha were added to the watershed. These are relatively insignificant inputs to the budgets for the deforested watershed. Bromide is equivalent to chloride with the analytical method we use for chloride (Iwasuki et al., 1952). Thus we have calculated a maximum possible input of 3.0 kg/ha to the chloride budget from Bromacil.

Approximately 87 liters of an ester (propylene glycol butyl ether) of 2, 4, 5-T were sprayed on the cutover watershed during the summer of 1967. This herbicide was mixed with water from the drainage stream of the watershed to produce the spray solution. Chemical analyses of the solution showed that negligible amounts of cations were added to the watershed-ecosystem from this application of herbicide. However we calculated a maximum input of 0.7 kg Cl$^-$/ha to the watershed from the herbicide solution. Inputs for other anions were negligible.

STREAMWATER PARAMETERS

Temperature

The drainage streams in the undisturbed watersheds during summer are in deep shade beneath the forest canopy; in winter they are under a deep snow pack. Thus, for most of the year, daily stream water temperatures are relatively constant and the annual temperature range is only about 16°C (Figs. 2 and 3).

The altitudinal temperature gradient in stream water varies somewhat in the undisturbed watersheds, and in Watershed 4 during midsummer it is about 5–7°C. At times it may be as much as 12°C, however the steepest portion of this thermal gradient is confined largely to the upper 50 m of the stream (McConnochie and Likens, 1969). Macan (1958) and others have shown that small streams usually warm and reach equilibrium temperatures rapidly, and that the average

water temperature approximates the average air temperature. In winter, the altitudinal temperature gradient in streams within the watersheds is not more than 1 or 2°C.

Streamwater temperatures in the deforested watershed (W2) were higher than in the undisturbed watersheds during both summer and winter. In the absence of shade, temperature varied by 3–4°C during the day in summer; the maximum temperature occurred about 1500 hours and the minimum at 0700–0800 hours (Fig. 3). The annual variation was about 18 to 20°C (Fig. 2).

Even though the mean January air temperature is −9°C (U. S. Forest Service, 1964), the streams do not freeze solid. In fact, they are relatively warm below the thick snow pack in winter (Figs. 2 and 3). A thicker snow depth in the stream channel of the deforested watershed could provide more insulation and may account for the somewhat higher streamwater temperature. As mentioned earlier, frost is uncommon in the soils of the watersheds.

The streams seem to have discrete summer and winter temperature regimes with rapid seasonal transitions. Warm-up during the spring is very rapid, occurring mostly in May in the undisturbed situation (Fig. 2).

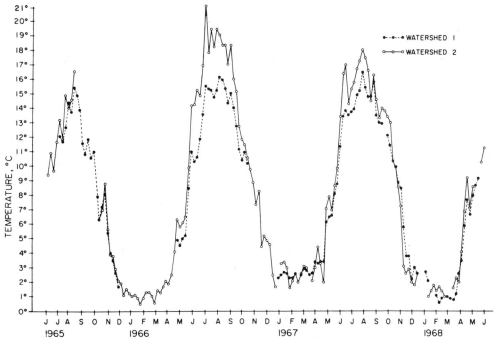

FIG. 2. Mean weekly stream water temperatures in Watersheds 1 and 2 (deforested) during 1965–68. The recording thermometers were located above the weirs in each stream. The points were not connected during periods when the stream discharge was negligible or when the thermograph functioned improperly.

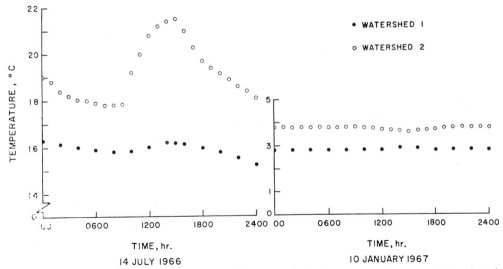

Fig. 3. Mean hourly stream water temperatures during one day in the summer and one day in the winter in Watersheds 1 and 2 (deforested).

DISSOLVED OXYGEN

The stream water from the Hubbard Brook watersheds is normally saturated or slightly supersaturated with dissolved oxygen, except during periods of very low flow, *e.g.,* late summer and early autumn (Fig. 4). There is no apparent

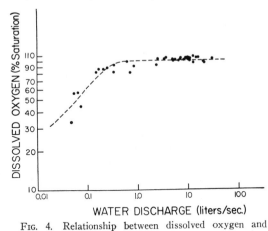

Fig. 4. Relationship between dissolved oxygen and stream water discharge in undisturbed Watershed 3 during 1965–66. The equation for the dashed line is $C = (1 - e^{-(D)(0.0693)})\ 79.26 + 22.59$, where $C =$ dissolved oxygen in % saturation and $D =$ stream discharge in liters/sec. The constants in this equation were determined by linear regression analysis and the F-ratio for the regression line is very highly significant (>0.001).

seasonal pattern separate from this dependence on discharge. The dissolved oxygen values are adjusted for altitude and streamwater temperature to calculate the percent saturation. Ruttner

(1953) indicates that the dissolved oxygen in streams, even in cascading water, should not exceed the saturation equilibrium with air. However, supersaturation of dissolved oxygen has been found by several other workers in natural streams and rivers (*e.g.,* Järnefelt, 1949; Harvey and Cooper, 1962; Minckley, 1963; Woods, 1960). The explanation of the supersaturated condition in streams tributary to Hubbard Brook is not readily apparent, although it may be "forced" to supersaturation by turbulence (Lindroth, 1957; Harvey and Cooper, 1962) or may be a function of the altitudinal temperature gradient in the streams. Photosynthetic organisms are not abundant in these tributary streams, and diel variations in dissolved oxygen are not apparent.

At low flows the water mostly seeps from one small pool to the next through gravel and organic debris dams. The biological oxygen demand from decaying leaves and other organic debris in the stream is apparently high enough to reduce significantly the dissolved oxygen concentration when flow and turbulence are low. Rapid depletion of dissolved oxygen resulting from decomposition of leaves in small streams has been demonstrated by several workers (see Minckley, 1963).

Increased water yield from the deforested watershed, particularly in the summer months (Fig. 1), results in greater discharge, more turbulence and a constant high level of dissolved oxygen in this stream. Thus, during the summer and autumn large differences in dissolved oxygen concentration may be anticipated between the streams

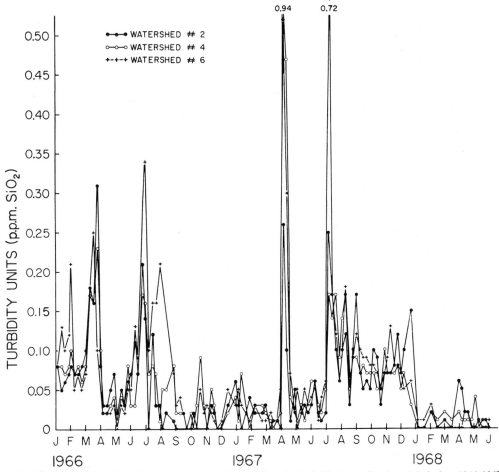

FIG. 5. Turbidity values for stream water from Watersheds 2 (deforested), 4 and 6 during 1966–1968.

in the deforested and undisturbed watersheds. With increased water temperature, sunlight and high concentrations of dissolved oxygen, more rapid decomposition of the organic debris would be expected in the stream of W2.

TURBIDITY

Following disturbance of the vegetation and litter in forested watersheds, drainage waters may become quite turbid as the result of erosion and transport of inorganic and organic matter from the watershed (*e.g.*, Lieberman and Hoover, 1948a; Tebo, 1955). For example, this frequently occurs as a consequence of unregulated commercial logging operations. However, with care and planning, turbidity and sedimentation in drainage streams may be minimized following commercial logging (Hewlett and Hibbert, 1961; Lieberman and Hoover, 1948b; Trimble and Sartz, 1957).

Stream water draining the Hubbard Brook Experimental Forest is very clear, and no obvious differences were noted in the turbidity of the stream water from W2 following the cutting and herbicide treatment of the vegetation (Fig. 5). In fact, the peak turbidity values seemed to be depressed in comparison with values for streams in the undisturbed watersheds. Of the three watersheds compared, W6 showed the greatest extremes in turbidity, and these extreme values were also slightly out-of-phase with changes in the other watersheds (Fig. 5). High values were not always correlated with high runoff values, *e.g.*, during June 1966 and July 1967. All in all, the measurements of turbidity were of little value in assessing the changes in water quality of the stream water of these forested and deforested ecosystems.

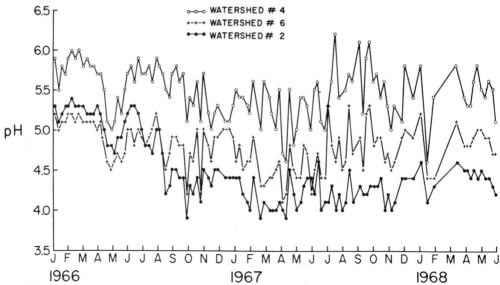

FIG. 6. pH values for stream water from Watersheds 2 (deforested), 4 and 6 during 1966–1968.

PARTICULAR MATTER

Undisturbed forest ecosystems lose relatively little organic and inorganic particulate matter. Average annual losses are about 25 kg/ha-yr., and are about equally divided between organic and inorganic particulate matter (Bormann, *et al.,* 1969). These minor particulate matter losses, about one-sixth of the dissolved substance losses, are attributed to the operation of biotic factors, which 1) decrease the erodability of the ecosystem, 2) decrease the amount of runoff and 3) tend to damp the frequency of high discharge rates.

The influence of these biotic factors has been severely limited by removal of the living vegetation in Watershed 2. Compared to undisturbed Watershed 6, the new conditions in Watershed 2 are reflected by a 4-fold increase in particulate matter in the settling basin above the V-notch weir for the period May 1966 to May 1968. Also, particulate matter output from Watershed 2 is becoming increasingly inorganic in content (76% inorganic, May 1966 to May 1968).

The increased output in particulate matter, and particularly the increased proportion of inorganic materials, from W2 occurred primarily from the unraveling of the stream channel. In many places the banks have been eroded and many of the small debris dams, composed of leaves, twigs, etc., that were common in the undisturbed stream, have worn away, with a subsequent release of trapped inorganic and organic materials. Because these dams no longer exist, and without the annual

replenishment of leaves and other organic debris to build new dams, stream water is able to transport much more particulate material downslope. Also, stream channel erosion is increased since the binding action of roots has been reduced, and leaves, which provide a protective cover on the banks, have been removed and not replenished.

PH

Acidic water characterizes the drainage streams from the undisturbed watersheds. This is typical for many streams in New England with podzolic soils (*e.g.,* Anderson and Hawkes, 1958). The pH values are variable throughout the year, but consistently show the same relationship between watersheds (Fig. 6). There was particularly good agreement in pattern between the undisturbed watersheds, W4 and W6, although stream water from W4 was relatively less acidic.

As a result of deforestation, and concurrently with other chemical changes in the drainage waters during June 1966, the hydrogen ion content (calculated from pH measurements) in stream water from W2 increased by 5-fold (Fig. 6). That is, the weighted average pH, decreased from 5.1 during 1965–66 to 4.3 during 1966–67 and 1967–68 (Likens, *et al.,* 1969). During the same period the weighted pH value remained relatively unchanged for W4 and W6.

ELECTRICAL CONDUCTIVITY

Electrical conductivity averages about 20 μmhos/cm^2 at 25°C in stream water of the undis-

TABLE 4. Weighted average concentration in stream water from watersheds 2, 4, and 6 of the Hubbard Brook Experimental Forest, expressed in mg/liter

Ion	W2			W4#		W6		
	1965-66	1966-67	1967-68	1965-66	1966-67	1965-66	1966-67	1967-68
Ca++	1.81*	6.45	7.55	1.82*	1.80	1.36*	1.27	1.28
Mg++	0.37	1.35	1.51	0.41	0.40	0.36	0.35	0.36
K+	0.19	1.92	2.96	0.22	0.24	0.18	0.20	0.26
Na+	0.87	1.51	1.54	1.13	1.10	0.83	0.80	0.93
Al+++	0.22	1.5	2.0	0.12	0.12	0.32	0.33	0.32
NH4+	0.14	0.07	0.05	0.12	0.06	0.12	0.05	0.02
NO3-	0.94	38.4	52.9	0.86	0.88	0.85	0.69	1.30
SO4=	6.8	3.8	3.7	6.4	6.2	6.2	6.0	6.1
Cl-	0.54	0.89	0.75	0.56**	0.58	0.57***	0.55	0.56
HCO3-	0.8	0.1	0	1.6	2.0	0.1	0.2	0.3
SiO2—aq	4.1	5.6	5.7	5.4	5.5	4.1	4.4	3.8
Total	16.8	61.6	78.7	18.6	18.9	15.0	14.8	15.2

*Values for the initial 7 months of the water-year have been increased by a factor of 1.6 to compensate for analytical interferences (Likens, et al., 1967; Johnson, et al., 1968)
**Juang and Johnson (1967) calculated a value of 0.51 for the period September through August.
***Juang and Johnson calculated a value of 0.50 for the period September through August.
#Data not available for 1967-68.

turbed watersheds and changes very little either on a daily or seasonal basis. This would be expected from the relative constancy of cation and anion concentrations in stream water at Hubbard Brook (Likens, et al., 1967; Fisher, et al., 1968; Johnson, et al., 1969).

In contrast, the conductivity of stream water from the deforested watershed is quite variable, ranging from 65 and 160 μmhos/cm² at 25°C. Usually the conductivity decreased during rain storms. However occasionally the conductivity increased or was unaffected. The hydrogen ion concentration of rain water added to the stream water is important in this regard, as well as other variables such as amount of rainfall, its duration, streamwater discharge, and water content of the soil.

IONS

Ammonium and Nitrate

The ammonium ion occurs in very low concentration in stream water from undisturbed watersheds of the Hubbard Brook Experimental Forest (Fisher, et al., 1968). Essentially no change was observed in the concentration of this ion in drainage water from the deforested watershed. In contrast the nitrate concentration increased from an average weighted value of 0.9 mg/liter prior to cutting of the vegetation, to 53 mg/liter, two years later (Table 4). Measured concentrations soared to 82 mg/liter in October 1967 (Fig. 7). It should be noted that the initial increase (June 1966) in streamwater nitrate concentration in W2 occurred 16 days before the application of the Bromacil and at the same time the nitrate

concentration in the stream water from the undisturbed watershed showed the normal late-spring decline (Fig. 7; Bormann, et al., 1968).

Nitrate concentrations in stream water from the undisturbed watersheds, show a pronounced, recurring seasonal pattern (Bormann, et al., 1968; Fisher, et al., 1968). Average monthly concentrations are low (<0.1 mg/1) throughout the summer growing season, increase in November, and reach values as high as 2 mg/liter during the spring (April) thaw (Fig. 8). This variation may be explained by a combination of mechanical and biological effects (Johnson, et al., 1969).

The decline of nitrate concentrations during May and the low concentrations throughout the summer correlate with heavy nutrient demands by the vegetation and increased heterotrophic activity associated with warming of the soil. The winter pattern of NO_3^- concentration may be explained in strictly physical terms, since the input of nitrate in precipitation from November through May largely accounts for nitrate lost in stream water during this period (Bormann, et al., 1968). Also, sublimation and evaporation of water from the snow pack could account for some 15–20% increase in concentration of NO_3^- in the stream water in the spring. Concentration of nitrate in stream water after deforestation show a pattern that is nearly the reciprocal of the undisturbed situation (Fig. 8).

Since yearly input of nitrate-nitrogen in precipitation exceeds losses in stream water in the undisturbed ecosystems (Table 5), the concentration of nitrate in stream water provides no conclusive evidence for nitrification in these un-

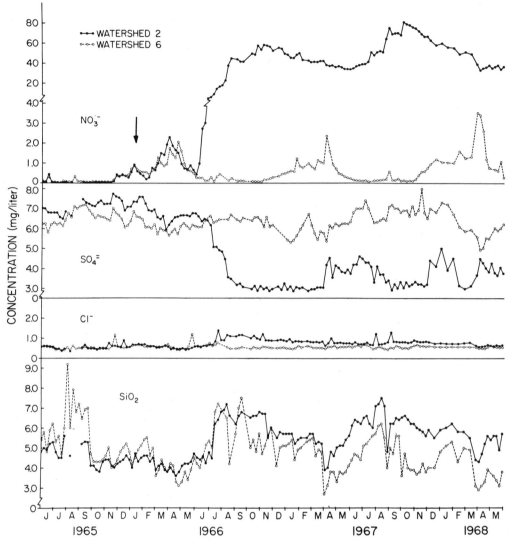

FIG. 7. Measured streamwater concentrations for nitrate, sulfate, chloride and dissolved silica in Watersheds 2 (deforested) and 6. Note the change in scale for the nitrate concentration. The arrow indicates the completion of cutting in W2.

disturbed acid soils. The low levels of ammonia and nitrate in the drainage water of the undisturbed ecosystem (W6) may attest to the efficiency of the oxidation of ammonia to nitrate, and to the efficiency of the vegetation in utilizing nitrate. However, Nye and Greenland (1960) state that growing, acidifying vegetation represses nitrification; thus the vegetation may draw directly on the ammonium pool, and little nitrate may be produced within the undisturbed ecosystem. However, in the absence of forest vegetation, the microflora of the deforested watershed apparently oxidize ammonia to nitrate, and the

nitrate is rapidly flushed from the watershed-ecosystem (Bormann, et al., 1968; Likens, et al., 1969).

Many microorganisms convert organic nitrogen into ammonia. The ammonia then may be oxidized to nitrite by bacteria of the genus *Nitrosomonas*. The nitrite may be further oxidized to nitrate by bacteria of the genus *Nitrobacter*. The important end products of these reactions in terms of nutrient losses from the watershed-ecosystem are the increased production of nitrate and hydrogen ions. Increased biological nitrification apparently is the principal factor responsible for

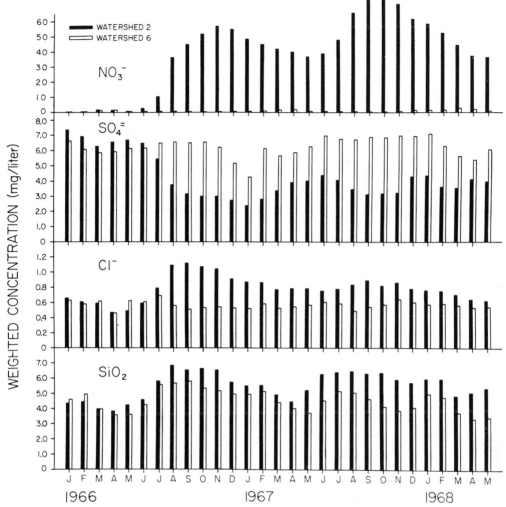

FIG. 8. Weighted average monthly concentrations of nitrate, sulfate, chloride and dissolved silica in stream water from Watersheds 2 (deforested) and 6.

the total flush of ions from the deforested watershed. The increase in the milliequivalent value for nitrate in the water years subsequent to cutting, balances (within 10–15% error limits for the watershed systems) the increased net losses of cations and the decreased net losses of anions (Likens, *et al.,* 1969). This will be discussed in more detail in a later section.

Alexander (1967) indicates that nitrification usually decreases greatly below a pH of 6.0 and becomes negligible at a pH of 5.0. Our data would indicate that this is not the case for our deforested watershed as shown by the increased concentration of nitrate in the drainage water during 1966–68 (Fig. 7), and the low pH of stream wa-

ters (Fig. 6) and soils. Also, in comparison with soils under undisturbed forest, bacteria of the genera *Nitrosomonas* and *Nitrobacter* have increased 18-fold and 34-fold, respectively, in the soil of Watershed 2 (Smith, *et al.,* 1968). Other workers have shown nitrification at similarly low pH's (Boswell, 1955; Weber and Gainey, 1962). It may be that we are dealing with relatively little known species of nitrifying bacteria adapted to more acid conditions (*e.g.,* Alexander, 1967).

Sulfate

Sulfate concentration in stream water from the undisturbed watersheds is relatively constant in relation to the highly variable stream discharge

TABLE 5. Nutrient budgets for undisturbed (W6) and cu over (W2) watersheds of the Hubbard Brook Experimental Forest. Values are expressed in kg/ha for the period 1 June to 31 May

Element	W6			1966-67	W2	
	Input	Output	Net loss or gain	Input	Output	Net loss or gain
Ca	2.4	10.7	−8.3	2.3	77.3	−75.0
Mg	0.4	2.9	−2.5	0	16.2	−15.7
K	0.6	1.7	−1.1	0.5	23.0	−22.5
Na	1.3	6.8	−5.5	1.3	18.1	−16.8
Al	1.4	2.8	−1.4	1.3	18.2	−16.9
NH_4-N	1.9	0.3	+1.6	1.8	0.7	+ 1.1
NO_3-N	4.6	1.3	+3.3	6.8	104	−97.2
SO_4-S	14.4	17.1	−2.7	13.6	15.3	− 1.7
Cl	6.9**	4.6	+2.3	9.5**	10.6	− 1.1
HCO_3-C	*	0.4	−0.4	*	0.2	− 0.2
SiO_2-Si	*	17.2	−17	*	31.3	−31
			1967-68			
Ca	3.0	12.2	−9.2	2.7	93.1	−90.4
Mg	0.8	3.4	−2.6	0.7	18.6	−17.9
K	0.8	2.4	−1.6	0.7	36.5	−35.8
Na	1.8	8.8	−7.0	1.7	19.0	−17.3
Al	*	3.1	−3.1	*	24.5	−24
NH_4-N	2.6	0.2	+2.4	2.4	0.5	+ 1.9
NO_3-N	5.2	2.8	+2.4	4.9	147	−142
SO_4-S	16.0	19.3	−3.3	15.16	15.15	0
Cl	5.2	5.3	−0.1	5.5	9.2	− 3.7
HCO_3-C	*	0.5	−0.5	*	0	0
SiO_2-Si	*	17.0	−17	*	32.6	−32

*Not determined, but very low.
**Based on data for 9 months.

(Figs. 7 and 8; Fisher, et al., 1968); and on close examination the sulfate concentration shows a very small and irregular volume concentration effect (Johnson, et al., 1969). In addition stream-water concentrations of sulfate seem to show some general, recurring seasonal patterns. There is a gradual rise to maximum values in the autumn with lower values during the late winter and early spring. This pattern is nearly the reciprocal of the nitrate pattern (Figs. 7 and 8).

The weighted concentration of $SO_4^=$ in drainage water from the deforested watershed decreased by about 45% in the first year subsequent to cutting and herbicide treatment (Table 4). The decrease in sulfate concentration was somewhat delayed in relation to the increase in nitrate concentration (Fig. 7). The explanation for this decrease in sulfate concentration is complicated and will be elaborated in the section on Nutrient Budgets.

Chloride

Since the quantity of chloride bearing rocks in the Hubbard Brook watershed-ecosystems is small (Juang and Johnson, 1967), and since biological immobilization is small, we expected the weighted concentration of Cl^- in precipitation after correction for evapotranspiration, to balance the concentration in drainage water from undis-

turbed ecosystems. However, during 1965–66 a higher concentration was found in the stream water of the undisturbed watersheds than could be accounted for by precipitation input (Juang and Johnson, 1967). The excess Cl^- was attributed to the accumulation of Cl^- in the ecosystem by dry removal of aerosols through impaction on the forest canopy. More recent comparisons of the chemical results from precipitation collectors that are continuously open and those that open only during periods of rain or snow are inconclusive, but do not support the dry fallout hypothesis. Data for years subsequent to 1965–66 show that the long-term, mean chloride concentration in stream water is about the same as the average concentration in precipitation, after adjustment for water loss by evapotranspiration (Johnson, et al., 1969). We feel that the problems encountered previously in the chemical analysis and collection of chloride samples (e.g., Fisher et al., 1968) were greatly minimized during 1967–68, and significantly, the chloride input in precipitation balanced the output in stream water in undisturbed W6 (Table 5).

After deforestation of W2, the streamwater concentration of chloride increased, but the increase was somewhat delayed relative to most of the other ions (Fig. 7). The weighted concentration of chloride in drainage waters from W2

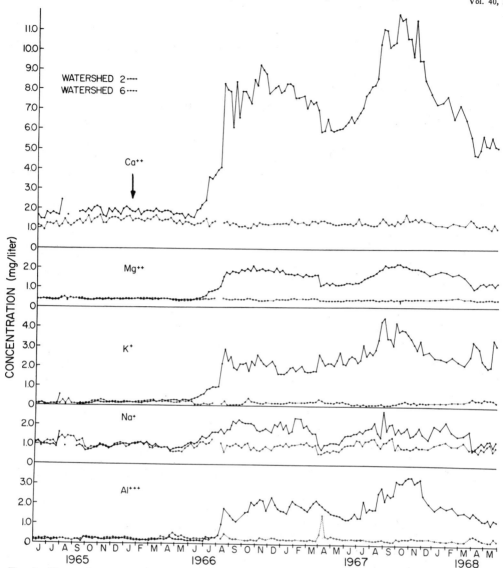

Fig. 9. Measured streamwater concentrations for Ca++, Mg++, K+, Na+, and Al+++ in Watersheds 2 (deforested) and 6. The points were not connected during periods when the stream discharge was negligible. The arrow indicates the completion of cutting in W2.

increased about 65% in the first year (Table 4), indicating the existence of a small chloride reservoir within the ecosystem. A maximum of about 70% of this increased chloride concentration could be explained by the addition of the Bromacil solution to the watershed, however not all of the Bromacil was lost from the watershed during the first year (Pierce, 1969). A continuing but relatively smaller increased chloride concentration also was observed in stream water from W2 during the second year, 1967–68 (Table 4; Fig. 7). Some 30% of the increase in streamwater con-

centration during 1967–68 may be attributed to the addition of the 2, 4, 5-T solution to the watershed.

Calcium, Magnesium, Potassium and Sodium

The concentration of these cations characteristically has been relatively constant in stream water of the undisturbed watersheds despite the highly variable discharge of water (Likens, *et al.*, 1967). The constancy of magnesium is phenomenal in this regard. The relatively small variations that do occur may be explained by appropriate equations for dilution and concentration, based on

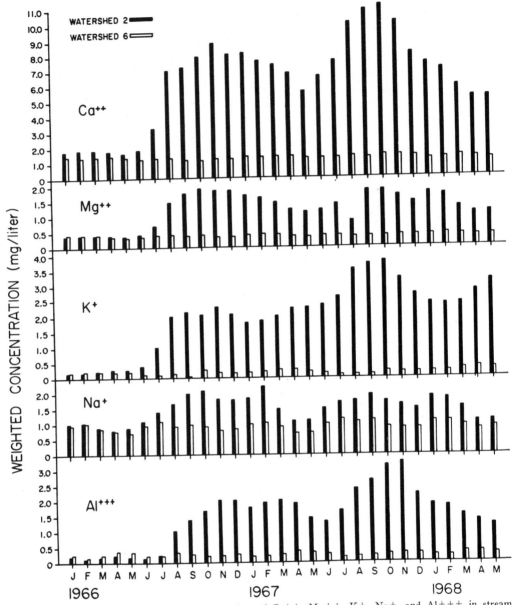

Fig. 10. Weighted average monthly concentrations of Ca++, Mg++, K+, Na+, and Al+++ in stream water from Watersheds 2 (deforested) and 6.

water discharge (Johnson, et al., 1969). In the deforested watershed the volume-concentration model still applies but is less important as a controlling factor than the effect of nitrification.

Remarkable increases in average stream water concentration (417% for Ca++, 408% for Mg++, 1558% for K+ and 177% for Na+) were observed during the two years subsequent to deforestation (Figs. 9 and 10; Table 4). The increased concentration and resultant net losses

of these cations in drainage waters is dependent upon the increased concentration of nitrate and hydrogen ions brought about by nitrification (Likens, et al., 1969). With decreasing pH and available nitrate these cations are more readily mobilized and leached from the watershed. This may occur 1) as the hydrogen ions replace the cations on the humic-clay ion exchange complexes of the soil, 2) as organic compounds are decomposed and 3) as chemical decomposition of bedrock and till is accelerated. However, since nitrification is occurring in the humic layers of the soil, the excess ions are most likely released from the decay of humic substances (Likens, et al., 1969). This is best shown by the changes in the ratio of Ca:Na in the drainage water of Watershed 2 before and after deforestation. Prior to cutting, the Ca:Na ratio in the stream water of Watershed 2 was 1.6/1.0. This ratio is consistent with that observed in adjacent undisturbed watersheds, and has been attributed to the steady-state chemical weathering of bedrock and till (Johnson, et al., 1968). However, after deforestation, the ratio climbed to an average of 4.8/1.0 for the period 1966–68. More significantly, the Ca:Na ratio for the "excess" ions of the stream (that is, the amount added above the undisturbed condition) is 7.4/1.0. Thus, the net chemical effect of the deforestation was to differentially produce soluble calcium within the ecosystem. Hardwood forest litter is rich in calcium relative to sodium (Ca:Na is >20/1.0; Scott 1955) in contrast to average bedrock for the Hubbard Brook watershed-ecosystems (Ca:Na is 0.9/1.0; Johnson, et al., 1968), suggesting that most of the "excess" ions produced are derived from the bulk decomposition of humic materials (Likens, et al., 1969).

Only a small portion of the 6- to 22-fold increase in net losses of calcium, magnesium and potassium were apparently derived from an accelerated rate of chemical weathering (Table 5). Based on the change in the sodium budget of Watershed 2, and attributing net sodium losses solely to chemical weathering (Johnson, et al., 1968), the chemical decomposition of bedrock and till within Watershed 2 has apparently accelerated no more than 3-fold.

Aluminum

Aluminum concentrations in stream water from the undisturbed watersheds show seasonal pattern and differences between watersheds (Fisher, et al., 1968). Concentrations of aluminum are highest in April during peak runoff (Figs. 9 and 10). In fact, there is a direct relationship between aluminum concentration and stream discharge (Johnson, et al., 1969). Concentration of aluminum in stream water varies with watershed, with W4 lower than the other streams (Table 4). The average stream water concentration for W4 is about 1/3 of the value for W6.

Deforestation resulted in an increased concentration of aluminum in stream water from W2, but the increase was delayed relative to other ions (Figs. 9 and 10). Aluminum concentration increased by 10-fold in two years following deforestation. The primary source for aluminum in the system is the bedrock and till. Hence, the increase in aluminum concentration indicates the importance of decreasing pH on the solubility of common minerals containing aluminum in the soils, especially the clay minerals such as kaolin and vermiculite. The decrease in pH from 5.2 to 4.3, we observed as a result of the deforestation, is critical to the solubility of alumina (e.g., Mason, 1966, p. 167).

Dissolved Silica

Seasonal variations in the concentration of dissolved silica in stream water are inversely correlated with runoff (Johnson, et al., 1969). Maximum streamwater concentrations occur in late summer, whereas minimum concentrations occur during peak flow periods associated with spring thaw. There is apparently a yearly recurring slump in the dissolved silica concentration in August preceding the seasonal decline in September (Fig. 7) for which the explanation is not clear.

After forest cutting the average weighted concentration of dissolved silica increased by about 37%. This increase probably reflects the increased solution or weathering of the geologic substrate, or both.

The relatively new finding that the solubility of amorphorus silica is inversely related to pH between about pH 3 and 7 (see Marshall, 1964, p. 81) partially explains the increase in dissolved silica in stream water from the deforested watershed.

Bicarbonate

Although the bicarbonate ion is normally the most abundant in freshwaters, it is a relatively minor constituent of the acidic runoff water from the undisturbed Hubbard Brook watersheds (Fisher, et al., 1968; Johnson, et al., 1968; Likens, et al., 1969; and Table 4). In fact, it is barely detectable with routine methods in the stream water from W6.

In W2 the measured concentration dropped to nearly zero by the middle of the first year after cutting (1966–67), and concurrently with the decrease in streamwater pH. Thus the weighted value for 1966–67 (Table 4) reflects the conditions during the first part of the water-year and is somewhat misleading as a temporal average for the year. The streamwater concentration is zero for 1967–68 as expected since the average pH was 4.3.

TABLE 6. Dependence of chemical concentration on the volume of water discharged from Watersheds 2 (cut-over) and 6 during 1967–68. Regression lines were fitted to the relationship y = ax + b, where y = chemical concentration in mg/liter, a = slope, x = a function of discharge $\left(\dfrac{1}{1 + \beta D}\right)$, β is a proportionality constant, and D is stream discharge in liters/ha-day, and b = y intercept (Johnson, et al., 1969). The F-ratio for the values given for slope are significant at the 0.01 level

		W2	W6
calcium	slope	ns*	ns
	Y-intercept	—	—
magnesium	slope	ns	ns
	Y-intercept	—	—
sodium	slope	1.08	0.66
	Y-intercept	1.23	0.69
potassium	slope	-1.79	-0.36
	Y-intercept	4.40	0.50
aluminum	slope	ns	-0.34
	Y-intercept	—	0.57
hydrogen	slope	ns	ns
	Y-intercept	—	—
nitrate	slope	ns	-1.63
	Y-intercept	—	1.49
sulfate	slope	ns	ns
	Y-intercept	—	—
chloride	slope	ns	ns
	Y-intercept	—	—
dissolved silica	slope	1.85	3.24
	Y-intercept	5.31	2.78

*Not significant

EFFECT OF NITRIFICATION ON CATION LOSSES

In the undisturbed watersheds, the relatively small variations in cation and anion chemistry may be explained almost entirely by the effects of dilution and concentration as brought about by changes in stream discharge and biological activity (Johnson, et al., 1969). Initially we thought that the large oscillations in streamwater chemistry after deforestation (Figs. 7, 8, 9 and 10) should be explained by this same model. However a regression analysis showed that only sodium, potassium, and dissolved silica were significantly (<0.01) related to discharge of water after deforestation (Table 6).[4] Moreover, the slope of the regression lines and the Y-intercepts for these three elements were changed greatly,

[4] The same pattern of results was obtained for 1966–67, however, the following discussion is based entirely on 1967–68 because of the complexities associated with the sharp chemical transition period that occurred during the water-year of 1966–67 (Figs. 7 and 9).

relative to the undisturbed situation (Table 6; Johnson, et al., 1969). The correlation coefficients for these regression lines were all less than 0.57, which indicates a great deal of scatter in the points and suggests either experimental error or other important controlling factors. In addition to large changes in the slope and Y-intercept for the solutes that show volume or concentration effects with changes in stream discharge after deforestation (Johnson, et al., 1969), aluminum and nitrate, which usually were related to discharge, no longer show any significant relationship (Table 6). Thus, the increased nitrification after deforestation has swamped the rather small volume and concentration effects of discharge, characteristic of these ions in the undisturbed watersheds.

Regression analyses for each ion on nitrate concentration in stream water showed very high significance (<0.001) for all ions except hydrogen, ammonium and dissolved silica (Table 7). The more important cations (Ca^{++}, Mg^{++}, and Al^{+++}), in terms of milliequivalent values, had extremely high correlation coefficients in the regression analysis; whereas the correlation coefficients were lower for Na^+ and K^+ (Table 7;

TABLE 7. Relationship between the nitrate concentration and other ions in stream water from the cut-over watershed (W2) during 1967–68. Regression lines fitted to the relationship y = ax + b, where y = chemical concentration in mg/liter, a = slope, x = concentration of nitrate in mg/liter, and b = y intercept. An F-ratio > 12.6 for the slope is significant at the 0.001 level for these regression analyses

	Slope	Correlation coefficient	F-ratio for slope
magnesium	0.024	0.97	673
calcium	0.137	0.95	424
aluminum	0.044	0.93	279
potassium	0.023	0.60	23.9
sodium	0.014	0.57	20.4
sulfate	-0.023	-0.01	30.4
chloride	0.005	-0.28	21.5
dissolved silica	0.017	0.37	6.8
ammonia	0.0006	-0.28	1.0
hydrogen	0.0003	0.08	0.3

Figs. 11 and 12). Thus, these data show the quantitative importance of nitrification as a major controlling factor in determining the quantity and quality of dissolved materials flushed from the cutover watershed in drainage waters. Thus, in the undisturbed watershed, dilution and concentration effects on water discharge were the principal mechanisms controlling ionic concentrations, while in the deforested watershed nitrification was the more important controlling factor.

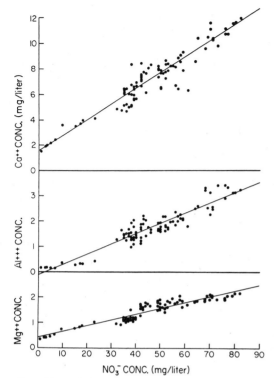

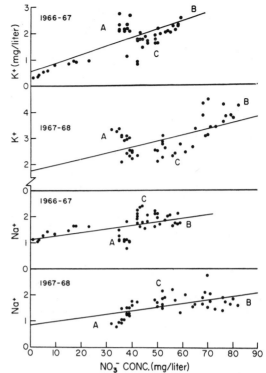

Fig. 11. Relationship between nitrate concentration and calcium, magnesium and aluminum concentrations in stream water from Watershed 2 during 1 June 1966 to 31 May 1968.

Fig. 12. Relationship between nitrate concentration and sodium and potassium concentrations in stream water from W2 during 1 June 1966 to 31 May 1968. All points in Area A were observed during the period of high spring runoff; points in Area B were observed in relatively high runoff periods in November; and Area C is a mixture of summer and winter points with lesser runoff. The points for nitrate concentrations <25 mg/liter represent the chemical transitional period between undisturbed and deforested conditions (1 June 1966 through 31 July 1966). See text for full explanation.

Plots of the actual points in the regression of Na^+ and K^+ on NO_3^- show an interesting relationship and indicate the relative importance of 1) nitrate concentration and 2) discharge in regulating ionic concentration of stream water (Fig. 12). That is, if the regression line is taken as the reference, points above the line represent concentration of Na^+ or K^+ and points below the line represent dilution of Na^+ or K^+ over that expected for the relationship between each of these cations and nitrate. An analysis of the temporal distribution for these points reveals that this is just the case. The points above the line for K^+ vs. NO_3^- (Area A) are all associated with periods of high spring runoff; Area B is associated with higher runoff periods in November; and Area C is a mixture of summer and winter points with lower runoff values. According to the volume-concentration model for undisturbed watersheds, K^+ concentrations are directly related to volume of water discharge (Table 6; Johnson, et al., 1969).

The reverse is shown for the relationship for Na^+ vs. NO_3^-. Area A, represents the spring

runoff period; B is the November period; and C is the summer and winter periods. Sodium is inversely related to the volume of discharge in the undisturbed watersheds (Johnson, et al., 1969; Table 6), and in the cutover watershed (Table 6).

The sodium and potassium were normalized for volume and concentration effects during 1967-68 according to the equations presented by Johnson, et al. (1969). As a result, the F-ratio and correlation coefficients for a regression analysis of nitrate and normalized sodium and potassium concentrations were improved. The residual scatter of points about this latter regression line indicates the presence of other minor controlling factors.

The significant slope but poor correlation coefficient for NO_3^- vs. $SO_4^=$ (Table 7) is the result of the marked curvilinear relationship after deforestation shown in Fig. 14.

NUTRIENT BUDGETS

Nutrient budgets for dissolved ions and dissolved silica for the Hubbard Brook watershed-ecosystems were determined from the difference between the meteorologic input per hectare and the geologic output per hectare (Bormann and Likens, 1967). Input was calculated from the product of the ionic concentration (mg/liter) and the volume (liters) of water as precipitation (Likens, et al., 1967 and Fisher, et al., 1968). Additional input from applications of herbicides was added to the precipitation input. Output was calculated as the product of the volume (liters) of water draining from the watershed-ecosystems and its ionic concentration (mg/liter). Budgets for all of the ions and substances measured are given in Table 5.

Net losses were greatly increased after deforestation and herbicide treatment for all ions except ammonium, sulfate, and bicarbonate. Two factors are involved in the removal of nutrients from the deforested watershed: 1) increased runoff and 2) increased ionic concentrations in stream water. If the concentrations had not increased from the undisturbed condition, increased runoff would have accounted for a 39% increase in gross export the first year and a 28% increase the second year after deforestation. However, the gross outputs for 1967–68 were greater than the undisturbed watershed, W6 (Table 5), by 7.6-fold for Ca^{++}, 5.5-fold for Mg^{++}, 15.2-fold for K^+, 2.2-fold for Na^+, 46-fold for NO_3-N, 1.8-fold for Cl^-, 7.9-fold for Al^{+++} and 1.9-fold for SiO_2-Si, clearly indicating that increased stream water concentrations are primarily responsible for the increased nutrient loss from the ecosystem.

Nitrogen losses from W2 after deforestation, although very large already, do not take into account volatilization. Allison (1955) reported volatilization losses averaging 12 percent of the total nitrogen losses from 106 fallow soils. However, denitrification is an anaerobic process and requires a nitrate substrate generated aerobically (Jansson, 1958); consequently, for substantial denitrification to occur in fields, aerobic and anaerobic conditions must exist in close proximity. The large increases in subsurface flow of water from the deforested watershed suggests that such conditions may have been more common than in the undisturbed ecosystem. Moreover, Alexander (1967) points out, "When ammonium oxidation takes place at a pH lower than 5.0 to 5.5, or where the acidity produced in nitrification increases the hydrogen ions to an equivalent ex-tent, the formation of nitrite can lead to a significant chemical volatilization of nitrogen."

Net losses of SO_4-S from the deforested ecosystem were about 40% lower in 1966–67, and 100% lower in 1967–68 than from undisturbed watersheds. In fact, the 1967–68 budget for SO_4-S in W2 was balanced in contrast to the undisturbed situation (Table 5). Precipitation is by far the major source of sulfate for the undisturbed watersheds (Fisher, et al., 1968). Although the amount of sulfate added by precipitation in 1967–68 was increased slightly relative to previous years, the net export of sulfate was zero, with sulfate input in precipitation exactly balancing streamwater export (Table 5). The decreases in streamwater sulfate concentration and gross export from the ecosystem occurred concurrently with the increases in streamwater nitrate concentration and gross export after forest cutting (Figs. 7 and 8).

Average sulfate concentrations in stream water were 3.8 and 3.7 mg/liter during 1966–67 and 1967–68 (Table 4), far below the 6.4 and 6.8 mg/liter values recorded in 1964–65 and 1965–66 before cutting (Fisher, et al., 1968). Much of this change can be explained by two facts, 1) the 39 to 28% increase in streamwater discharge from 1966 to 1968, which resulted from the elimination of transpiration by deforestation, and 2) the elimination of sulfate generation by sources internal to the ecosystem. If the decreases in sulfate concentrations were wholly due to increased runoff after deforestation, concentrations calculated on the basis of expected runoff (i.e., normal for the undisturbed system, Table 2) and measured gross sulfate lost from W2 during 1966–67 and 1967–68 (Table 5), should approximate the weighted streamwater concentrations for the undisturbed period, 1964–66. However, these calculated concentrations (5.3 and 4.7 mg/liter) equal only 79 and 70% respectively of the average weighted concentrations for 1964–66. These differences in concentration may be due to some year-to-year variation, but are largely explained by a sharp reduction in the internal release of sulfate from the ecosystem, which we earlier attributed to chemical weathering and biological activity (Fisher, et al., 1968, Likens, et al., 1969). The average annual internal release of sulfate (i.e., an amount equivalent to net loss) supplies about 10 kg/ha in the undisturbed watersheds (Table 5). Removal of this source of sulfate would account for the lower than expected adjusted sulfate concentrations mentioned above.

Thus, apparently, the normal, relatively small release of $SO_4^=$ from the ecosystem by chemical weathering and microbial activity (Table 5) probably became negligible following forest cutting. There are at least two possible mechanisms, operating simultaneously or separately, which may account for this:

i) There may be decreased oxidation of various sulfur compounds to $SO_4^=$. Waksman (1932) has suggested that a high concentration of nitrate is very toxic to sulfur oxidizing bacteria, such as *Thiobacillus thiooxidans*. This species may be important in sulfate oxidation in the deforested watershed since *T. thiooxidans* is capable of active growth at low pH (Alexander 1967). In the undisturbed watersheds we have observed a highly significant inverse linear relationship between the concentration of nitrate and sulfate in drainage water. This relationship is particularly clear in plots of sulfate concentrations against nitrate concentration using data from November through April, when the vegetation is dormant (Fig. 13). The inverse relationship between NO_3^- and

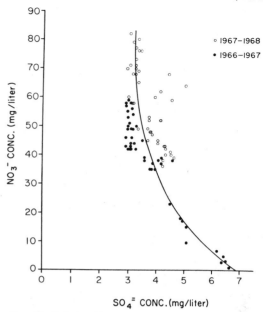

FIG. 14. Relationship between nitrate and sulfate concentrations in stream water from Watershed 2 during 1966–67 and 1967–68. Nitrate values less than 25 mg/liter indicate the chemical transition period (1 June 1966 through 31 July 1966, Fig. 7) between undisturbed and deforested conditions.

ter-year after cutting, 1967–68, the nitrate concentration in stream water from W2 increased even more, whereas the sulfate concentration decreased very little and coincided with the concentration of sulfate in precipitation after adjustment for water loss by evaporation. Perhaps there is an intricate feedback mechanism between the toxicity of the nitrate concentrations and microbial oxidation of sulfur compounds within the soil. Another possibility is that the number of sulfur oxidizing bacteria have been selectively reduced by the herbicides in the deforested watershed.

ii) Although somewhat unlikely, there may be increased sulfate reduction brought about by more anaerobic conditions, particularly in the lower more inorganic horizons of the soil (*e.g.,* Waksman, 1932). That is, an increased zone of water saturation in the deeper layers and in topographic lows on the cutover watershed probably has less free oxygen than in the undisturbed situation, promoting sulfur reduction. One difficulty, however, is that the growth of the most important sulfur reducing bacteria (*Desulfovibrio* spp.) is greatly retarded by acid conditions (Alexander, 1967). Also, molecular hydrogen released by anaerobic bacterial decomposition of organic matter may be used for the reduction of sulfate (Postgate, 1949; Rankama and Sahama, 1950).

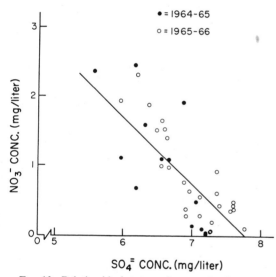

FIG. 13. Relationship between nitrate and sulfate concentrations in stream water from Watershed 2. Data were obtained during November through April of 1964–65 and 1965–66, which was prior to the increase in nitrate concentration resulting from clearing of the forest vegetation (Fig. 7). The F-ratio for this regression line is very highly significant ($p < 0.001$) and the correlation coefficient is 0.79.

$SO_4^=$ concentrations is very obvious in the first water-year after deforestation, 1966–67, when nitrate concentrations in stream water increased from normal (undisturbed) values to very high concentrations (Fig. 14). During the second wa-

The chloride budget for the undisturbed watershed during 1966–67 showed that input in precipitation exceeded the gross output, whereas the budget was essentially balanced during the 1967–68 water-year (Table 5). However after deforestation, significant net losses of chloride were observed (Table 5). The application of Bromacil in 1966–67 potentially added the equivalent of 3.0 kg Cl/ha or about 50% of the chloride input as precipitation. From the pattern of chloride changes in stream water following the addition of this herbicide (Fig. 7), it would appear that the herbicide and/or its degradation products were lost from the watershed quite gradually throughout the year. Measurements of Bromacil in stream water seemed to confirm this (Pierce, 1969). Since the Bromacil (1966–67) and possibly 2, 4, 5–T (1967–68) were not all flushed from the ecosystem within a year, then the internal release of chloride from the ecosystem probably represented an even greater percentage of the gross annual output (Table 5). Based upon streamwater concentrations in W2 and W6 (Fig. 7), it would appear that the internal reservoir (plus external inputs from herbicides) of chloride within the ecosystem has been essentially exhausted in two years following deforestation.

GENERAL DISCUSSION AND SIGNIFICANCE

The intrasystem cycle of a terrestrial ecosystem links the organic, available nutrient, and soil and rock mineral compartments through rate processes including decomposition of organic matter, leaching and exudate from the biota, nutrient uptake by the biota, weathering of primary minerals, and formation of new secondary minerals (Fig. 15). The deforestation experiment was designed to test the effects of blockage on a major ecosystem pathway, i.e., nutrient and water uptake by vegetation, on other components of the intrasystem cycle and on the export behavior of the system as a whole. The block was imposed by cutting all of the forest vegetation and subsequently preventing regrowth with herbicides. We hoped that this experimental procedure would provide information about the nature of the homeostatic capacity of the ecosystem. The deforested condition has been maintained since 1 January 1966.

Forest clearing and herbicide treatment had a profound effect on the hydrologic and nutrient relationships of our northern hardwood ecosystem. Annual runoff (water export) increased by some 33 cm or 39% in the first year and 27 cm or 28% in the second year over that expected. Moreover, the discharge pattern was altered so

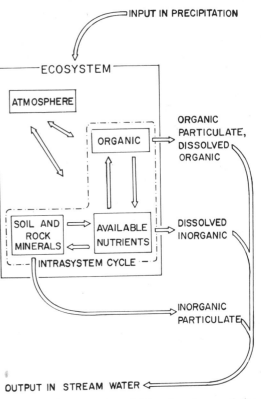

Fig. 15. Diagrammatic model for sites of accumulation and pathways of nutrients in the Hubbard Brook ecosystem (after Bormann, et al., 1969).

that sustained, higher flows occurred in the summer months and the snow pack melted earlier in the spring. This overall increase in stream runoff is large compared to the average increase (about 20 cm) found for other such experiments throughout the world (Hibbert, 1967), but is less than the maximum increase of 41 cm found for clearcut watersheds in North Carolina (Hoover, 1944).

No previous comprehensive measurements have been made of the homeostatic ability of a watershed-ecosystem to retain nutrients despite major shifts in the hydrologic cycle, including increased discharge, following deforestation (Odum, 1969).

Our results showed that cation and anion export did not change for the first 5 months (winter and spring) after deforestation, but then the ionic concentrations increased spectacularly, and remained at high levels for the 2 years of observation (Tables 4 and 5, Figs. 7, 8, 9 and 10). Annual net losses in kg/ha amounted to about 142 for nitrate-nitrogen, 90 for calcium, 36 for potassium, 32 for dissolved silica, 24 for aluminum, 18 for magnesium, 17 for sodium, and 4 for chlorine during 1967–68. These losses are much

TABLE 8. Comparative net gains or losses of dissolved solids in runoff following clear-cutting of Watershed 2 in the Hubbard Brook Experimental Forest for the period 1 June to 31 May. In metric tons/km²-yr

	1966-67		1967-68	
	W2	W6	W2	W6
Ca........	−7.5	−0.8	−9.0	−0.9
K.........	−2.3	−0.1	−3.6	−0.2
Al........	−1.7	−0.1	−2.4	−0.3
Mg........	−1.6	−0.3	−1.8	−0.3
Na........	−1.7	−0.6	−1.7	−0.7
NH₄.......	+0.1	+0.2	+0.2	+0.3
NO₃.......	−43.0	+1.5	−62.8	+1.1
SO₄.......	−0.5	−0.8	0	−1.0
HCO₃......	−0.1	−0.2	0	−0.3
Cl........	−0.1	+0.2	−0.4	0
SiO₂—aq...	−6.6	−3.6	−6.9	−3.6
Total.....	−65.0	−4.6	−88.4	−5.9

greater than for adjacent undisturbed ecosystems (Table 5). Ammonium-nitrogen was essentially unchanged relative to the undisturbed condition during this period and showed an annual net gain of about 1 to 2 kg/ha. In comparison with the undisturbed watershed-ecosystems the greatest changes occurred in nitrate-nitrogen and potassium export. Nitrate-nitrogen is normally accumulated in the undisturbed ecosystem in contrast to this very large export, and the net potassium output increased about 18-fold. The total net export of dissolved inorganic substances from the deforested ecosystem is 14–15 times greater than from the undisturbed ecosystem (Table 8).

The terrestrial ecosystem is one of the ultimate sources of dissolved substances in surface water. The contribution of dissolved solids (gross export) by our undisturbed forest ecosystems, 13.9 metric tons/km² (Bormann, et al., 1969) is only about 25% of the dissolved load predicted by Langbein and Dawdy (1964) for regions with 75 cm of annual runoff. Their estimates were based on data from watersheds of the north Atlantic slope, which probably include areas disturbed by agriculture or logging. The difference between our undisturbed forest ecosystems and the regional prediction is credited in part to the operation of various regulating biotic factors associated with mature undisturbed forest (Bormann, et al., 1969).

Deforestation markedly altered the ecosystem's contribution of dissolved solids to the drainage waters. Total gross export, exclusive of dissolved organic matter, was about 75 metric tons/km² in 1966–67 and 97 metric tons/km² in 1967–68. These figures exceed the regional prediction of Langbein and Dawdy (1964). However, it should be noted that the accelerated export of

dissolved substances results primarily from mining the nutrient capital of the ecosystem and cannot be sustained indefinitely.

Surprisingly, the net export of dissolved inorganic substances from the cutover watershed is about double the annual value estimated for particulate matter removed by debris avalanches in the White Mountains (Bormann, et al., 1969). Thus, the effects of deforestation may have almost twice the importance of avalanches in short-term catastrophic transport of inorganic materials downslope in the White Mountains.

Coupled with this increase in gross and net export of dissolved substances, there has been at least a 4-fold increase in the export of inorganic and organic particulate matter from the deforested ecosystem. This increase indicates that the biotic mechanisms that normally minimize erosion and transport (Bormann, et al., 1969) are also becoming less effective.

The greatly increased export of nutrients from the deforested ecosystem resulted primarily from an alteration of the nitrogen cycle within the ecosystem. Whereas in the undisturbed system, nitrogen is cycled conservatively between the living and decaying organic components, in the deforested watershed, nitrate produced by microbial nitrification from decaying organic matter, is rapidly flushed from the system in drainage waters. In fact, the increased nitrate output accounts for the net increase in total cation and anion export from the ecosystem (Likens, et al., 1969). With the increased availability of nitrate ions and hydrogen ions from nitrification, cations are readily leached from the system. Cations are mobilized as hydrogen ions replace them on the various exchange complexes of the soil and as organic and inorganic materials decompose. Based upon the increased output for sodium, chemical decomposition of inorganic materials in the deforested ecosystem is also accelerated by about 3-fold.

If the streamwater concentrations had remained constant after deforestation, increased water output alone would have accounted for 39% of the increased nutrient export the first year and 28% of the increased nutrient export the second year. However, the very large increase in annual export of dissolved solids from the deforested ecosystem occurred primarily because the streamwater concentrations were vastly increased, mostly as a direct result of the increased nitrification. The increased output of nutrients originated predominantly from the organic compartment of the watershed-ecosystem (Fig. 15).

Our study shows that the retention of nutrients within the ecosystem is dependent on constant and efficient cycling between the various components of the intrasystem cycle, i.e., organic, available nutrients, and soil and rock mineral compartments (Fig. 15). Blocking of the pathway of nutrient uptake by destruction of one subcomponent of the organic compartment, i.e., vegetation, leads to greatly accelerated export of the nutrient capital of the ecosystem. From this we may conclude that one aspect of homeostatis of the ecosystem, i.e., maintenance of nutrient capital, is dependent upon the undisturbed functioning of the intrasystem nutrient cycle, and that in this ecosystem no mechanism acts to greatly delay loss of nutrients following sustained destruction of the vegetation.

The increased output of water from the deforested watershed was readily visible during the summer months, however the increased ion and particulate matter concentrations were not. The stream water from the deforested watershed appeared to be just as clear and potable as that from adjacent, undisturbed watersheds. However this was not the case. By August, 1966, the nitrate concentration in stream water exceeded (at times almost doubled) the concentration recommended for drinking water (Public Health Service, 1962).

The high nutrient concentrations, plus the increased amount of solar radiation (absence of forest canopy) and higher temperature in the stream, resulted in significant eutrophication. A dense bloom of *Ulothrix zonata* (Weber and Mohr) Kütz, has been observed during the summers of 1966 and 1967 in the stream of W2. In contrast the undisturbed watershed streams are essentially devoid of algae of any kind. This represents a good example of how an overt change in one component of an ecosystem, alters the structure and function, often unexpectedly, in another part of the same ecosystem or in another interrelated ecosystem. Unless these ecological interrelationships are understood, naive management practices can produce unexpected and possibly widespread deleterious results.

CONCLUSIONS

1. The quantity and quality of drainage waters were significantly altered subsequent to deforestation of a northern hardwoods watershed-ecosystem. All vegetation on Watershed 2 of the Hubbard Brook Experimental Forest was cut, but not removed, during November and December of 1965; and vegetation regrowth was inhibited by periodic application of herbicides.

2. Annual runoff of water exceeded the expected value, if the watershed were undisturbed, by 33 cm or 39% during the first water-year after deforestation and 27 cm or 28% during the second water-year. The greatest increase in water discharge, relative to an undisturbed situation, occurred during June through September, when runoff was 414% (1966–67) and 380% (1967–68) greater than the estimate for the untreated condition.

3. Deforestation resulted in large increases in streamwater concentrations of all major ions except NH_4^+, $SO_4^=$ and HCO_3^-. The increases did not occur until 5 months after the deforestation. The greatest increase in streamwater ionic concentration after deforestation was observed for nitrate, which increased by 41-fold the first year and 56-fold the second year above the undisturbed condition.

4. Sulfate was the only major ion in stream water from Watershed 2 that decreased in concentration after deforestation. The 45% decrease the first year (1966–67) resulted mostly from increased runoff of water and by eliminating the generation of sulfate within the ecosystem. The concentration of sulfate in stream water during 1967–68 equalled the concentration in precipitation after adjustment for water loss by evaporation. Sulfate concentrations were inversely related to nitrate concentrations in stream water in both undisturbed and deforested watersheds.

5. In the undisturbed watersheds the stream water can be characterized as a very dilute solution of sulfuric acid (pH about 5.1 for W2); whereas after deforestation the stream water from Watershed 2 became a relatively stronger nitric acid solution (pH 4.3), considerably enriched in metallic ions and dissolved silica.

6. The increase in average nitrate concentration in precipitation for the Hubbard Brook area compared to data from 1955–56, as well as the consistent annual increase observed from 1964–1968, may be some measure of a general increase in air pollution.

7. The greatly increased export of dissolved nutrients from the deforested ecosystem was due to an alteration of the nitrogen cycle within the ecosystem. Whereas nitrogen is normally conserved in the undisturbed ecosystem, in the deforested ecosystem nitrate is rapidly flushed from the system in drainage water. The mobilization of nitrate from decaying organic matter, presumably by increased microbial nitrification, quantitatively accounted for the net increase in

total cation and anion export from the deforested ecosystem.

8. Increased availability of nitrate and hydrogen ions resulted from nitrification. Cations were mobilized as hydrogen ions replaced them on various exchange complexes of the soil and as organic and inorganic materials were decomposed. Chemical decomposition of inorganic materials in the deforested ecosystem was accelerated about 3–fold. However, the bulk of the nutrient export from the deforested watershed originated from the organic compartment of the ecosystem.

9. The total net export of dissolved inorganic substances from the deforested ecosystem was 14–15 times greater than from undisturbed ecosystems. The increased export occurred because the streamwater concentrations were vastly increased, primarily as a direct result of the increased nitrification, and to a much lesser extent because the amount of stream water was increased.

10. The deforestation experiment resulted in significant pollution of the drainage stream from the ecosystem. Since August, 1966, the nitrate concentration in stream water has exceeded, almost continuously, the maximum concentration recommended for drinking water. As a result of the increased temperature, light and nutrient concentrations, and in sharp contrast to the undisturbed watersheds, a dense bloom of algae has appeared each year during the summer in the stream from Watershed 2.

11. Nutrient cycling is closely geared to all components of the ecosystem; decomposition is adjusted to nutrient uptake, uptake is adjusted to decomposition, and both influence chemical weathering. Conservation of nutrients within the ecosystem depends upon a functional balance within the intrasystem cycle of the ecosystem. The uptake of water and nutrients by vegetation is critical to this balance.

LITERATURE CITED

Alexander, M. 1967. Introduction to Soil Microbiology. John Wiley and Sons, Inc., New York, 472 pp.

Allison, F. E. 1955. The enigma of soil nitrogen balance sheets. Advan. Agron. 7: 213–250.

Anderson, D. H., and H. E. Hawkes. 1958. Relative mobility of the common elements in weathering of some schist and granite areas. Geochim. Cosmochim. Acta 14(3): 204–210.

Bormann, F. H., and G. E. Likens. 1967. Nutrient cycling. Science 155(3761): 424–429.

Bormann, F. H., G. E. Likens, D. W. Fisher, and R. S. Pierce. 1968. Nutrient loss accelerated by clear-cutting of a forest ecosystem. Science 159: 882–884.

Bormann, F. H., G. E. Likens, and J. S. Eaton. 1969. Biotic regulation of particulate and solution losses from a forest ecosystem. BioScience 19(7): 600–610.

Boswell, J. G. 1955. The microbiology of acid soils. IV. Selected sites in Northern England and Southern Scotland. New Phytol. 54(2): 311–319.

Federer, C. A. 1969. Radiation and snowmelt on a clear-cut watershed. E. Snow Conf. Proc., Boston, Mass. (1968) pp. 28–41.

Fisher, D. W., A. W. Gambell, G. E. Likens, and F. H. Bormann. 1968. Atmospheric contributions to water quality of streams in the Hubbard Brook Experimental Forest, New Hampshire. Water Resources Res. 4(5): 1115–1126.

Gambell, A. W., and D. W. Fisher. 1966. Chemical composition of rainfall, eastern North Carolina and southeastern Virginia, U.S. Geol. Survey Water-Supply Paper 1535K: 1–41.

Harvey, H. H., and A. C. Cooper. 1962. Origin and treatment of a supersaturated river water. Internat. Pacific Salmon Fish. Comm., Prog. Rept. No. 9: 1–19.

Hart, G., R. E. Leonard, and R. S. Pierce. 1962. Leaf fall, humus depth, and soil frost in a northern hardwood forest. Forest Res. Note 131, Northeastern For. Exp. Sta., Durham, N. H.

Hibbert, A. R. 1967. Forest treatment effects on water yield. pp. 527–543. In: Proc. Internat. Symposium on Forest Hydrology, ed. by W. E. Sopper and H. W. Lull, Pergamon Press, N. Y.

Hoover, M. D. 1944. Effect of removal of forest vegetation upon water yields. Trans. Amer. Geophys. Union, Part 6: 969–975.

Hornbeck, J. W., and R. S. Pierce. 1969. Changes in snowmelt run-off after forest clearing on a New England watershed. E. Snow Conf. Proc., Portland, Maine (1969). (In press).

Hornbeck, J. W., R. S. Pierce, and C. A. Federer. Streamflow changes after forest clearing in New England. (In preparation).

Iwasaki, I., S. Utsumi, and T. Ozawa. 1952. Determination of chloride with mercuric thiocyanate and ferric ions. Chem. Soc. Japan Bull. 25: 226.

Hewlett, J. D., and A. R. Hibbert. 1961. Increases in water yield after several types of forest cutting. Quart. Bull. Internatl. Assoc. Sci. Hydrol. Louvain, Belgium, pp. 5–17.

Jansson, S. L. 1958. Tracer studies on nitrogen transformations in soil with special attention to mineralisation-immobilization relationships. Kungl. Lantbrukshögskolans Annaler. 24: 105–361.

Järnefelt, H. 1949. Der Einfluss der Stromschnellen auf den Sauerstoff-und Kohlensäuregehalt und das pH des Wassers im Flusse Vuoksi. Verh. Internat. Ver. Limnol. 10: 210–215.

Johnson, N. M., G. E. Likens, F. H. Bormann, and R. S. Pierce. 1968. Rate of chemical weathering of silicate minerals in New Hampshire. Geochim. Cosmochim. Acta. 32: 531–545.

Johnson, N. M., G. E. Likens, F. H. Bormann, D. W. Fisher, and R. S. Pierce. 1969. A working model for the variation in streamwater chemistry at the Hubbard Brook Experimental Forest, New Hampshire. Water Resources Res. 5(6): 1353–1363.

Juang, F. H. F., and N. M. Johnson. 1967. Cycling of chlorine through a forested watershed in New England. J. Geophys. Research 72(22): 5641–5647.

Junge, C. E. 1958. The distribution of ammonia and nitrate in rain water over the United States. Trans.

Amer. Geophys. Union **39**: 241–248.

———. 1963. Air Chemistry and Radioactivity. Academic Press, N. Y. 382 pp.

———, and **R. T. Werby.** 1958. The concentration of chloride, sodium, potassium, calcium and sulphate in rain water over the United States. J. Meteorol. **15**: 417–425.

Langbein, W. B., and D. R. Dawdy. 1964. Occurrence of dissolved solids in surface waters in the United States. U. S. Geol. Survey Prof. Paper **501–D**: D115–D117.

Lieberman, J. A., and M. D. Hoover. 1948a. The effect of uncontrolled logging on stream turbidity. Water and Sewage Works **95**(7) : 255–258.

———. 1948b. Protecting quality of stream flow by better logging. Southern Lumberman : 236–240.

Likens, G. E., F. H. Bormann, N. M. Johnson, and R. S. Pierce. 1967. The calcium, magnesium, potassium, and sodium budgets for a small forested ecosystem. Ecology **48**(5) : 772–785.

Likens, G. E., F. H. Bormann and N. M. Johnson. 1969. Nitrification: Importance to nutrient losses from a cutover forested ecosystem. Science **163**(3872) : 1205–1206.

Lindroth, A. 1957. Abiogenic gas supersaturation of river water. Arch. für Hydrobiol. **53**: 589–597.

Macan, T. T. 1958. The temperature of a small stony stream. Hydrobiologia **12**: 89–106.

Marshall, C. E. 1964. The physical chemistry and minerology of soils. Vol. I. Soil materials. Wiley and Sons, N. Y. 388 pp.

Mason, B. 1966. Principles of geochemistry. 3rd ed. Wiley and Sons, N. Y. 329 pp.

McConnochie, K., and G. E. Likens. 1969. Some Trichoptera of the Hubbard Brook Experimental Forest in central New Hampshire. Canadian Field Naturalist. **83**(2) : 147–154.

Minckley, W. L. 1963. The ecology of a spring stream, Doe Run, Meade County, Kentucky. Wildlife Monogr. **11**: 1–124.

Nye, R. H., and D. J. Greenland. 1960. The soil under shifting cultivation. Commonwealth Bureau of Soils, Harpenden, England, Tech. Bull. No. **51**, 156 pp.

Odum, E. P. 1969. The strategy of ecosystem development. Science **164**(3877) : 262–270.

Pierce, R. S. 1969. Forest transpiration reduction by clearcutting and chemical treatment. Proc. Northeastern Weed Control Conference. **23**: 344–349.

Postgate, J. R. 1949. Competitive inhibition of sulphate reduction by selenate. Nature (**4172**) : 670–671.

Rankama, K., and T. G. Sahama. 1950. Geochemistry. Chicago Univ. Press, 912 pp.

Ruttner, F. 1953. Fundamentals of Limnology. Univ. of Toronto Press. (Transl. by D. G. Frey and F. E. J. Fry). 242 pp.

Scott, D. R. M. 1955. Amount and chemical composition of the organic matter contributed by overstory and understory vegetation to forest soil. Yale Univ. School of Forestry. Bull. No. **62**, 73 pp.

Smith, W., F. H. Bormann, and G. E. Likens. 1968. Response of chemoautotrophic nitrifiers to forest cutting. Soil Science **106**(6) : 471–473.

Tebo, L. D. 1955. Effects of siltation, resulting from improper logging, on the bottom fauna of a small trout stream in the southern Appalachians. Prog. Fish Culturist **12**(2) : 64–70.

Thiessen, A. H. 1923. Precipitation for large areas. Monthly Weather Review **51**: 348–353.

Trimble, G. R., and R. S. Sartz. 1957. How far from a stream should a logging road be located? J. Forestry **55**(5) : 339–341.

U. S. Forest Service. Northeastern Forest Experiment Station. 1964. Hubbard Brook Experimental Forest. Northeast. For. Exp. Sta., Upper Darby, Penn. 13 pp.

U. S. Public Health Service. 1962. Drinking water standard. U.S. Public Health Service Publ. **956**. Washington, D. C.

Waksman, S. A. 1932. Principles of soil microbiology. 2nd ed. Williams and Wilkins Co., Baltimore. 894 pp.

Weber, D. F., and P. L. Gainey. 1962. Relative sensitivity of nitrifying organisms to hydrogen ions in soils and in solutions. Soil Sci. **94**: 138–145.

Whitehead, H. C., and J. G. Feth. 1964. Chemical composition of rain, dry fallout, and bulk precipitation at Menlo Park, California, 1957-1959. J. Geophys. Res. **69**(16) : 3319–3333.

Woods, W. J. 1960. An ecological study of Stony Brook, New Jersey. Ph.D. Thesis, Rutgers Univ. New Brunswick. 307 pp.

16

Reprinted from *Ecology*, **54**(1), 70–80 (1973) with permission of the publisher, Duke University Press, Durham, North Carolina

STUDIES OF CATION BUDGETS IN THE SOUTHERN APPALACHIANS ON FOUR EXPERIMENTAL WATERSHEDS WITH CONTRASTING VEGETATION[1]

Philip L. Johnson[2] and Wayne T. Swank[3]

Abstract. Nutrient fluxes within and through watershed ecosystems at the Coweeta Hydrologic Laboratory are under study. This paper describes the annual budgets and seasonal fluctuations for selected cations. Concentrations and flux of cations moving through a hardwood forest stand, a weed to forest succession, a hardwood coppice stand and an eastern white pine stand on steep mountain topography are compared. Stream discharge was greater by 6% for the successional weed stand, and 10% for the second hardwood coppice, but 15% less for the young pine stand in contrast to pretreatment levels. Although concentrations for Ca^{++}, Mg^{++}, K^+ and Na^+ combined were usually less than 3.5 ppm, over 98% of the loss of each cation was in dissolved form on all four watersheds. Regression analysis showed that 50 to 60% of the variation in monthly weighted average concentration was accounted for by monthly discharge amounts. Annual losses of the four cations from the mature hardwood stand were in the amounts of approximately 7, 3, 5, and 10 kg/ha respectively for the Ca^{++}, Mg^{++}, K^+, and Na^+. Annual budgets showed net changes to be −0.8, −1.8, −2.0, and −4.3 kg/ha, respectively, for this mature hardwood ecosystem. In contrast, the weed stand lost significantly greater amounts, and the young pine and hardwood coppice watersheds showed a net gain in Ca^{++} and significantly lower losses than the mature ecosystem for the other three ions. These budgets show that major alterations to these forest ecosystems are not now producing a substantial out-flux for these cations.

INTRODUCTION

Steady state or stable natural ecosystems are generally agreed to be the standard against which to evaluate man's impact on the landscape, and the processes involved in nutrient cycling and biological productivity are keys to the understanding of such complex systems. Although ecological theory is fundamental to food and fiber production, cycling of materials through large ecosystems must be quantified before ecological principles can be effectively applied to ecosystem management for maximization of a variety of products. Moreover, the results of ecosystem studies are perhaps best applied within the framework of predictive models which are based on empirical data.

Both forested and agricultural watersheds have been studied for decades to specify the variables and flux rates of the hydrologic cycle. Bormann and Likens (1967) concluded that small watersheds can be eminently suitable for studies of mineral flux through ecosystems because (1) drainage basins are definable, natural ecosystems, (2) gaged watersheds minimize gains and losses due to deep seepage, (3) budgets of mineral flux for whole ecosystems may be calculated, including estimates of weathering input and erosional losses, and (4) land-water inter-

actions can be evaluated in the context of responses to various environmental pollutants or land management manipulations. The utility of watersheds as units for ecosystem research as been further illustrated through a first generation model for predicting chemical output from catchments (Johnson et al. 1969). On the other hand, concern is sometimes expressed in regard to how readily mineral budgets from small experimental watersheds can be interpreted for larger basins. We agree that the application of results must proceed with caution in light of a possible scalar dependence, but we also recognize that experimental catchments are in many cases comparable in size (8–30 ha) to current forest management units.

The natural and manipulated watersheds at the Coweeta Hydrologic Laboratory in North Carolina offer several unique advantages as sites for mineral cycling research. These advantages include a 35-year history of hydrologic and climatological monitoring, a well-documented history of the vegetation on the watersheds and a variety of experimental manipulations designed for water yield improvement studies. Furthermore, the research strategy at Coweeta during the past decade of studying hydrologic processes on catchments provides an essential base for interpreting the results of mineral flux studies. Thus, a cooperative effort between the Institute of Ecology, University of Georgia, and the Coweeta Hydrologic Laboratory, U. S. Forest Service, was initiated in 1968 with the primary objective of measuring the annual and seasonal fluxes of selected nutrients in four ecosystems as defined by watersheds and represented by four different vegetation types. A second, and long-term, objective of the study is to inventory

[1] Based partially on a paper presented in Section II, Forest Influences and Watershed Management, XVth Congress of the International Union of Forest Research Organizations, March 15–20, 1971, U. of Florida, Gainesville, Fla. Manuscript received March 22, 1972, accepted August 7, 1972. Contribution No. 55 from the Eastern Deciduous Forest Biome, US-IBP.

[2] National Science Foundation, Washington, D. C.

[3] Coweeta Hydrologic Laboratory, Franklin, N. C.

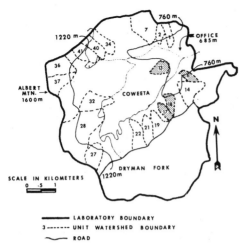

FIG. 1. The 2,270 ha. Coweeta Hydrologic Laboratory, Macon Co., N.C. The two major drainages are Coweeta and Dryman Fork. The four shaded experimental watersheds are the sites of nutrient cycling studies.

the biotic and abiotic compartments as a basis for estimating the standing pool of nutrients and net turnover for each ecosystem. As studies progress, quantitative models will be developed in an effort to describe and account for the behavior of these systems. The purpose of this paper is to describe some of the current research effort and to present initial estimates of selected nutrient fluxes in four contrasting ecosystems.

The Experimental Watersheds

The 2,270 ha Coweeta Hydrologic Laboratory is located in the southwest corner of North Carolina in the southern Appalachians and is comprised of two basins, Coweeta and Dryman Fork (Fig. 1). Prior to about 1842 when the white man settled the basin, the area was occupied by the Cherokee Indians. Light semiannual burning of the woods and grazing were the principal land practices until about the turn of the century. Controlled selection logging was conducted in the period 1909 to 1923; cutting

operations were concentrated in the valleys and in the more accessible coves and lower slopes. The Forest Service acquired all rights to the area in 1924 and except for experimental treatments, there have been no fires or cutting operations on the Laboratory.

The climate of the region is classified as marine with cool summers, mild winters and adequate rainfall in all seasons (Trewartha 1954). Precipitation is frequently cyclonic in origin with air masses coming from the Gulf of Mexico, the Atlantic Coast or from the subarctic region of North America. However, the steep mountains of the southern Appalachians sometimes produce orographic precipitation and frequent thunderstorms during the summer. Within the experimental site, average annual precipitation varies from 2,500 mm on the upper slopes to 1,700 mm at the lower elevations with snow typically contributing less than 2% to total precipitation.

The basin lies within the Blue Ridge province of the Appalachian Highlands physiographic division. The underlying rock is classified as Carolina gneiss of pre-Cambrian origin and includes granite, diorite, mica gneiss, and mica schist. In response to climate, the formation is deeply weathered and the derived soils are classified as Porter loam, steep phase with the regolith over the basin averaging about 7 meters in depth (Devereux et al. 1929).

The four experimental catchments selected for ecosystem study are among 22 currently gaged at Coweeta. Specific physical characteristics of the four study areas are given in Table 1 and their relative location within the Laboratory are shown in Fig. 1. The catchments range in size from 8 to 16 ha and have well-defined topographic boundaries. Streamflow is perennial and the flow regime is typical of that exhibited by Watershed 18 (Table 2). Although the range of flow is rather narrow for all months, discharge is highest and most variable during the late winter months of February and March. In the late summer and early fall months, flow tends to be lowest and most stable. Quickflow (or direct runoff) usually comprises less than 10% of the total runoff, and there is essentially no overland flow on the catchments.

TABLE 1. Summary of physical characteristics on four Coweeta Experimental Watersheds

Watershed no.	Area	Vegetation type	Precipitation (30 year avg.)	[1]Runoff P-RO (30 year avg.)	Elevation avg.			Aspect	Channel length
					Max.	Min.	Slope		
	ha.		mm	mm	mm	m	m	%		m
6	8.86	Weeds, grass, shrubs	1823	833	990	905	699	54	NW	241
13	16.11	7-year-old hardwood coppice	1851	888	963	912	725	49	NE	604
17	13.48	13-year-old white pine	1916	800	1116	1021	742	57	NW	352
18	12.46	Mature hardwood	1813	955	858	1006	721	52	NW	292

[1] Runoff for treated watersheds was normalized for post-treatment period by regression on control watershed.

TABLE 2. Mean daily minimum and maximum mean monthly discharge; quarterly quickflow based on 33 years of record on Coweeta Watershed 18

Month	Discharge (1/sec/km²)			Quarterly quickflow (cm)		
	Minimum	Daily Maximum	Mean	Minimum	Maximum	Mean
May	16	61	37			
June	10	54	25			
July	11	45	19			
Qtr. Quickflow				0.31	3.91	1.18
August	8	40	17			
September	5	41	14			
October	4	62	14			
Qtr. Quickflow				0.14	8.72	1.22
November	4	54	16			
December	5	70	25			
January	6	95	38			
Qtr. Quickflow				0.22	9.54	2.76
February	17	132	53			
March	18	121	56			
April	26	98	54			
Qtr. Quickflow				0.50	10.28	3.83
Annual, monthly mean			31			8.99

Hewlett et al. (1969) have shown how the paired catchment experiment offers control over leakage problems in assessing *changes* in water yield due to treatment, but water balance or nutrient budget studies require a complete accounting of input and loss of water. Leakage into or out of the watershed is usually assumed to be negligible but in fact, the extent of leakage is unknown. Lee (1970) used theoretical and empirical models to estimate the long term water balance of small catchments as a check on the observed water balance. He found that theoretical and empirical models were, in general, less accurate than water balance studies and were not helpful in confirming or challenging the accuracy of water balance measurements. Thus, he concluded that there is no satisfactory method of estimating watershed leakage. The best safeguards against leakage are selection of watersheds of tight bedrock and well-defined boundaries; construction of weir cutoff walls which force all water to leave the catchment via the weir blade; and a rigorous program of weir checks and maintenance. Based on available geologic information, it is reasonable to assume that the bedrock at Coweeta is generally impermeable, and weirs are constructed and maintained to strict standards. Thus, watershed leakage problems are probably negligible for most of these experimental watersheds. This belief is supported by the constancy of annual differences between total precipitation (P) and total runoff (RO) over a wide range of values for Watershed 18 (Dils 1957) and by the similarity of water balances for Watersheds 6, 13, and 18 (Table 1). By comparison, the annual P-RO for Watershed 17 is substantially larger than for other watersheds. Lee

(1970) notes that there is a genuine discrepancy between observed and theoretical water balances for the watershed which amounts to about 10% of annual flow. Thus, it is highly probable that some flow from Watershed 17 is bypassing the weir. While nutrient budgets for Watersheds 6, 13, and 18 are felt to be accurate (within measurement errors), interpretation of the budget for Watershed 17 must take into account the probability of unmeasured flow.

Except for Watershed 18, each of the study catchments has a long and varied treatment history. Watershed 18 has remained relatively undisturbed since at least 1924 and has served as a reference catchment beginning in 1934. The area supports an unevenaged mature hardwood forest containing 3,042 stems/ha and 25.6 m² of basal area/ha of which 70% is comprised of *Quercus* species, *Carya* species, and red maple (*Acer rubrum* L.). American chestnut (*Castanea dentata* (Marsh.) Borkh.) was originally a major constituent of the forest but was reduced to a minor component by the chestnut blight in the middle 1930's. According to Day (1971), there is a total above-ground biomass of approximately 137.5 metric tons/ha.

In 1942, 12% of Watershed 6 along the stream was cut to determine the effect of cutting streamside vegetation on water yield (Dunford and Fletcher 1947). The streamside strip was allowed to regrow until 1958 when all merchantable timber over the entire catchment was cut and removed and the cover was converted to Kentucky 31 fescue grass (*Festuca elatior* var. *arundinacea* Schreb.). The objective of the experiment was to compare water use of the grass and the original mixed hardwood forest. A seedbed was prepared by a combination of burning, grubbing, and harrowing. In March, 1959, 6.7 metric tons of ground dolomitic limestone containing a minimum of 56% $CaCO_3$ and 32% $MgCO_3$; 2.2 metric tons of 2-12-12 fertilizer; and 22 kg of fescue seed were applied to each hectare. In May, 1960, a nitrogen deficiency was observed on about 3 hectares and was corrected with an application of 33.5% ammonium nitrate at a rate of 224 kg/ha. During a subsequent 5-year period, the only treatment applied was occasional control of laurel, rhododendron, and other hardwood sprouts with herbicides. Then in March, 1965, an additional 672 kg of 30–10–0 fertilizer and 170 kg of 60% potash were applied to each hectare of the catchment. The effects of the hardwood-to-grass conversion on the quantity of streamflow have been described by Hibbert (1969). Beginning in May, 1966, and continuing for a two-year period, the grass cover on the catchment was killed with herbicides to determine streamflow response in the absence of a plant cover. Details of herbicide types, quantities, and application schedule have been reported by Douglas et al. (1969). No treatment was applied in the spring of 1968 and subsequently the catchment has been allowed to revert to successional vegetation. In the first year of succession, horseweed (*Erigeron canadensis* L.) and cottonweed (*Erechtites hieracifolia* (L.) Raf.) were the dominant species on the watershed and comprised 50% of the total vegetative productivity (460 g/m²). Horseweed and cottonweed were also present during the second year of succession in 1969, but dropped to 20% of the aboveground biomass of 283 g/m² while grass species added on additional 45%. In 1970, woody shrubs were abundant on the watershed and contributed 50% of the aboveground total productivity of 396 g/m².

The treatment on Watershed 13 represents one of the earliest examples of the effect of complete forest clearfelling on streamflow in the Eastern United States. From September 1939 through January 1940, all woody vegetation on the catchment was cut and left in place. A coppice forest regrew until 1962 when the watershed was again clearfelled. As in the first treatment, no products were removed and the forest was allowed to regrow. The response of streamflow to the cutting treatments has been described by Kovner (1956) and Swank and Helvey (1970). In 1969, when the coppice forest was 7 years old, there were 9,659 stems/ha; 7.0 m² basal area/ha. *Quercus* species, *Carya* species, yellow-poplar (*Liriodendron tulipifera* L.) and dogwood (*Cornus florida* L.) accounted for 70% of the total basal area.

All shrub and forest vegetation on Watershed 17 was cut in 1942 (Hoover 1944). In contrast to Watershed 13, annual sprout growth was cut back most years between 1943 to 1955 and a low herbaceous and woody cover developed. Eastern white pine (*Pinus strobus* L.) was planted in 1956 to compare water use by pine and the original hardwood forest. In subsequent years, competing hardwood sprouts were cut or sprayed with herbicides as required to prevent competition with the pine. Some early effects of the cover type conversion on streamflow were described by Swank and Miner (1968). By 1969, the 13-year-old stand averaged 15.3 m² per ha basal area with 1,668 stem per ha.

PROCEDURES

By monitoring quantities of precipitation, streamflow, and the concentration of nutrients received and leaving the catchment, the mass flux and response to treatment are being evaluated for the four ecosystems. A number of nutrients are under study at Coweeta, but this report deals only with calcium, magnesium, potassium, and sodium.

Since 1934, precipitation records have been collected at 130 different sites within the Coweeta drainage with a combination of standard and recording rain gages. Gaging density ranged from 1 gage per 24 ha over the entire drainage to 1 per 4 ha for some experimental catchments and 1 per 0.2 ha for detailed rainfall distribution studies. Based on these records, an isohyetal weighting system was developed to estimate the areal precipitation received by each of the experimental watersheds (Swift 1970). Sources of nutrient input to the catchments are measured in both precipitation and dry fallout. Precipitation samples are collected weekly from gages of the water trap type described by Likens et al. (1967) which eliminate concentration of solutes by evaporation. Several sampling studies have been conducted to determine variation in precipitation chemistry over a watershed and between watersheds. Dry fallout is collected in two Wong Model ARC Mark V collectors which automatically close to exclude rain or snow. One gage is installed at a permanent weather station near the Coweeta office and the other is located at the base of Watershed 6. Dry fallout is recovered weekly in 500 ml of deionized water by rinsing the collection vessel. The weekly samples of precipitation and dry fallout are returned to the laboratory for chemical analysis. Infrequently samples are contaminated by birds, insects, or other debris and such samples are rejected. Total nutrient input to a watershed is taken as the product of precipitation volumes and chemical concentrations.

Streamflow is measured continuously at the base of each catchment using sharp-crested, V-notch blades set in concrete cutoff walls (weir). Watersheds 6 and 17 are instrumented with 90° weir blades and Watersheds 13 and 18 have 120° blades. The depth of water flowing over the weir blade is recorded on analog-to-digital water level recorders, and the digital output is machine-translated to IBM cards. Hibbert and Cunningham (1967) have described the steps of streamflow data processing and the various types of flow summaries routinely used at Coweeta. Samples for stream chemistry are collected weekly in the stream mid-channel directly above each weir installation. Samples are collected in 125 ml polyethylene bottles which are first acid washed with 10% HCl and then rinsed with demineralized water. At the collection site, bottles are flushed with stream water prior to sample collection.

The relationship of concentration to discharge is an important consideration in sampling the stream regime for chemical concentration. Therefore, a sampler which accumulates sample aliquots proportional to stream discharge rates is needed to contrast proportional and weekly samples. A modified Brailsford Model EV-1 effluent sampler was installed and tested for proportionality. The unit utilizes a separate head detector in the stilling pond to energize a pump at intervals proportional to discharge to achieve a proportional composite sample. The water sample intake line is located in the flowing stream and automatically backflushes between sample collections. Only minor maintenance problems during winter temperatures have been experienced in 18 months of operation. All weekly stream water samples for cation analysis are returned to the laboratory within 24 hours after collection and held near freezing until analyzed. Total dissolved nutrient output is calculated as the product of streamflow volumes and chemical concentrations.

Suspended materials which are deposited in the settling basins behind the weir are removed quarterly in September, December, March, and June. The settling basins contain solid concrete and rock bottoms; after diverting the stream temporarily, the accumulated sediments are removed and placed in a plywood box (1.2 by 1.2 by 1.8 meters) located at the weir site. The sediments are stirred to achieve homogeneity; then three pipe sections, each 15 cm in diameter, are inserted and their contents later recovered for analysis. Samples are sieved and the < 2 mm fraction is used for cation analysis. Ion concentration is determined for 5 g samples extracted in 20 ml of mixed acid extracting solution (0.05 N HCl, 0.025 N H_2SO_4) by the University of Georgia Soil Testing Laboratory. Fractions larger than 2 mm in diameter were not considered of biological interest for this study. The erosional loss of the entire collection is calculated from the volume on a dry weight basis, but chemical concentration is based on only the <2 mm fraction of the three subsamples to calculate chemical losses via sediments.

Cation concentrations for water samples are determined with a Bausch and Lomb Model AC2–20 Atomic Absorption Spectrophotometer. Since certain substances interfere with Ca^{++} and Mg^{++} determination, samples are buffered with appropriate amounts of lanthanum. However, lanthanum interferes with Na^+ determinations and thus Na^+ and K^+ are analyzed separately. A comparison of immediate analysis of water samples and subsamples held for up to 10 weeks revealed no change in cation concentration levels with storage time.

RESULTS AND DISCUSSION

A primary objective of this study was to measure the annual and seasonal fluxes of Ca^{++}, Mg^{++}, K^+, and Na^+. These results must be considered preliminary for two reasons. First, two years of record are suggestive but not conclusive with respect to the mean annual and seasonal chemical budgets. Second, final interpretation must await completion of the studies on internal components of the cycle, because the rate of input and loss of minerals from most natural ecosystems is much smaller than the rate of internal biological recycling. It may be for this reason that fewer values have been reported on flux through ecosystems than of recycling within them.

Hydrology

Although only two complete years of nutrient budget data are reported in this study, the hydrologic components are typical of average conditions. For Watershed 18 (the undisturbed catchment), precipitation and streamflow for water years 1970 and 1971 were only 5 to 10 percent higher than the 30-year average shown in Table 1. The seasonal distribution of precipitation and streamflow during 1971 was almost identical to the mean but in 1970, values tended to be above normal during August through November and below normal in February through April (Table 2).

Both evapotranspiration and streamflow have been altered in the three treated ecosystems as a result of changes in vegetative type and structure. The paired, or reference, watershed method of analysis is used to assess the effect of vegetative changes on water yield (Wilm 1943, Reinhart 1967). Using this technique, streamflow is carefully gaged over a period of years for two closely located watersheds similar in size, geology and cover conditions, and the relationship of flow between the watersheds for the calibration (pretreatment) period is determined by regression analysis. In the subsequent treatment period, one watershed serves as a reference (undisturbed) while some treatment is applied to the other watershed. The effect of treatment on streamflow is described as the difference between observed flow and flow predicted for the treated watershed based on the calibration regression.

For the two water years covered by cation data (1970 and 1971), annual flow on Watershed 6 averaged 5.3 cm or 6% greater than flow expected from the hardwood forest which originally covered the watershed. Annual flow from the young coppice forest on Watershed 13 was 9.4 cm (10%) above pretreatment levels. Evapotranspiration from the young pine forest on Watershed 17 is much greater compared to the hardwood forest it replaced, and thus, average annual streamflow was reduced by 12.7 cm (15%). Error terms for the calibration regressions for the watersheds are about ± 2 cm (95% confidence intervals) and therefore, the quantitative changes in water yield are regarded as significant. The impact of these hydrologic changes on nutrient fluxes in the system could also be important as discussed later in this paper.

Precipitation chemistry

Collection of precipitation samples for chemical analysis was not begun until October, 1969. Initially samplers for bulk precipitation were located near the weir of Watershed 6 where samples were collected for a seven-month period. In the early spring of 1970, a study of variability in cation concentration with elevation was conducted on Watershed 6. Two

gages were placed at each of three different elevations over the watershed to represent the lower, middle and upper third of the catchment and samples
were collected for a four-month period. The mean
concentration of the four cations over the period
was relatively constant between sample elevations
for any given weekly collection. For example, the
mean Na+ concentration of one collection was 135,
132 and 140 ppb, respectively, at the lower, middle
and upper elevations; but storm to storm concentration varied over an order of magnitude for this ion.
Equal precipitation concentrations with elevation is
not surprising because isohyets for the watersheds
run perpendicular to the slope. Thus, average cation
values can be used to describe the precipitation chem-

istry for a small experimental watershed at Coweeta.
Differences in cation amounts are more likely to
exist between watersheds because of the variability in
total precipitation received by each catchment. Therefore, two chemistry gages were located at the weir
of each watershed and samples from 21 storms were
collected for analysis. For most storms, K+ showed
the largest variability of the cations between watersheds and the mean K+ concentration for all storms
differed most between Watershed 6 and Watershed
17. However, an analysis of variance showed that the
difference between mean K+ concentrations (35
ppb) of the two watersheds was not significant at the
0.05 level. Therefore, cation values for all gages were
averaged to determine bulk precipitation chemistry.

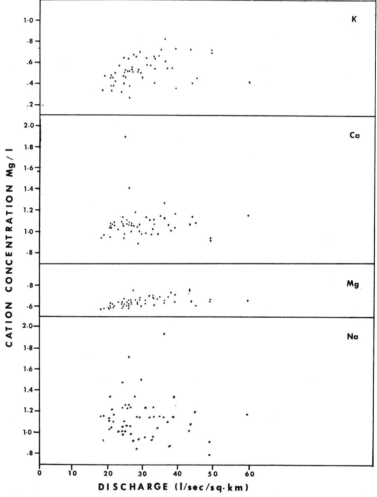

Fig. 2. Relationship between cation concentration and streamflow discharge rate for
Watershed 6, June–May, 1969–70.

Cation concentrations for the two dryfall samplers were also averaged since mean periodic values differed by only 5% or less for most cations.

The average annual weighted concentrations for the four cations in bulk precipitation (1969–71) were Ca^{++}, 0.323; Mg^{++}, 0.066; K^+, 0.165 and Na^+, 0.283 ppm. The dryfall contributions were 12, 10, 27 and 14% of bulk precipitation respectively for these cations. Annual values are very similar to those reported for bulk precipitation at Hubbard Brook in New Hampshire (Likens et al. 1967). Both Coweeta and Hubbard Brook are semi-isolated from major industrial or urban activity and therefore, dryfall is a minor constituent of the total precipitation chemistry. Precipitation chemistry at Coweeta is quite variable from month to month and several years of additional data will be required to determine temporal trends.

Concentration and discharge

The mature hardwood forest, Watershed 18, is a relatively stable ecosystem characteristic of lower slopes in the southern Appalachians. The remaining catchments have been significantly altered, particularly Watershed 6 which was heavily limed and fertilized in 1960, and fertilized again in 1965. Moreover, this catchment was completely denuded of vegetation with herbicide in 1967 and 1968. As runoff from this denuded basin increased, it was anticipated that both the amount and the concentration of chemical outflux would increase accompanied by greater seasonal variation. Thus, Watershed 6 was expected to have the greatest change of cation concentration with discharge contrasted to the relatively constant chemical concentration reported for hardwood watersheds at Hubbard Brook, New Hampshire (Likens et al. 1967). Actually, as shown in Fig. 2, cation concentration, when considered throughout the year, appears to be independent of discharge rate. Concentration levels for Mg^{++} tended to be quite stable while Ca^{++}, Na^+, and K^+ fluctuated widely over the range of flow rates. Additional data, particularly throughout the interval of a major storm event are needed to further define the relationship between concentration and discharge.

On Watershed 6, estimates of average cation concentrations from continuous proportional sampling for a twelve-month period were only different by +8%, +6%, +2%, and +0.7%, respectively, for K^+, Na^+, Ca^{++}, and Mg^{++} of the amounts obtained by weekly hand samples. Thus, weekly samples appears to be sufficiently frequent to provide confidence in annual budgets for these cations.

Cation concentrations and seasonal losses

The general levels of cation concentrations in streams of undisturbed and treated watersheds were

TABLE 3. Average annual weighted concentrations of four cations (dissolved form) in stream water of four watersheds at Coweeta (June–May, 1969–71)

Vegetation type, watershed no.		Cation concentration ppm			
		Ca	Mg	K	Na
		1969–70			
Grass-to-forest succession,	W.S.6	1.081	.651	.540	1.135
Coppice,	W.S.13	.489	.256	.390	.653
White pine,	W.S.17	.531	.212	.404	.752
Mature hardwood,	W.S.18	.678	.297	.444	.948
		1970–71			
Grass-to-forest succession,	W.S.6	1.037	.644	.691	1.114
Coppice,	W.S.13	.451	.247	.472	.624
White pine,	W.S.17	.500	.213	.488	.763
Mature hardwood,	W.S.18	.619	.281	.516	.877

low (Table 3); in most cases, the cations were present in quantities less than 1 ppm. Nevertheless, there were some large differences in ionic concentrations between ecosystems. Using the stream chemistry data from Watershed 18 as baseline information for a mature ecosystem, the concentration levels of Mg^{++} on the grass-to-forest succession catchment (Watershed 6) were twice as large, and Ca^{++}, and Na^+ and K^+ were 20–50% greater. Past application of lime and fertilizer five years earlier is probably a principal reason for the higher concentrations in Watershed 6. It can also be postulated that greater concentrations may be partially attributed to the life cycle of the vegetation on the catchment which was primarily annual, and nutrients incorporated annually into the standing vegetative crop may be relatively small during most of the year. Cation concentrations leaving the coppice and pine-covered watersheds were considerably below those of Watershed 18 except for K^+ which was only slightly lower. Although possible reasons could be given as to why concentrations might be lower in streams of these ecosystems, supporting data are incomplete and arguments would be largely speculative at this time.

The monthly weighted average concentrations for the four cations in the undisturbed hardwood forest showed a distinct seasonal trend which was repeated over a two-year period (Figure 3). Maximum concentrations occurred during August and September followed by a sharp decline in November with minimum values occurring in the winter months. Thereafter, concentrations increased during the early spring and into the summer months until a maximum was again apparent in the fall season.

The influence of streamflow on the seasonal efflux of cations is shown by total monthly dissolved losses for the mature hardwood forest during both water years (Table 4). Because precipitation and hence streamflow are distributed rather uniformly throughout the year, significant losses of cations occurred during all months. Although losses do not show the

TABLE 4. Total monthly streamflow and dissolved cation losses for a mature hardwood forest (Watershed 18) at Coweeta, g/ha

	Total monthly streamflow (mm)		Ca++		Mg++		K+		Na+	
Month	1970	1971	1970	1971	1970	1971	1970	1971	1970	1971
June	90	73	746	593	311	246	474	440	1153	703
July	59	45	505	376	208	163	318	273	678	485
Aug.	83	39	737	332	295	144	456	247	917	424
Sept.	59	34	518	281	206	116	319	216	826	349
Oct.	55	57	510	397	192	190	292	388	501	529
Nov.	79	80	433	477	212	230	321	463	576	698
Dec.	95	68	474	385	234	184	315	320	621	581
Jan.	112	126	503	717	243	317	353	585	733	1076
Feb.	94	195	594	1064	250	474	399	912	1014	1555
Mar.	103	186	623	981	293	462	438	864	909	1459
Apr.	108	130	668	718	312	339	432	630	978	1167
May	70	104	516	694	231	323	358	518	645	914

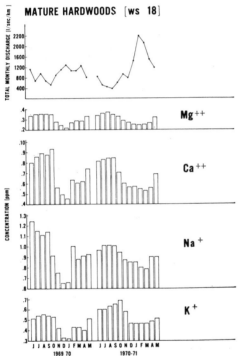

FIG. 3. Average monthly weighted concentration for four cations in stream water of watershed covered with mature hardwood.

distinct seasonal trends exhibited by concentrations, losses appeared to be greatest during the early spring and summer months when biologic functions are active and when stream flows are still relatively high. During the winter months when biological activity is low, streamflow is generally high, but losses tended to be lower for all cations considered.

The association between monthly concentration and discharge is somewhat variable within a year for the two-year period shown in Fig. 3, but can be summarized by regression analysis where monthly discharge accounts for 50–60% of the variation in monthly cation concentrations. Therefore, it appears that seasonal variation in discharge is partially responsible for seasonal variation in cation outflux, but the biotic components involved in mineral recycling within the ecosystem must also be a major factor affecting the seasonal outflux of ions.

Almost identical seasonal trends in monthly concentrations of all four cations were also found in the streams draining the coppice and white pine-covered catchments although the concentration amplitudes were dampened in both systems. The importance of biotic processes is thought to be demonstrated by the seasonal trends of K+ and Mg++ for the grass-to-forest succession watershed (Fig. 4), where the hydrology is essentially the same as for the other watersheds. As with the other three watersheds, these two cations showed a distinct seasonal trend which can be described by a sine wave function. That is, minimum concentrations appear during the late summer months with the occurrence of maximum concentrations in late winter. Calcium and Na+ showed no consistent pattern over the two-year span.

Annual budget and treatment response

Comparison of results among the four catchments is primarily a contrast in management histories. Annual total sediment losses were 33, 81, 258 and 152 kg/ha oven-dry sediments from the mature hardwood, pine, coppice and successional ecosystems, respectively, for the year 1969–70. Thus, sedimentation losses were greater by factors of about 3 for conversion to pine, 8 for second cycle hardwood coppice and 5 for the two-year-old successional stand com-

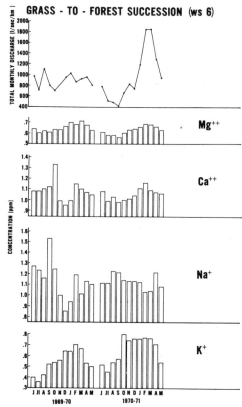

FIG. 4. Average monthly weighted concentration for four cations in stream water of a watershed in a grass-to-forest succession stage.

TABLE 5. Annual cation loss as sediment and the percent contribution to total loss for four Coweeta watersheds

WS #	Annual cation loss as sediment <2 mm, kg/ha				Sediment cation loss as % of dissolved & sediment losses			
	Ca	Mg	K	Na	Ca	Mg	K	Na
6	.14	.021	.006	.003	1.4	0.4	0.1	0.0
13	.09	.013	.004	.002	1.8	0.7	0.1	0.0
17	.05	.007	.003	.002	1.3	0.5	0.1	0.0
18	.05	.004	.002	.001	0.7	0.2	0.0	0.0

pared with the mature hardwood stand. Although treatment responses are partially obscured by different times since alteration, absolute losses were quite small in any case; particularly considering the degree of perturbation imposed on these three ecosystems in a relatively high rainfall regime. The export of cations by sediments comprised only about 1% or less of the total annual cation loss that can be considered biologically available (Table 5). The high infiltration rate characteristic of forest soils in the basin is a significant factor in reducing these losses.

The annual budgets for the four cations for each watershed (Table 6) reflect differences in the treatment history, hydrology and biology of the four ecosystems which cannot be adequately separated at this time. If the hardwood forest (Watershed 18) is taken as a mature ecosystem, that is in steady state with respect to inorganic chemical cycling, it is apparent from comparison that there was a net loss of all elements from the weedy, successional catchment. This observation is most evident for Ca^{++} and Mg^{++}. Conversely, data for the pine and coppice watersheds suggest a net gain (accumulation) in all cations; particularly for Ca^{++}, Mg^{++} and Na^+.

Arguments why greater concentrations and hence losses might be expected from the grass-to-forest succession catchment have already been briefly stated. The apparent accumulation of cations for the other two altered ecosystems is at least partially caused by a rapid biomass accretion. Because total loss is strongly dependent on total water flow, the net gain for the pine catchment is partly attributable to the significantly smaller amount of water passing the weir blade. A 15% greater evapotranspiration loss from the young pine compared to mature hardwood has been measured from the ecosystem; moreover, there may be some flow bypassing the weir. Although unmeasured flow for Watershed 17 may contribute to unmeasured cation loss, the difference in annual loss between Watersheds 17 and 18 is about 40%, a difference which cannot be attributed entirely to unmeasured flow. Thus, hydrologic factors, combined with lower concentrations in flow from the pine-covered catchment, produce relatively small cation outputs.

Total outputs from the coppice are somewhat larger than outputs from the pine, but still considerably less than for the mature hardwood forest. Differences in losses from the two hardwood ecosystems are partially due to lower average cation concentration differences (Table 3) from the coppice forest. However, concentration differences are partially offset by a 10% greater flow from the coppice stand than the flow expected from the original hardwood forest on this catchment.

It is clear even at this early stage in the study that the interaction of hydrologic and biological processes must be examined closely. Furthermore, it seems probable that an understanding of the flux of moisture and nutrients through the soil profile to the stream is an important key to the interpretation of mineral budget studies. It is significant that annual budgets for the young successional forest ecosystems suggest net gains in cations compared to the mature forest. The interaction of cation flux through these ecosystems with cycling within them is under study. Whether successional accretion of biomass is a mineral conserving mechanism, or if mature forest eco-

TABLE 6. Average annual cation budgets for one undisturbed and three manipulated Coweeta watersheds during two water years (June–May, 1969–77), kg/ha

Vegetation type and watershed number	Ca^{++}			Mg^{++}			K$^+$			Na$^+$		
	In-put	Out-put	Net difference	In-put	Out-put	Net difference	In-put	Out-put	Net difference	In-put	Out-put	Net difference
Grass-to-forest succession (6)	5.73	10.40	−4.68	1.20	6.26	−5.06	3.02	5.98	−2.96	5.11	10.86	−5.75
Coppice (13)	5.76	5.01	+0.75	1.34	2.68	−1.34	3.25	4.62	−1.38	5.40	6.82	−1.42
White pine (17)	6.51	4.10	+2.42	1.34	1.69	−0.35	3.32	3.56	−0.24	5.70	6.06	−0.36
Mature hardwoods (18)	6.16	6.92	−0.76	1.26	3.09	−1.82	3.16	5.17	−2.02	5.40	9.74	−4.34

systems develop tighter, more closed biogeochemical cycles than successional stages (Odum 1969) is not yet shown by these studies. Decoupling the physical and biological processes that account for the observed behavior must await completion of biota and mineral studies now in progress within these ecosystems.

Clearly, there may have been accelerated losses immediately following treatments which are not reflected in these data. If so, then readjustment toward present rates within a few years is suggested. It is also likely that large pulses of materials may leave the watersheds during especially severe storms of intensities greater than occurred during this study. Considering the implications of forest management on water quality, perhaps the most significant overall result suggested by the study is the fact that drastic alterations of these forest ecosystems have not resulted in substantial current rates of loss of cations by erosion or drainage waters. It is reasonable to expect our results to approximate long term response to these perturbations.

Comparison with other selected eastern watersheds

Precipitation in the southern Appalachians and at Coweeta is very high compared to most areas in the Eastern United States; but the cation concentrations and flux into and out of the mature hardwood-covered watershed at Coweeta are similar to values reported elsewhere. Rainfall data from eastern North Carolina (Gambell and Fisher 1966) show that input values were quite similar to those at Coweeta but output values were approximately 2.5 times greater for Ca^{++} and Na$^+$. Although the two sites are at about the same latitude, they are grossly different in geology and land use history so that agreement in annual cation budgets would be unexpected. The Hubbard Brook Experimental Forest (Likens et al. 1970) is on similar igneous bedrock, although soils were derived from glacial till and are shallow. In contrast to Coweeta, snow is a major form of precipitation at Hubbard Brook and dominates stream hydrology and nutrient flux for half of the year. Comparable annual nutrient budget studies for mature hardwood forests at Hubbard Brook (Table 7) show a somewhat higher net loss of Ca^{++}, lower net

loss of K$^+$, but budgets are similar to Coweeta results for Mg^{++}, and Na$^+$. Comparable measurements at Oak Ridge National Laboratory (Swank and Elwood 1971) are similar for K$^+$ and Na$^+$, but very much greater for Ca$^+$ and Mg$^+$. The Oak Ridge drainage, while only 80 km northwest of Coweeta, is underlain by dolomitic limestone, a substrate which clearly contributes to large losses of Ca^{++} and Mg^{++}.

These three sites offer comparisons on similar igneous substrate but over 9 degrees of latitude, versus similar latitude but contrasting substrates. Such comparisons provide an insight into biogeochemical cycling and, when coupled with studies of biological and physical processes, should contribute to improved resource management.

ACKNOWLEDGMENTS

The authors express their appreciation for financial support for this research in part by the Southeastern Forest Experiment Station and the National Science Foundation (GB-7918) and in part by the Eastern Deciduous Forest Biome, US-IBP, funded by the National Science Foundation under Interagency Agreement AG-199, 40-193-69 with the Atomic Energy Commission–Oak Ridge National Laboratory.

LITERATURE CITED

Bormann, F. H., and G. E. Likens. 1967. Nutrient cycling. Science 155: 424–429.

Day, F. P. 1971. Vegetation of a hardwood watershed at Coweeta. M.S. Thesis. University of Georgia, 145 p.

Devereux, R. E., E. F. Goldston, and W. A. Davis. 1929. Soil survey of Macon County, North Carolina. U. S. Dept. Agr. Bur. Chem. and Soils, No. 16. Series 1929, 21 p.

Dils, R. E. 1957. A guide to the Coweeta Hydrologic Laboratory. SE Forest Exp. Sta., Forest Serv., U. S. Dept. of Agr. 40 p.

Douglass, J. E., D. R. Cochrane, G. W. Bailey, J. I. Teasley, and D. W. Hill. 1969. Low herbicide concentration found in streamflow after a grass cover is killed. U.S.D.A. Forest Serv. Res. Note SE-108, 3 p.

Dunford, E. G., and P. W. Fletcher. 1947. Effect of removal of streambank vegetation upon water yield. Trans. Amer. Geophys. Union 28: 105–110.

Gambell, A. W., and D. W. Fisher. 1966. Chemical composition of rainfall, eastern North Carolina and southeastern Virginia. U. S. Geol. Surv. Water Supply Pap., 1535K: 1–41.

Hewlett, J. D., H. W. Lull, and K. G. Reinhart. 1969.

TABLE 7. Average annual cation budgets for undisturbed watersheds at three sites in the Appalachian Highlands. Oak Ridge values based on one water year (June–May, 1969–70), Coweeta values based on two water years (1969–71), and Hubbard Brook values based on four water years (1963–67). All values in kg/ha

Site	Vegetation type	Ca++			Mg++			K+			Na+		
		Input	Output	Net dif-ference	Input	Output	Net dif-ference	Input	Output	Net dif-ference	Input	Output	Net dif-ference
Coweeta	Mature Hardwoods	6.2	6.9	−0.8	1.23	3.1	−1.8	3.2	5.2	−2.0	5.4	9.7	−4.3
Hubbard Brook[a]	Mature Hardwoods	2.6	10.6	−8.0	0.7	2.5	−1.8	1.4	1.5	−0.1	1.5	6.1	−4.6
Oak Ridge[b]	Mature Hardwoods	41.4	100.0	−58.6	3.8	50.5	−46.7	5.6	4.0	1.6	6.1	5.0	1.1

[a] Johnson, N. M., et al. (1968).
[b] Swank, W. J. and J. W. Elwood (1971).

In defense of experimental watersheds. Water Resources Res. **5**: 306–316.

Hibbert, A. R., and G. B. Cunningham. 1967. Streamflow data processing opportunities and application. Int. Symp. on Forest Hydrol. Proc. 1965: 725–736. Pergamon Press, Inc., New York.

Hibbert, A. R. 1969. Water yield changes after converting a forested catchment to grass. Water Resources Res. **5**: 634–640.

Hoover, M. D. 1944. Effect of removal of forest vegetation upon water yields. Trans. Amer. Geophys. Union **6**: 969–975.

Johnson, N. M., G. E. Likens, F. H. Bormann, D. W. Fisher, R. S. Pierce. 1969. A working model for the variation in stream water chemistry at the Hubbard Brook Experimental Forest, New Hampshire. Water Resources Res. **5**: 1353–1363.

Kovner, J. L. 1956. Evapotranspiration and water yields following forest cutting and natural regrowth. Soc. Amer. Forest. Proc. 106–110.

Lee, R. 1970. Theoretical estimates versus forest water yield. Water Resour. Res. 6: **5**: 1327–1334.

Likens, G. E., F. H. Bormann, N. M. Johnson, and R. S. Pierce. 1967. The calcium, magnesium, potassium and sodium budgets for a small forested ecosystem. Ecology **48**: 772–785.

Likens, G. E., F. H. Bormann, N. M. Johnson, D. W. Fisher, and R. S. Pierce. 1970. Effects of forest cutting and herbicide treatment on nutrient budgets in the Hubbard Brook watershed-ecosystem. Ecol. Monogr. **40**: 23–47.

Odum, E. P. 1969. The strategy of ecosystem development. Science **164**: 262–270.

Reinhart, K. G. 1967. Watershed calibration methods. Int. Symp. on Forest Hydrol. Proc. 1965: 715–723, Pergamon Press, Inc., New York.

Swank, W. T., and J. W. Elwood. 1971. Seasonal and annual flux of cations for forested ecosystems in the Appalachian Highlands. Unpublished paper, Second National Biological Congress, Miami, Florida 23–26.

Swank, W. T., and J. D. Helvey. 1970. Reduction of streamflow increases following regrowth of clearcut hardwood forests. In Symposium on the results of research on representative and experimental basins. UNESCO-AIHS Pub. (Assoc. Int. Hydrol. Sci.) **96**: 346–360.

Swank, W. T., and N. H. Miner. 1968. Conversion of hardwood-covered watersheds to white pine reduces water yield. Water Resources Res. **4**: 947–954.

Swift, L. W., Jr. 1970. Comparison of methods for estimating areal precipitation totals for a mountain watershed. Coweeta Hydrologic Laboratory, Franklin, North Carolina, unpublished data and report.

Trewartha, G. T. 1954. An introduction to climate. McGraw-Hill Book Co., New York. 402 p.

Wilm, H. G. 1943. Statistical control of hydrologic data from experimental watersheds. Amer. Geophys. Union Trans., Part 2: 618–624.

17

Reprinted from *The Ecosystem Concept in Natural Resource Management,* George Van Dyne, ed., Academic Press, Inc., New York, 1969, pp. 77–93

Chapter V A Study of an Ecosystem: The Arctic Tundra

ARNOLD M. SCHULTZ

I. INTRODUCTION

A few years ago some ecologists predicted that the ecosystem bubble would soon burst, after which investigators would "go back to tried and true methods" for probing nature. Others have said there is nothing new in the ecosystem concept at all except a fancy name and another language for students to learn. Like any other science, ecology has had its share of fads, some of which turned out to be nothing but old ideas dressed in new semantics.

The concept of system is indeed very old but not so the area we have come to call loosely systems analysis. The way in which modern systems analysis has been applied to the ecosystem—just since 1962—is truly fantastic. It comes closer to a breakthrough than any other event in the history of ecology.

We now have a conceptual tool which allows us to look at big chunks of nature as integrated systems (Schultz, 1967). Also we now have the technical tools to handle the information obtained in this framework. Ecologists no longer fear complexity. These innovations have come along not any too soon. The alarm of "Silent Spring" is still ringing in our ears. Finally we realize that nature is not as piecemeal as science is.

Where does one begin, to study an ecosystem? The first part of this chapter explains some concepts fundamental to ecosystem study. In a discussion of models, it tells how the author decided to look at the tundra. The second part gives our first-hand experience in studying a tundra ecosystem in northern Alaska. In a discussion of results, it tells what we decided to look for.

II. THE REAL SYSTEM AND THE MODEL

A. The Language of Systems

The concept of system dates back to the very dawn of thought although its language is quite modern. From the beginning man has perceived only wholes; his penchant for taking them apart is rather a recent development but his ability to put the parts together again has scarcely developed at all. Our language of systems derives from these three ways of looking at things: taking things apart (analysis), assembling parts into wholes (synthesis), and seeing things only as wholes.

Therefore a system is a whole thing; it has three kinds of components (Fig. 1).

The elements of a system are the physical objects, often thought to be the "real" parts. In an ecosystem the elements are space–time units in that they occupy some volume in space for a certain length of time. Rain-

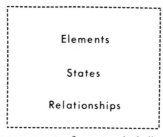

Fig. 1. The components of a system including boundary.

drops, sand grains, and mosquito larvae are examples of elements in a system. Each element has a set of properties or states, e.g., number, size, temperature, color, age, or value. Between two or more elements or between two or more states there are relationships which can be expressed as mathematical functions or less formally, with plain English verbs.

A system can now be defined as a set of elements together with relations among the elements and among their states (Hall and Fagen, 1956).

The term "set" implies that the components can be bounded. In the diagram the boundary is permeable to indicate an open system into which elements can enter from outside.

We might think that the elements, being physical entities of some kind such as nitrogen ions or living organisms, are the real and important components while the others are mere abstractions. In system thinking, however, we put more emphasis on the state. In a thermostatic control system, for example, it is not the air in the room but its temperature that is important; nor is it the hardware of the furnace but its state of being off or on, low or high that we consider. The elements of most interest to us in a system are capable of taking on two or more alternative states. In other words, the element is a variable and over time its difference in state is what we observe, measure, and record. We cannot record the element itself.

There would be no need to invoke the systems concept were it not for the crucial component relationship. A system becomes a whole thing only because its elements and states are connected together in some way. Thus, by understanding the linkages we see how the whole system works.

We like to use this very simple model of a system (Fig. 1) as a reminder of what is real, what is abstraction, and why we have the systems concept in the first place.

How different is this approach from the one used in science before systems analysis? Scientists have always studied systems but they have shied away from complex ones. Early physicists learned that the mathematics needed to describe the attractions of more than two bodies at a time were beyond their powers of calculation or too time-consuming to carry out. Ecology is the science of relationships; yet ecologists, though awed by the complexity of nature, have long used methods which treat one factor at a time. How can one unravel the many interrelationships in a species-rich temperate forest, for example, or for that matter, in the arctic tundra?

B. Homomorphic Models

It is improbable that any ecosystem will ever be studied in its full complexity. In some systems there may be thousands of kinds of organisms and billions of interacting individuals. If we had spectroscopic X-ray

. . . .   **Atoms, Molecules**
                                                     Cells, etc.

(a) (b) (c) (d) (e) (f) (g) (h) (i) (j) (k) (l) (m) (n) (o) (p)   Individuals

(a + b) (c + d) (e + f) (g + h) (i + j) (k + l) (m + n) (o + p)   Species, Genera, etc.

(a + b + c + d) (e + f + g + h) (i + j + k + l) (m + n + o + p)   Trophic Levels

                                                     Living & Non-living
(a + b + c + d + e + f + g + h) (i + j + k + l + m + n + o + p)   **Biomass**

(a + b + c + d + e + f + g + h + i + j + k + l + m + n + o + p)   Entire System

FIG. 2. Homomorphic models showing six levels of discrimination of system parts.

eyes we could see yet finer subdivisions: cells, molecules, atoms, perhaps even electrons. Obviously, to study the many states and relations in such a complex system would be too much for our best computers. The system must be simplified.

In the past ecologists would select a certain few parts of a larger system for study, for example, a species population or a relationship such as plant succession or competition. This kind of simplification falls short of studying the whole system; the other parts and the other relations that occur are completely neglected.

A complex system can be simplified by making a homomorphic model of it. Here the system remains intact. Its parts are discriminated at some level that can be handled conveniently. Figure 2 shows the ABC's of homomorphic models.

In the row second from the top are sixteen distinct individuals. Not everyone can see them, however, so my statement is only a point of view. Suppose the letters *a* to *p* represent grass plants in a dense sward. You would not be able to distinguish individuals at all. The next row down shows the units combined by twos into superunits. We can think of these as populations of taxonomic groups (species, genera, families) or of physiognomic ones (herbs, shrubs, trees). All of the finer units are still present but undiscriminated at the next higher level and so on through the hierarchy—trophic levels, then living and nonliving biomass, until the entire system emerges as a Gestalt and there is no discrimination of parts whatsoever.

Homomorphic models are not designed just for lazy taxonomists. There is a very practical reason for looking at systems this way. Let us look at the top and bottom lines of Fig. 2. At the top, every possible state has been distinguished but the sheer bulk of information is so overwhelming that we can make little use of it. At the other end, all the states are fused into one grand but platitudinous expression and all you can say for it is "There it is!" In between are a number of handy simplifications. On the fine end some realism is retained. On the coarse end we gain generality. We now have a set of models which allows us to coordinate all the discoveries made by specialists on the separate parts of the system. Any part can be handled as a black box coupled into the system at the appropriate level of organization.

C. The Tundra as a Simple Ecosystem

The history of scientific investigation of arctic tundra follows a pattern different from that of other regions of the world. Because of the sparse indigenous human population, pressure for agricultural research, as we know it in lower latitudes, has not occurred. Basic studies from diverse disciplines and with broad objectives were initiated; they did not assume the single-minded purpose of maximizing crop yields.

Many ecologists were lured to the tundra because it was supposed to be simple. Here in the arctic could be found a paucity of growth forms and species, shallow soil, a short growing season, an extreme climate, and essentially no disturbance of the landscape by humans. If ever the total processes of nature could be put in order, it would be done for the tundra. Let us see how simple the tundra really is.

I can cite numerous detailed descriptions of coastal tundra in the vicinity of Point Barrow, Alaska. These include my own investigations and those of my students which began in 1958 (Schultz, 1964; Pieper, 1964; Van Cleve, 1967) and the excellent work on microtines and their predators by F. A. Pitelka and others begun in 1952 (Pitelka et al., 1955; Pitelka, 1958; Maher, 1960). Intensive studies on soils, meteorological phenomena, bacteriology, and other aspects of the tundra have been going on at Point Barrow since 1950.

It is not my intent here to redescribe the tundra. Rather I shall point out the proximity between the real ecosystem and its simplified model.

It would be possible to print on two pages of this book a list of all the species of plants, animals, and microorganisms known to occur in the massive ecosystem under study. Perhaps it would go on one page. The list could be reduced to about ten species, and still include 90% or more or the biomass in each major group. It would include sphagnum moss among the lower plants, several grasses and sedges among the higher plants, the brown lemming and pomarine jaeger among herbivores and carnivores, respectively. For quantitative studies of energy and major nutrients, an analysis of samples from these few predominant groups would give essentially the same results as would a total ecosystem study. To put it another way, the properties or states of the trophic levels are at least 90% predictable from just one or two of their component parts.

We have already seen how the complexity of a system is determined by the number of distinguishable parts. At the species population level of discrimination, the tundra is fairly simple compared to other ecosystems of the world. But there is another determinant of complexity: the number of recognizable states which the parts can assume. Some examples are depth of active soil layer, exchangeable calcium in the soil, phosphorus level of forage plants, population density of lemmings, jaegers, and owls, and decomposition rate of organic matter. The number of states depends entirely on the yardstick and stopwatch used to measure. The investigator can make it as simple or complex as he wishes.

We must come to the conclusion that the easiest way to analyze the tundra or any other ecosystem is to lump all populations of organisms into trophic levels and the resources (atmosphere and soil) into convenient compartments. For each compartment a sample, in exactly the proportion in which the various populations occur, can be digested for nutrients and bombed for energy. Egler (1964) calls this the "meatgrinder" approach. A wealth of information inside each box is conveniently ignored. Our primary interest is directed to relationships between boxes. How much phosphorus flows from one tropic level to another? In this framework, one system is as simple as another—tundra or temperate zone grassland.

D. Isomorphic Models

We can now return to the third component of systems, relationships. The homomorphic model does not help us here; it is concerned only with the power of resolution used for the elements and their states. We need an isomorphic model for studying relationships.

An isomorphic model is a map. A road map of New Mexico is a model of the real geographical area of the state. It shows among other things the distance and direction between Albuquerque and Santa Fe. By inference from the kind of highway, it also shows the rate of traffic flow between the two cities. So a map can be a flow chart.

Figure 3 is a map of a homomorphic model of an ecosystem—two kinds of models combined. This one has been proposed by the subcommittee on Terrestrial Productivity for the International Biological Program as a tentative model for studying all terrestrial ecosystems. The arrows indicate paths of nutrient or energy flow from one box to another. Research on ecosystems can be standardized and routinized by using

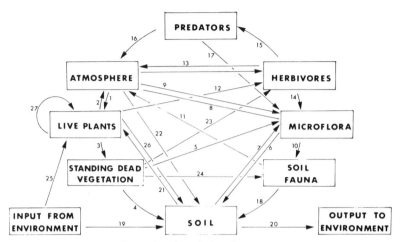

FIG. 3. Isomorphic model mapping the relations between ecosystem compartments.

such a scheme. One can use the numbers on the arrows as a checklist to see which relations of Fig. 3 one has forgotten to measure.

Let us consider what kinds of relationships there are and how they are measured.

Arrow No. 12 represents a process with a familiar name: grazing. The model suggests that something travels physically along the path from *live plants* to *herbivores*. We measure this something in units of chemical compounds, biomass, or energy. It is measured as a rate—so many pounds of dry matter per unit time. Rarely is the transfer measured directly. The *live plants* box is weighed at time one (T_1) and again at time two (T_2). The difference in state between T_1 and T_2 is read as a gain or loss. A series of exclosures and enclosures will indicate what proportion has been gained or lost via pathways No. 1, 25, 2, 3, 12, or 21. The same is done for *herbivores* and every other box in the model. We should note here that enclosures and exclosures are experimental tools not to be pre-empted by range managers. The bacteriologists' agar plates and the radiobiologists' isotope tracers are based on the same principle as a fenced plot.

Isomorphic models can be constructed for many kinds of relationships. A correlation matrix is such a model; the coefficients show spatial or phenetic distances. Regression coefficients and ratios may be used to express causation, conjunction, or succession of events. Finally, a model can be set up with vernacular expressions like "boy meets girl," or more appropriately, as in André Voisin's definition of grazing, "cow meets grass." You will recognize pathway 15 as owl eats lemming, 5 as bacterium rots straw, and 20 as ocean takes soil. I am not being trite. If we could clearly establish verbally expressible relationships between all of the parts we would indeed understand ecosystems very well and could develop valuable theorems about them. The numbers would then be superfluous.

E. System Boundary and Environment

Earlier in this chapter an ecosystem was described as a space-time unit—a volume that exists in time. But you cannot see an ecosystem in the same way that you can see, for instance, a lemming. The skin clearly marks the boundary of the animal. It is simply a matter of perception and all observers would agree as to what is lemming and what is not lemming.

Not all ecosystems have a skin you can touch. Defining the boundaries of his ecosystem is one of the biggest problems an ecologist has to face. Some insight into the problem can be obtained by considering the nature of boundary for a generalized system (Fig. 4).

The system has within it all the elements the observer is interested in (the set, by definition). These elements have a certain density or concentration. Outside, the elements are different either in kind or in concentration. The observer is not interested in these to the same degree. If they had been of the same kind of elements or had the same density as inside, he would have included them in the system. Thus, the environment (e) (outside) is different from the system (s) itself. It is not entirely true that the observer is disinterested in the environment; he cares about what effect it has on the system. There is pressure for some of the elements to cross over the boundary.

Think of the boundary as a membrane with a certain thickness Δx. It has a texture which determines how easily any of the elements can flow across it. This can be thought of as a permeability factor (m). Since there are different densities of elements inside and outside, there must be a concentration gradient across the boundary. If Δx is narrow, the gradient will

FIG. 4. The boundary between system and environment.

be steep. The flow of elements across the membrane is governed by the
following three factors (Jenny, 1961):

$$\text{Flow} = -\left(\frac{\text{concentration}_e - \text{concentration}_s}{\Delta x}\right) m$$

This can be illustrated with an exclosure in a pasture. The fence, of
course, represents the boundary. The exclosed plot has no animals, but
outside are 10 steers per acre. Some of these are pressing against the
fence. The gradient is sharp—from 10 to 0 over a distance the diameter
of the wire. If the fence is strong ($m = 0$), the flow will be zero no matter
what the steer pressure is.

In most ecosystems there is no actual membrane separating system
from environment. The boundary is imaginary and is located at the convenience of the observer. One choice is to set the boundary in a zone
where there is no gradient (where concentration$_e$ − concentration$_s = 0$).
Now any crucial variations in density are trapped within the system and
must be measured as state transformations. The other approach is to use
a natural boundary (where $m \rightarrow 0$) such as the shore line of an island.
Here the problem is that the permeability coefficient may be low for
only one kind of element, and as we have already seen, a complex may
have a thousand kinds.

F. The Role of the Observer

One of the tenets of systems research is that the observer is always a
part of the system. Refer back to the four diagrams and picture the role
of the observer. He is an element of the system, with definite properties
and unique relations to the other elements and states (Fig. 1). The observer decides the level of discrimination to be used in the study (Fig. 2).
He selects from the large number of possible relations just those he wants
to measure (Fig. 3). He fixes the boundary of his system according to his
resources and his interests (Fig. 4). Apart from the observer there can be
no unique ecosystem. No one can go to Point Barrow and see the same
system that I see. It follows that I cannot possibly describe to you the
real or absolute ecosystem, only my model of it.

Within the system he has circumscribed, the observer looks for time-
invariant relationships. During the investigation he records the activity
of the system. This includes haphazard events which may happen only
once, activities which occur frequently, and those which occur every
time. These represent, respectively, the temporary, the hypothetical, and
the permanent or real behavior of the system. In graphical form, the three
kinds of activity can be shown as scatter diagram, prediction curve, and
equation for an absolute law.

It is within this framework that I present some of the results of my research.

III. THE TUNDRA AS A HOMEOSTATIC SYSTEM

A. Cyclic Phenomena

Many of the activities of the tundra ecosystem under study are cyclic, with a periodicity of three or four years, but with varying amplitudes between cycles. We can think of a cycle as a series of transformations of state (Ashby, 1963). Thus, if a subsystem (compartment) has four clearly recognizable alternative states, a, b, c, and d, and the transformation always goes a → b → c → d → a → b, etc., then the sequence of states is a cycle. This can be shown kinematically:

$$
\begin{array}{ccc}
a & \to & b \\
\uparrow & & \downarrow \\
d & \leftarrow & c
\end{array}
$$

or, when put on a time scale, as a sine wave:

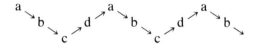

and so on.

At Point Barrow we have good records of yearly lemming population densities, starting from 1946 (Fig. 5). Lemming peaks occurred in 1946, 1949, 1953, 1956, 1960, and 1965. Neither the amplitude nor the wave-length of these cycles is always the same. Yet there are some striking similarities. Using the systems language given above with a 1-year time interval, generally we can recognize four states: a, high density; b, very low density; c, low density; and d, medium density. The states can be named by reading the histogram, without any knowledge of lemming population dynamics or life histories. Sometimes c is missing and once c and d are transposed. Always the high year was immediately followed by a very low year. But even during the low years, there were found local "pockets" supporting a denser population; for example, on the outskirts of the Eskimo village of Barrow, the fluctuations in lemming numbers have never been as pronounced as on the open tundra. Also of significance, some areas are out of phase with Point Barrow. At a point 100 miles east of Barrow, the population peaked in 1957, a year after the Point Barrow high. By 1960, it was in phase again.

What can be said about lemming population cycles? By Ashby's criterion, we have definitely observed cycles; but an engineer would say,

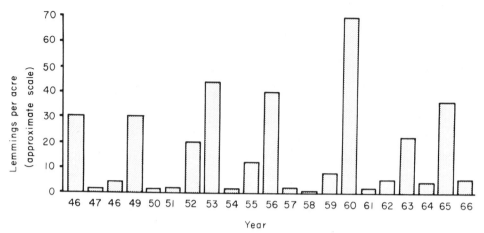

FIG. 5. Lemming population cycles at Point Barrow, Alaska, from 1946 to 1966.

if he saw the waves on an oscilloscope, that there was a lot of noise on the channel.

We have data on standing crop of plants (tops only), starting from 1958. Clippings were made in the vicinity of traplines used for the lemming census. Ninety percent of the dry matter was contributed by three species (*Dupontia fischeri, Eriophorum angustifolium,* and *Carex aquatilis*); these same species constitute the bulk of the lemming diet. When a graph is constructed for standing crop each year at phenologically equivalent dates, a cyclic pattern appears. Moreover, the pattern is in synchrony with that of the lemming population cycle, noise and all. A short lag develops between the two curves as the high point approaches. I do not want to explain the facts at this time, but I should say, in passing, that the high correlation between forage yield and lemming stocking rate would not come as a surprise to a range management audience.

Samples of the material clipped for production records were analyzed for total nitrogen, phosphorus, calcium, and several other elements. Concentrations of nutrients in the herbage, at any given phenological stage (i.e., date) increased through the year corresponding to the peak lemming year, then dropped to low values the year after, only to increase again. Figure 6 shows the activity for phosphorus superimposed on the histogram of lemming population density. Within a season, nitrogen, phosphorus, and potassium decrease percentagewise as grasses mature, while calcium increases. The line in Fig. 6 should not be construed as a continuous increase in phosphorus level from 1957 through early 1960.

Calcium, potassium, and nitrogen show the same trends as phosphorus. Due to greater plot-to-plot variability, the nitrogen data are not as sig-

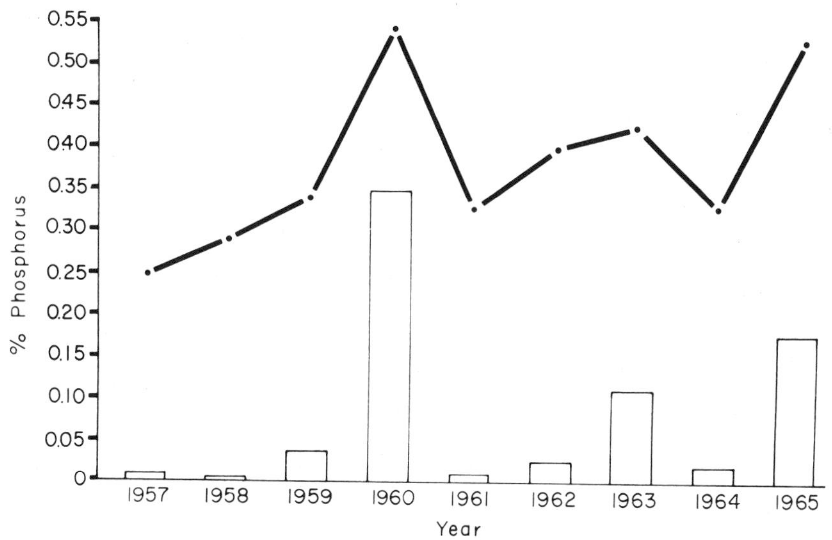

FIG. 6. Phosphorus levels in forage at Point Barrow, Alaska. Bars represent relative lemming numbers.

nificant as are those of the other three elements, but the trend is nevertheless the same. Magnesium and sodium show no relationship to lemming numbers at all, nor were the data cyclic.

Still another activity studied was decomposition of organic matter on the soil surface. This, too, turned out to be a cyclic phenomenon, and correlation with the activities already mentioned is high.

I have given a rough sketch of the behavior of the tundra, as discovered by survey techniques. The observations seem to fit closely a hypothesis of synchronous cycles. But at this stage, the results could be spurious. The close fit might result from artifacts of sampling. The transformation sequence a → b → c and/or d → a might occur frequently just by chance.

The next step is experimental: to introduce a disturbance at any one of the compartments and watch for reverberations throughout the system.

B. Experiments in Stressing the System

If the fluctuations in herbage production and nutrient level are related to immediate grazing history, then the cyclic aspect should disappear when grazing is eliminated. A simple exclosure, in effect, removes the herbivore from the system.

In 1950, a series of exclosed plots was established, alongside paired plots open to normal grazing (Thompson, 1955). Records kept for 13 years show cyclic variation on the outside paired plots, while the fenced plots show a constant decline. Since 1958, percentages of phosphorus, calcium, potassium, and nitrogen in the herbage from the grazed plots show the same marked cycles that occur elsewhere on the tundra (see Fig. 6). By comparison, year-to-year fluctuations inside the exclosures are slight and not cyclic.

With regard to decomposition of litter, the outside plots responded as did the tundra on the whole; inside the exclosures, decomposition rates were low and constantly decreasing.

An unexpected bonus came from the exclosure experiment. It gave an opportunity to assess the effect of lemming activity on the depth of thaw. By comparing, at the time of maximum thaw, soil depths inside and outside exclosures, I could separate the lemming-caused (within-system) effects from the summer temperature (environmental) effects. The results were most interesting. During a peak lemming year, the thickness of the active soil layer was maximum and it gradually diminished to the shallowest point the year before the next peak.

A second experiment was to stabilize artificially the fluctuating nutrient levels in the soil. This was done by fertilizing annually 6 acres of tundra with nitrogen, phosphorus, potassium, and calcium. Heavy applications were made to make sure that the variations in native soil nutrients were completely masked. What effect would this kind of disturbance have on primary and secondary production?

Net primary production, for the 4 years studied, was stabilized at a level 3–4 times that of the control plot. Annual variation was obliterated. Herbage quality was also stabilized. Protein levels, for example, were 4–5 times those of the vegetation of the control plot. Percent of calcium and phosphorus in the green tissue at equivalent dates remained high and constant in the four years.

The first fertilization was applied in 1961. No animals were seen either on fertilized or on control plots in 1961 or 1962. In 1963, animals were abundant all over the tundra (see Fig. 5 or 6), while in 1964, they were generally sparse. Immediately after the snowmelt in 1964, 30 winter nests per acre were counted on the fertilized area, none on the control plot, and less than 1 per 10 acres on the tundra in general. However, jaegers had found this 6-acre pantry and picked it clean. The few survivors observed at the time of the winter nest survey were large and fat. In 1965, a year of high lemming density all over the tundra, lemmings were abundant both on the fertilized and unfertilized control plot.

C. Hypotheses of Ecosystem Cycling

Only a fraction of the information collected so far can be presented in this paper. For the sake of brevity, I have shorn away all evidence of

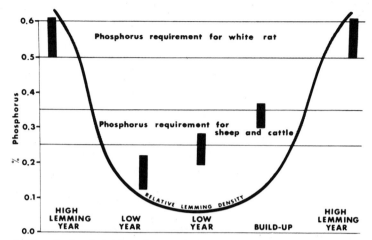

FIG. 7. Hypothesis relating lemming populations and nutritional quality of forage. Lemming density curve is generalized, as are the black bars showing phosphorus in forage.

"noise" and shown, so to speak, only the slick regression lines. These represent the hypothetical behavior of the system compartments.

With the evidence at hand, let us develop a more general hypothesis to explain the synchronous cycles apparent in our ecosystem. Tentatively it might be called the nutritional threshold hypothesis (Fig. 7) but such a name places undue emphasis on just one part of the system.

Let us review a generalized 4-year cycle.

1. Early in summer of the high lemming year, the forage is calcium- and phosphorus-rich. Because of high production and consumption, much of the available calcium and phosphorus is tied up in organic matter. The soil has thawed down deep into the mineral layer because grazing, burrowing, and nest-building has altered the albedo and insulation of the surface.

2. The next year, not only is forage production low, but also the percent of calcium and phosphorus in the diet is below that which would be required for lactation by sheep or cattle. Nutrients in organic matter have not yet been released by way of decomposition.

3. The following year production is up, the plants are recovering from the severe grazing two years earlier, and dead grass from the previous year insulates the soil surface. At the same time, decomposition of that dead material is speeding up. Forage quality is still quite low. Whether there is enough calcium and phosphorus in the diet to support lemming reproduction and lactation depends on how closely the species resembles domestic livestock, on the one hand, or the laboratory rat, on the other, in mineral requirements.

4. In the fourth year plants have fully recovered from grazing. Forage species accumulate minerals in their stem bases. Freezing action concentrates solutes in upper soil layers. Dead grass from several years has accumulated and plant cover is high; soil surface is well insulated and the thawed layer is very shallow. Decomposition rate is high. Calcium and phosphorus (and also potassium and nitrogen) content of forage is satisfactory for reproduction. There is enough food to support a large population of herbivores.

Next, the sequence is repeated.

Not until a nutritional threshold has been reached can a large lemming population build up. But the population does not keep getting bigger and bigger. This would be disastrous to the vegetation. So a deferred-rotation grazing scheme is built into the system. No grazing at all would also be disastrous to the vegetation and to the soil as well. Predators play a role at the time of herbivore decline. Indeed, all parts of the system play a role. It is a homeostatically controlled system.

This is only a hypothesis. It can be tested in the framework of the ecosystem concept: First, by showing that all parts bear some relationship to all others; second, by experimentally stressing the system to see how it adapts to disturbance; third, by opening the black box and studying its physiology—that is, explanation of a phenomenon at a lower level of organization.

D. Contemporary Hypotheses on Cycling

Needless to say, the nutritional threshold hypothesis is at variance with several prominent hypotheses that have been advanced in recent years. The hypotheses of Christian and Chitty minimize the role played by energy and nutrition in controlling animal populations. The stress hypothesis of Christian (1950) associates population declines with shock disease and changes in adrenal–pituitary functions. The increase in adrenal activity at high population densities lowers reproduction and raises mortality. The hypothesis involving genetic behavior (Chitty, 1960) suggests that when animal numbers fluctuate, the populations change in quality. This is brought about through selection resulting from mutual antagonisms at high breeding densities.

All hypotheses concerning animal population cycles have in common the notion of feedback. There are two kinds of feedback, negative and positive. The kind generally involved in control mechanisms is negative or deviation-counteracting while the "vicious circle" kind is positive or deviation-amplifying (Maruyama, 1963). Most ecosystems have both kinds. We can think of loops running through a series of compartments

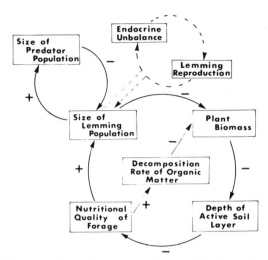

FIG. 8. Feedback-loop model showing homeostatic controls in the arctic tundra ecosystem.

so that the state of each compartment either counteracts (−) or amplifies (+) the change of state of the next (Fig. 8).

It is probable, in fact common, that any given element will be stationed on several loops. It may be checked via one loop and amplified via another. Consider the herbivore compartment of any ecosystem. Amount of forage, its quality, and availability of space are all positive; predators and pathogens are negatively related. In some cases, a density control mechanism operates within the compartment itself—as described by the stress theory or the genetic selection theory. This is simply an additional loop in the system. There is no reason why a control system should have but one governor.

The idea of one cause–one effect is left over from the nineteenth century when physics dominated science. The whole notion of causality is under question in the ecosystem framework. Does it make sense to say that high primary production causes a rich organic soil and a rich organic soil causes high production? This kind of reasoning leads up a blind alley. We are dealing with the different dependent properties of the same system. Only things outside the system can cause something to happen inside. For the same reason, we cannot say that the lemmings are the driving force, any more than the vegetation, the soil, or the microflora, in making the ecosystem tick.

ACKNOWLEDGMENTS

Work on the coastal tundra near Point Barrow, Alaska, was done under grants from the Arctic Institute of North America, the National Science Foundation, and the Office of

Naval Research. R. D. Pieper, who participated in some of these studies, is thanked for presenting this paper at the symposium while the author was in Great Britain.

REFERENCES

Ashby, W. R. 1963. "An Introduction to Cybernetics." Wiley, New York. 295 pp.

Chitty, D. 1960. Population processes in the vole and their relevance to general theory. *Can. J. Zool.* **38**, 99–113.

Christian, J. J. 1950. The adreno-pituitary system and population cycles in mammals. *J. Mammal.* **31**, 247–260.

Egler, F. E. 1964. Pesticides—in our ecosystem. *Am. Scientist* **52**, 110–136.

Hall, A. D., and R. E. Fagen. 1956. Definition of system. *Gen. Systems Yearbook* **1**, 18–28.

Jenny, H. 1961. Derivation of state factor equations of soils and ecosystems. *Soil Sci. Soc. Am. Proc.* **25**, 385–388.

Maher, W. J. 1960. The relationship of the nesting density and breeding success of the pomarine jaeger to the population level of the brown lemming at Barrow, Alaska. *Alaskan Sci. Conf., Proc.* **11**, 24–25.

Maruyama, M. 1963. The second cybernetics: Deviation-amplifying mutual causal processes. *Am. Scientist* **51**, 164–179.

Pieper, R. D. 1964. Production and chemical composition of arctic tundra vegetation and their relation to the lemming cycle. Ph.D. thesis, University of California, Berkeley, California.

Pitelka, F. A. 1958. Some characteristics of microtine cycles in the arctic. *Ann. Biol. Colloq.* **18**, 73–78.

Pitelka, F. A., P. Q. Tomich, and G. W. Treichel. 1955. Ecological relations of jaegers and owls as lemming predators near Barrow, Alaska. *Ecol. Monographs* **25**, 85–117.

Schultz, A. M. 1964. The nutrient-recovery hypothesis for arctic microtine cycles. II. Ecosystem variables in relation to arctic microtine cycles. *In* "Grazing in Terrestrial and Marine Environments" (D. J. Crisp, ed.), pp. 57–68. Blackwell, Oxford.

Schultz, A. M. 1967. The ecosystem as a conceptual tool in the management of natural resources. *In* "Natural Resources: Quality and Quantity" (S. V. Ciriancy-Wantrup and J. J. Parsons, eds.), pp. 139–161. Univ. of California Press, Berkeley, California.

Thompson, D. Q. 1955. The role of food and cover in population fluctuations of the brown lemming at Point Barrow, Alaska. *Trans. 20th N. Am. Wildlife Conf.*, pp. 166–175.

Van Cleve, K. 1967. Nutrient loss from organic matter placed in soil in different geographic regions. Ph.D. thesis, University of California, Berkeley, California.

18

Reprinted with permission from *Ann. Rev. Ecol. Systems*, **3**, 33–50 (1972)

MINERAL CYCLING: SOME BASIC CONCEPTS AND THEIR APPLICATION IN A TROPICAL RAIN FOREST

Carl F. Jordan

AND

Jerry R. Kline[1]

Radiological and Environmental Research Division
Argonne National Laboratory
Argonne, Illinois

While the importance of mineral nutrition for plants and animals has been recognized for centuries, only recently has there been a systematic approach to mineral element cycling in entire ecosystems (47). By a systematic approach to ecosystem cycles we mean studies that yield estimates of the total amount of mineral elements and rate of element cycling within a complete landscape unit during a period of time, or under certain atmospheric or other physical conditions. Recent systematic studies (2, 43, 45, 50) have resulted in the formulation of several basic mineral-cycling concepts. In this paper we discuss several of these concepts in relation to a mineral-cycling study of a Puerto Rican tropical rain forest.

Worldwide cycles of mineral elements have recently been discussed by Deevey (6). He focused primarily upon carbon, hydrogen, oxygen, and nitrogen and their reactions in the biosphere, lithosphere, hydrosphere, and atmosphere. It is our purpose to discuss principles of mineral cycles at the terrestrial ecosystem level. Because carbon, oxygen, and nitrogen are in a gaseous form for a large proportion of their cycling time, it is difficult to relate their cycles to the processes of a single ecosystem. Likewise, hydrogen, which moves in water vapor in the air is also difficult to relate to a single ecosystem. In contrast, cycles of elements such as calcium, potassium, and phosphorus that do not ordinarily volatilize can be more easily viewed in the context of one ecosystem. For this reason we have limited our discussion to this type of mineral element.

The tropical rain forest mineral-cycling study on which this discussion is based was conducted between 1965 and 1969 and was part of the Atomic Energy Commission sponsored Ecology Project near El Verde, Puerto Rico. Background of the project and the results of a forest irradiation experiment have been compiled in a single volume (41), and methods used in the mineral-cycling study will be published elsewhere (20).

[1] Work performed under the auspices of the US Atomic Energy Commission.

33

Input-output balance.—The input-output balance of mineral elements in an ecosystem can be positive or negative; that is, elements can be accumulating in an ecosystem, or the system can be in the process of depletion. Accumulation of elements in the biotic portion of the ecosystem often occurs as the result of successional processes (38) as the amount of biomass in a system increases. A positive balance may be an indication that succession is still occurring (65). A small negative balance may just be part of the normal weathering of the landscape (30), but a large deficit indicates a disruption of the usual ecosystem processes (31).

For the nonvolatile elements, the most important input routes into terrestrial systems are via rainfall and from weathering of parent rock, although wind-blown dust may constitute an important input under certain circumstances. The most important output route in undisturbed systems is via soil-water runoff into streams.

Rate of element input into ecosystems via rainfall is easily quantified, and this measurement has been included in most mineral-cycling studies. Rate of output via soil-water and stream flow is less easily measured, but there have been at least two important approaches to this problem: 1. In a study of mineral cycling in a Douglas-fir forest in Washington (4), it was assumed that the amount of dissolved mineral elements collected by a tension-plate lysimeter located below the rooting zone was proportional to the top area of the lysimeter. Loss from the system was calculated on the basis of the rate of movement of elements through this lysimeter. 2. At Hubbard Brook, New Hampshire, an ecosystem located on top of solid bedrock was studied. There was no ground-water percolation through the bedrock, and the rate of element movement in the drainage stream represented the entire loss of nutrient elements from the system (30).

In both these studies, output rate for most elements was greater than input rate via rainfall. In the Hubbard Brook study, it was assumed that the system was in steady state and that the difference between input and output represented weathering of parent material. In the Douglas-Fir ecosystem study it was suggested that the soil was being depleted of nutrient elements as a result of the net loss.

In an oak-pine forest on Long Island, an approach similar to the Douglas-fir study was used (65). Here input of potassium, calcium, magnesium, and sodium was greater than output of these elements. The gain was attributed to a net storage of incoming elements by the increasing biomass of the forest.

In the tropical rain-forest mineral-cycling study, we used another approach to the problem of determining rate of loss of elements from the ecosystem by movement of soil water. Throughfall and stem flow were measured directly, and evapotranspiration was estimated by the tritium method (28). The difference between throughfall plus stem flow, and evapotranspiration, was assumed to be the rate of water runoff. Concentration of mineral elements in runoff water was determined from soil-water collections in zero-tension lysimeters (16) located below the rooting zone. Rate of loss of nutrient elements was calculated by multiplying rate of water runoff times the nutrient element concentration in that water.

Rates of gain or loss of three important elements are shown in Figure 1. These rates represent the difference between rates of input into the system via rain and loss rates via runoff. The pattern results from a complex interaction of yearly cycles of leaf fall, wind direction during rain storms, and location of the study site. As in the Washington and New Hampshire studies, nutrient loss from the system exceeded rainfall input on a yearly basis. We had no reason to suspect that the difference was not being made up by weathering of parent material.

Element concentration and volume of drainage water.—The relationship between the concentration of mineral elements in ecosystem drainage water and the volume of that drainage water indicates certain properties of the ecosystem from which the water drains. For example, if the concentration of elements is independent of the volume of water (that is, the concentration remains constant while the volume varies), this may indicate that the soil through which the water flows has a high exchange capacity and that the exchange capacity is near saturation with nutrient elements. Under these conditions, as the exchangeable ions on the soil surfaces are removed by roots or

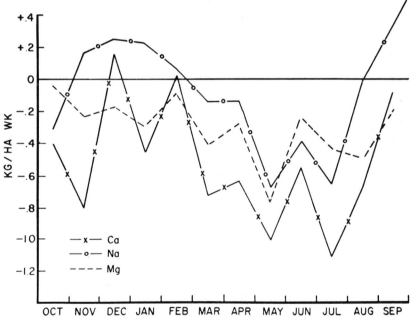

FIGURE 1. Net flux of calcuim, sodium, and magnesium into or out of the Puerto Rican tropical rain forest ecosystem, from October 1967 through September 1968.

soil water, other ions of the same kind become available for exchange into the soil solution. Forests growing on high exchange-capacity soils that are saturated with nutrient elements do not suffer deficiencies of these elements. Ecosystem drainage water in which the element concentration does not vary with volume indicates that the plants growing in the ecosystem do not suffer from a lack of these elements.

Lack of variation in element concentration during large variations in volume of soil water was shown for calcium and magnesium in a northern hardwoods forest (30), for six major cations in a Costa Rican tropical wet forest (36), and for sodium in the Puerto Rican tropical rain forest (17). In the Puerto Rican forest, concentrations of calcium and magnesium were influenced only slightly by the volume of runoff (17). While the volume of runoff increased by 100 times or more during storms, concentrations of calcium and magnesium went down by less than 25%.

In contrast, calcium and potassium concentrations decreased with increasing volumes of runoff in a New Jersey pine forest (16). A doubling of the volume almost halved the concentrations. Concentrations of calcium, magnesium, sodium, and potassium also were negatively correlated with stream volume in a California conifer forest (14). The lack of nutrient buffering ability in the soils of the pine and conifer forests may be related to the generally more acid conditions of the forest litter in these forests compared to hardwood forests (32). Acid litter results in relatively impoverished soil because the hydrogen ions replace the nutrient ions on the soil surfaces.

The relationship between nutrient content of the soil and concentration of nutrient elements in runoff has implications regarding two of man's recent activities, pollution of the air and weather modification. Man is putting sulfur compounds into the air where they may ultimately form sulfuric acid (2). When precipitation carries this acid into the soil, hydrogen ions replace the nutrient elements on the soil surfaces, and the nutrients are leached away (2). Ecosystems with nutrient-saturated, high exchange-capacity soils would lose relatively more of their nutrients than those with nutrient-unsaturated and/or low exchange-capacity soils. For example, the northern hardwood forest ecosystem (2), in which a large percentage of the total ecosystem nutrients is in the soil in exchangeable form, would lose a relatively large proportion of nutrients were the precipitation to become more acidic. In contrast, tropical rain forests in the Amazon basin would lose a relatively small quantity of nutrients. In these forests, most of the nutrient elements are tied up in the biomass (39), and the elements are recycled mainly through mycorrhiza (62). The mineral soil has a very low nutrient content (62). The first flush of runoff would leach away the bulk of available nutrients, and later runoff would be much more dilute. The percent of total ecosystem nutrients in the Amazon forest lost by increasing acidity of rainfall would be relatively low compared to the percent lost in the northern hardwoods.

If precipitation increases, regardless of the cause, and results in increases in volume of runoff, the ecosystems with nutrient-saturated, high exchange-capacity soils will lose more nutrients than those with nutrient-unsaturated, low exchange-capacity soils. In the former systems, nutrient loss is directly proportional to volume of runoff, because nutrient concentration in runoff water stays constant. In the latter systems, an increase in runoff results in dilution of the elements, and losses from the soil decrease as a function of the amount of runoff.

SYSTEMS ANALYSIS AND THE CYCLING OF ELEMENTS WITHIN ECOSYSTEMS

Attempts to generalize about current or future states of ecosystems based on individual studies of the parts are very difficult because of the complexity of most systems. The systems analysis approach appears to offer a way of understanding these complicated systems. We have used systems analysis in studying the mineral cycles of the Puerto Rican tropical rain forest. The analyses have resulted in several generalizations and predictions, which will be presented in the following sections. Here we present a very brief outline of systems analysis methods as applied to mineral cycles. A detailed explanation of systems analysis techniques is given by Patten (44).

The ecosystem is divided into a series of compartments, the number being determined by the degree of complexity desired. Compartmental content and flow rates into and out of each compartment are measured for periods of time compatible with budget and time requirements. Rates of change of compartmental contents are represented by a series of differential equations. For each compartment there is an equation which states that the rate of change of material in that compartment is equal to the flow into the compartment minus the flow out of the compartment. The equations are solved simultaneously by computer, and compartment content as a function of time is predicted.

Although there are seasonal variations in the mineral cycles of the Puerto Rican tropical rain forest, we represented the system by average yearly compartmental contents and flow rates so that the differential equations could be of linear form. Linear systems are easier to work with than are nonlinear systems, but predictions of states of the system represent only average yearly conditions and not the situation during any given season.

Differential chemical element accumulation by species, and ecological modeling.—In constructing models of element cycles, it is common practice to lump all species of a certain category, for example primary producers, into a single compartment (5). In constructing a calcium budget of a forest, for example, the calcium concentration in all trees might be determined and weighted according to the abundance of each species and an average value taken to represent the calcium content of the compartment representing trees. This procedure simplifies the modeling and probably is necessary because of computer limitations. Our experience illustrates such limitations. We have run a 50 compartment model (52) on the Argonne 360–50/75, a relatively large and fast computer. Before certain simplifying assumptions were made, it required 9 hours of computer time to simulate 3 months of real time, with solutions for every hour of real time. Even after simplification, simulation of one summer still required 1½ hours.

Even though ecological modelers often lump species together, they are generally aware that the differences in element concentration between different species may not be the result of random variation, but may represent significant differences. For example, most ecologists are aware of the tendency of flowering dogwood (*Cornus florida*) to accumulate and recycle calcium (56). Other reports of unusual element concentrations appear regularly in the literature. Recently Lyon et al (33) reported unusual accumulations of chromium and nickel in a New Zealand serpentine endemic, *Pimela suteri*.

To test the difference in element concentrations between tree species in the tropical rain forest at El Verde, we made the following systematic study (17): We analyzed four samples of roots, four of stems, four of clean leaves, and four of epiphyll (algae, mosses, liverworts) covered leaves for nine elements in five species at each of five similar sites in the same soil type in both July and December of 1968. Results of an analysis of variance showed not only that each species had a unique complement of element concentrations which was independent of site and season, but that each within-species division (roots, stems, clean and epiphyll-covered leaves) often had a unique complement also. As would be expected in light of the recent realization that epiphylls can be important in a nutrient budget (64), the epiphyll-covered leaves were significantly higher than clean leaves for most elements.

This discussion is not meant to imply that mineral-cycling models should include all species or all parts of species. In fact increasing the complexity of models does not necessarily increase the predictive power of the model (1). The discussion is intended to emphasize that the predictive power of a model is good only for those parameters that are explicitly modeled. For example, if an average calcium concentration of all tree species in a forest is used in a model, the model will predict only average concentrations, and the model cannot be used to predict calcium concentrations in any one species.

Relative loss of cations from the ecosystem in relation to their replacing power.—The tightness of an element cycle, that is, the resistance of a cycle to loss, is especially important when environmental disturbances threaten the fertility of ecosystems. Because loss of elements from ecosystems occurs primarily from leaching of the soil, it seems reasonable that the tightness of an element cycle in an ecosystem could be predicted by the element's replacing power on soil surfaces, except in ecosystems having direct mycorrhizal recycling such as those of the Amazon basin (62). Lists of relative replacing power have been compiled for various types of clay minerals (35).

To test this idea, we determined the half time of eight elements in the rain forest using a four-compartment model (19) consisting of soil, roots and

TABLE 1. Relative rate of element loss from the tropical forest ecosystem compared to the relative strengths with which cations are bound on exchange surfaces

Half time of element in the rain forest ecosystem		Relative replacing power of cations on a kaolinitic clay[a] ranked in an increasing series (35)
Element	Half time (yrs.)	
Cu	1.78	
Na	7.10	Na
K	7.28	K
Ca	12.91	Mg, Ca
Mg	14.56	
Sr	53.48	Sr
Mn	419.08	
Fe	$8.3 (10^6)$	

[a] Clay common in wet subtropical regions (13).

stems, leaves, and litter. We eliminated the input in the otherwise steady-state systems models, and from the computer output data we calculated the time at which only half of the original amount of material remained in the system The results (Table 1) show that for the mono- and divalent cations investigated, relative replacing power was a good indicator of the relative tightness of an element's cycle. The relationships of elements which have two or more possible valences such as copper, manganese, and iron are more complex.

Because the half times of manganese and iron in the rain-forest system are extremely long (Table 1), a word of explanation is appropriate. Half time of iron is very long because of the large amount of this element in the soil (10% of the soil is iron) (27). Manganese has a long half time because only a very small proportion of the total in the system moves into the soil where it would be susceptible to removal by leaching. Transfer from litter to roots may take place through mycorrhiza, as hypothesized by Went & Stark (62).

Stability of mineral cycling systems.—A system can be unstable, bounded, or asymptotically stable (51). If a system is unstable, it will deviate after a perturbation from an original steady-state condition. If it is bounded, it will also deviate from a steady-state condition but only within limits. If it is stable, it will return to steady-state conditions.

When an ecosystem is totally destroyed, as when one watershed of the Hubbard Brook ecosystem was clear-cut, the mineral cycles are unstable. When the perturbation is relatively small, as when a radioactive isotope of an element is introduced into a system via fallout, the cycle will be stable, provided there is no radiation damage to the ecosystem structure. In this section we discuss an investigation into the stabilities of systems of various elements in response to minor perturbations. This investigation, reported in detail elsewhere (19), gives an understanding of basic differences between cycles of essential and nonessential elements.

When all systems under investigation are basically stable, they can be compared by calculating their relative stability. Relative stability is basically related to the length of time required for a system to recover from a perturbation (45). The shorter the recovery time, the higher the stability.

Using computer models previously described, we compared the stability of the cycles of calcium, potassium, magnesium, sodium, iron, copper, strontium, and cesium in the Puerto Rican tropical rain forest (17); the cycles of calcium and potassium in the Douglas-fir forest in Washington (4); and of calcium in the Hubbard Brook forest, New Hampshire (2). Stabilities of these element systems in the three ecosystems all had equal values, except for sodium and cesium. Stability of sodium was slightly higher, and that of cesium much higher, then that of all the other elements.

It is basically the ratio of system input of an element to the recycling of that element which determines the stability of the element-cycling system (19). If a great deal of recycling takes place relative to input, as in essential element systems, the recovery time from a perturbation is relatively long, and the system has a low relative stability. If little recycling takes place, as in most nonessential systems, recovery time is short and stability is high. If a nonessential element such as strontium behaves in a manner similar to an essential element such as calcium, the nonessential element has a low stability.

The high stability of the cesium cycle means that the time required for the cycle to return to steady-state conditions is relatively short compared to the time required for return among essential elements. The rapid return to steady state by the cesium cycle is due to the relatively fast rate at which cesium leaves the active cycling system. This predicted behavior of cesium was confirmed in the Puerto Rican forest (29) and in the experimental forest at Oak Ridge (46) when it was found that ^{137}Cs virtually did not recycle through the trees and canopy after it reached the forest floor. Apparently, cesium becomes fixed in the soil in the clay minerals and is unavailable for further recycling (26).

The high stability of the sodium cycle in the Puerto Rican forest may occur because sodium is usually a nonessential element for plants (12) or because there is a superabundance of sodium in the Puerto Rican forest due to the proximity of the forest to the ocean. In the systems model, nonessential elements and elements which are in superabundance act in basically the same way. Both recycle very little through the plants, compared to the rate of input of the element into the system.

Prediction of pollutant concentrations in ecological systems.—Using the systems analysis techniques previously described, it is possible to predict future concentrations of pollutants in ecosystems if the transfer coefficients for the pollutant are known for every important compartment in the ecosystem and if the system behaves in the future in the same way it behaved while the transfer coefficients were being determined.

As a result of the atmospheric testing of nuclear devices in the late 1950s and early 1960s, ^{90}Sr was deposited throughout the world. Using a systems model, we have predicted future concentrations of this radioisotope in the Puerto Rican tropical rain forest (20).

To obtain transfer coefficients for ^{90}Sr, we measured amounts of stable strontium in major ecosystem compartments and rates of transfer of strontium between compartments. Our assumption was that ^{90}Sr behaves in the system in the same way as stable strontium. For input into the system model, we used actual ^{90}Sr fallout rates determined near the experimental site during the years of maximum fallout (59). The model predicts peak amounts of activity in the litter, canopy, and soil shortly after the maximum rate of fallout which occurred in 1963 (20) (Figure 2). Secondary peaks in the canopy and litter, resulting from recycling, are predicted to occur 10 to 15 years after the initial peaks, and maximum concentration in the wood is predicted to

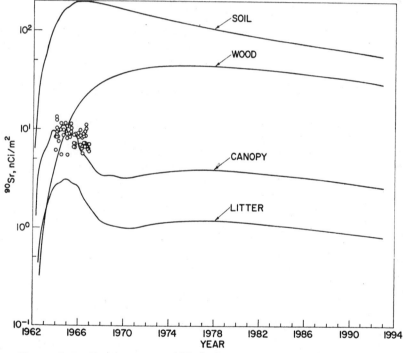

FIGURE 2. Predicted amounts of ^{90}Sr in four compartments of the Puerto Rican tropical rain forest as a function of time (lines) and measured amounts of canopy burden (circles).

occur in about 1975 (Figure 2). Half time of all the ^{90}Sr in the ecosystem is predicted to be 18 years (20). This half time includes physical decay of the isotopes as well as loss of the isotopes from the system through runoff.

We, of course, do not know whether the system will continue to behave in the future as it did while we were measuring it. We do, however, have a partial validation of the model. For a period of three years, actual levels of ^{90}Sr in the canopy leaves were measured at the study site in species used for the modeling. The data (circles, Figure 2) show good correlation between predicted and observed radioactivity.

The same approach was used for predicting radioisotope concentrations that would occur in Central American ecosystems if thermonuclear devices were to be used for excavation of a new transisthmian canal (34). Predicted concentrations were used to calculate doses of radioisotopes to indigenous populations (48).

Compartment size and mineral cycle variations.—Burns (3) has made a mathematical model of the mineral-element and energy flows in the tropical rain forest at El Verde. On the basis of computer experiments with the models, he has concluded that relatively large storage compartments, such as the large amounts of wood in tropical rain-forest trees buffer the cycles of minerals and flows of energy against variations in the physical environment. Jordan et al (22) have shown that the moose population on Isle Royale acts as an energy-buffering compartment that stabilized the wolf population. The presence of large compartments in an ecosystem may generally diminish the effects of large variations in the physical environment or of irregularities in the cycles and flows themselves. The converse of this principle may also be true. Lack of large compartments in an ecosystem may result in high sensitivity of the system to changes and variations.

To test the latter hypothesis, we have modeled the phosphorus cycle in the arctic tundra. This is an ecosystem in which the vegetation compartment is relatively small (50) and in which the active soil layer is shallow, because of perma-frost (7), and does not constitute an important storage compartment, at least for phosphorus (53). For purposes of modeling, the phosphorus cycle of the arctic tundra was abstracted into three compartments with transfers between them as follows: litter → roots → leaves (forage) → litter. Quantities of phosphorus in roots and leaves were calculated from data presented by Schultz (53) and Rodin & Bazilevich (50). Transfer from litter to roots (in years when it occurred) and from roots to leaves was assumed to occur during a 2-month growing season. Transfer from leaves to litter was assumed to occur gradually over the year. Release of phosphorus from litter was modeled to occur only once every 4 years, because Schultz (53) hypothesized that for 3 out of every 4 years most of the total pool of phosphorus in the tundra is bound in unavailable form in the litter. After decomposition of the litter during the fourth year, phosphorus is released and is transferred quickly to roots.

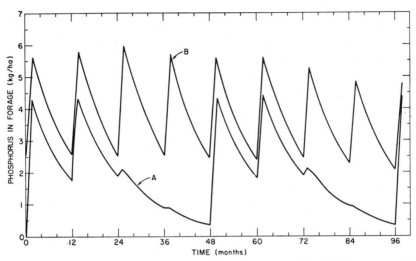

FIGURE 3. Total amount of phosphorus as a function of time in forage (leaves) of a tundra ecosystem (A), and in forage of an ecosystem differing from the tundra only in that it has a wood compartment equal in size to that of a tropical rain forest (B). Lines connect monthly values.

The amount of phosphorus in the leaves of tundra plants as a function of time during two 4-year cycles is shown in Figure 3, line A. The relatively small root compartment appears to buffer the leaf compartment against a large decrease in phosphorus for only 1 year. One year after the decrease in the litter to root transfer, the phosphorus content of the leaves shows a large decrease from the previous year.

To show the effect of a large buffering compartment on this ecosystem, the size of the phosphorus root compartment in the model was changed so it was the same size as the total phosphorus in the wood compartment in a tropical forest (50). The large buffer compartment eliminated the large decrease in phosphorus which occurred in the leaf compartment during the third and fourth years of each cycle in the unbuffered system (Figure 3, line B). It appears then that the lack of a large buffering compartment in an ecosystem can result in ecosystem variability much greater than occurs in a system buffered by large storage compartments. The stabilizing effect of a large compartment is brought out even more dramatically in Figure 4, where phosphorus contents of the leaves at the end of each growing season are connected with a smooth line, for each of the two contrasting systems.

If the vertical axis of Figure 4 were changed to designate population density of lemmings, line A, which represents lemming forage as well as leaves (53), would bear a striking resemblance to a graph of lemming populations at Point Barrow, Alaska (53). It also would match the cycles of lemmings and voles for Norway, presented by Elton (10). In fact, Schultz (53) has

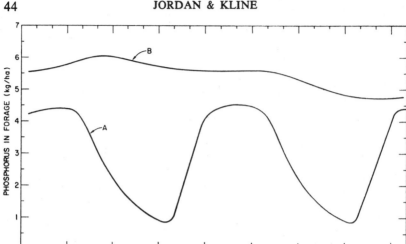

FIGURE 4. Total amount of phosphorus as a function of time in forage (leaves) of a tundra ecosystem (A), and in forage of an ecosystem differing from the tundra only in that it has a wood compartment equal in size to that of a tropical rain forest (B). Lines connect values for the end of each growing season.

hypothesized that it is the changes in the phosphorus content of the food of lemmings that is responsible for their 4-year cycle. In light of the above phosphorus models, we can hypothesize that the structure or lack of structure of an ecosystem can influence certain of the ecosystem's cyclic properties, such as oscillations of small mammal populations.

This is not to say that population oscillations cannot occur in ecosystems with large compartments. The essential point is that there are no large storage compartments in the food chain of animals exhibiting population oscillations. Two pertinent examples are the ruffed grouse and the snowshoe hare, which inhabit coniferous forests of the north and exhibit 10-year population cycles (23). Although coniferous forest soils are not large reservoirs of nutrient elements because of the soil's acidic nature, the coniferous trees constitute a large potential reservoir for energy and nutrient elements. However, the conifers are not part of the grouses' and hares' diets. These vegetarians eat solely angiosperms (58), which comprise a very small proportion of the coniferous forests.

These considerations, then, prohibit any broad statements about overall ecosystem resistance to change based on the presence or absence of large storage compartments. Rather, each variable of interest must be considered in light of the parameters which affect that particular variable.

Sensitivity analysis and cycling models.—Sensitivity analysis of a mineral-cycling or other type of model consists of a set of experiments on a computer. In each experiment, one compartment or flow of the system model is changed by a fixed percentage or amount, and the effect on the system output is then observed and compared with other changes. The purpose of the analysis is to determine which part of the model is most sensitive to error, so that future research efforts can be directed toward strengthening the weakest part of the model. Smith (55) describes sensitivity analysis procedures in detail.

In the course of modeling the Puerto Rican mineral-cycling data, we performed a sensitivity analysis on the strontium model and looked at changes in the predicted environmental half time of ^{90}Sr that would result from errors in estimates of quantities of strontium in the system and from errors in delineation of transfer routes of strontium. Our results (20) were in agreement with a general principal which has begun to emerge from the many ecological modeling efforts (61). That principle is, "hypothetical simple systems are far more sensitive to changes in the organizational structure of relations between components than to changes in the values of the components themselves" (5).

This point is of special significance to students interested in ecological modeling who have been schooled in the more traditional methods of biology. Such students often expend a relatively large effort in determining compartmental contents as accurately as possible and devote relatively little time to untangling the web of inter-relationships which can exist between compartments in ecological models. Smith (54) has suggested that field research for the purpose of modeling should be directed first at definition of system components and of links between them, second at the processes that influence transfer rates between components, and only third toward measurement of the actual quantities of the various components.

OTHER MINERAL CYCLING COMPARISONS

Mineral cycles in tropical forests as affected by burning and irradiation.— Slash and burn agriculture, practiced in many tropical areas, has a severe effect on mineral-element cycles. In this practice, forests are cut and burned, and crops are planted in the cut-over area. Yields are generally good the first year, but they decrease quickly thereafter (49). The reason for the decrease in yields, in contrast to the relatively sustained yields which are possible in temperate areas, is not yet completely understood.

One theory for the decline in productivity of the crop land is as follows: A large proportion of the nutrients in the undisturbed forests is held in the trees (39). The nutrients recycle directly from the litter of the trees into the roots via mycorrhiza (62). When the forest is cut and burned, not only are the nutrients in the trees released, but the recycling mechanisms—that is, the litter and mycorrhiza—are also destroyed. The released nutrients are available for the first season's crop, but they are rapidly leached downward so that succeeding crops have smaller quantities available to them. This theory is given credence by recent evidence of podsol soils over large areas of South America, Central America, Southeast Asia, and subtropical Australia (25).

In podsol soils, nutrient elements are readily leached out of the upper soil horizons (32).

Another reason for the decline in agricultural crops following cutting and burning may be the invasion of the crop land by fast-growing, succulent secondary successional species (40). These weed species compete with desirable crop species for light and nutrients. Unlike the forest trees, they cannot be cut and burned, but must be weeded, usually by hand. It may be easier to cut and burn a new area of forest than to pull these weeds from an already cleared area.

It was anticipated that the destruction of a small portion of the tropical rain forest at El Verde following irradiation might result in a decrease of nutrient elements in that part of the ecosystem. No decrease in nutrient levels in the soil occurred, however, either immediately following irradiation (9) or 3 years later (21). The secondary successional vegetation which invaded the irradiated area had, in fact, higher concentrations of certain nutrient elements than the undisturbed forest (15).

The retention of nutrient elements in the forest ecosystem at El Verde following irradiation, as contrasted to the loss of these elements following burning in other tropical systems, may have resulted from several factors. First, burning of a forest releases the nutrients in a single large pulse. Their availability may exceed the exchange capacity of the soils, and a large portion may be leached away before they can be incorporated into plants. In contrast, irradiation of the forest resulted in a slow destruction of the trees and consequent slow release of the nutrient elements. It required almost 4 years after cessation of the irradiation for most of the trees that ultimately died to fall down. At this writing, 7 years after cessation of irradiation, one large tree near the spot where the radiation source was located is still living but may ultimately die from the effects of irradiation. Even after falling over, some logs require many years to decompose because the very rocky terrain at the irradiation site suspends them well above the soil. Slow release of the nutrients may prevent the exchange capacity of the soil from being exceeded and allow time for successional vegetation to incorporate the nutrients as they are released from fallen logs.

Another factor in the retention of nutrient elements at the irradiated site is the relatively high exchange capacity of the soil. It has values between 25 and 50 meq/100 g soil for upper horizons (9). This is the same range as for major agricultural soils of the temperate zone (63).

A world pattern in mineral cycles.—Rates of mineral element uptake by natural plant communities and rates of return of elements to the litter are highest in the tropics, intermediate in temperate zones, and lowest in tundra regions (50). This pattern may be a result of the world pattern of leaf and litter production. Rates of leaf and litter production are highest in the tropics and decrease along a longitudinal line towards the poles (42).

In order to determine if a world pattern of ecosystem mineral cycles exists, and if so, whether it follows the same pattern as rates of mineral uptake and loss by plants, we constructed a model of a calcium cycle into which we put data from a series of ecosystems from the tropics to the taiga. The model has four compartments: wood, canopy (primarily leaves, but also flowers and fruits), litter (decomposing organic matter on the forest floor), and soil (exchangeable fraction of the calcium only). For each compartment we determined the turnover time by dividing the quantity of calcium in each compartment by the rate at which calcium leaves that compartment. Average yearly values were used. The flow of calcium was from soil to wood to canopy to soil via leaf fall, and leaf and stem leaching. Cycle time for each eocystem is the sum of turnover times for each compartment (Table 2, column 5).

Cycle times show a tendency to increase with increasing latitude, but there are exceptions, notably the Belgium forests. These forests are on calcareous soil (8), and consequently the soils have a high calcium content. Although rates of flow of calcium into and out of these soils are not much differ-

TABLE 2. Compartmental turnover times and cycling times for calcium

1 Ecosystem location	2 References	3 Type of ecosystem	4 Turnover time, years				5 Total cycle time, years	6 Total minus soil, years	7 Latitude degrees North	8 Length of growing season days (57)
			Soil	Wood	Canopy	Litter				
Puerto Rico	17	montane tropical rain forest	3.0	6.4	0.9	0.2	10.5	7.5	18	365
Ghana	11, 37	moist tropical forest	8.2	6.8	1.5	0.2	16.7	8.5	7	240 (24)
England	43	Scotch pine plantation	11.2	6.1	0.8	3.4	21.5	10.3	53	200
Virelles, Belgium	8	mixed oak	108.8	12.6	0.4	0.9	122.7	13.9	51	200
Wéve, Belgium	8	oak-ash	184.9	21.5	0.4	0.6	207.4	22.5	51	200
Washington state, US	4	Douglas-fir	57.4	20.2	5.5	10.2	93.3	35.9	48	200
Southern taiga, Russia	50	spruce	5.1	22.2	1.3	9.0	37.6	32.5	55	120
New Hampshire, US	2	northern hardwoods	14.0	10.8	0.8	34.8	60.4	46.4	44	120
Central taiga, Russia	50	spruce	7.6	18.3	3.2	13.6	42.7	35.1	65	105

ent from other ecosystems, the high calcium content of these soils results in long turnover times in the soils and consequent long cycling times in the systems. The pattern of increasing cycle time with increasing latitude is confounded by element availability patterns in soils, which are unrelated to distance from the equator.

If the turnover time in the soil compartment is subtracted from the sum of all the turnover times (Table 2, column 6), the pattern of decreasing rates of uptake and return of elements with increasing latitude described by Rodin & Bazilevich (50) becomes clearer. If time for uptake and return is compared to length of growing season (Table 2, column 8), an even better correlation exists. The growing season was taken to be the same as the frost-free season, given by Trewartha (57), in all cases except that of Ghana. There the growing season is shorter than the frost-free season, because of a dry season (24).

Cycling times for elements in ecosystems are influenced primarily by rates of element uptake and release by plants. Because uptake is closely related to biomass production, the world pattern of mineral cycles resembles the world pattern of vegetation production (18). Leaf and litter production, but not necessarily wood production, is highest in the wet tropics and decreases along lines of decreasing solar radiation available during the growing season and increasing lengths of dry season. Mineral cycles do not follow these worldwide trends in areas where nutrient availability in soils is unusually high or low.

CONCLUSION

We have attempted to place the results of our mineral-cycling studies in Puerto Rico into a general perspective by comparing them to other mineral-cycling studies. We hope that the principles that have begun to emerge from these comparisons will serve as building blocks for larger, more encompassing principles resulting from the analysis of ecosystems now being undertaken as part of the International Biological Program (60).

LITERATURE CITED

1. Anderson, H. W. 1966. Watershed modeling approach to evaluation of the hydrological potential of unit areas. Cited in Ref. 5
2. Bormann, F. H., Likens, G. E. 1970. The nutrient cycles of an ecosystem. *Sci. Am.* 223:No. 4, 92–101
3. Burns, L. A. 1970. Analog simulation of a rain forest with high-low pass filters and a programmatic spring pulse. *A Tropical Rain Forest*, ed. H. T. Odum, R. F. Pigeon, I-284–I-189. US

At. Energ. Comm., Tech. Inform. Ext., Oak Ridge, Tenn.
4. Cole, D. W., Gessel, S. P., Dice, S. F. 1967. Distribution and cycling of nitrogen, phosphorus, potassium, and calcium in a second-growth Douglas-fir ecosystem. *Symp. Primary Productivity and Mineral Cycling in Natural Ecosystems,* 196–232. Univ. Maine Press. 245 pp.
5. Cooper, C. F. 1969. Ecosystem models in watershed management. *The Ecosystem Concept in*

Natural Resource Management, ed. G. M. Van Dyne, 309–24. New York: Academic. 383 pp.

6. Deevey, E. S. 1970. Mineral cycles. *Sci. Am.* 223:No. 3, 149–58

7. Douglas, L. A., Tedrow, J. C. F. 1960. Tundra soils of arctic Alaska. *Trans. Int. Congr. of Soil Science, 7th,* Madison, Wisc. 4:291–304

8. Duvigneaud, P., Denaeyer-de-Smet, S. 1970. Biological cycling of minerals in temperate deciduous forests. *Analysis of Temperate Forest Ecosystems,* ed. D. E. Reichle, 199–225. New York: Springer-Verlag. 304 pp.

9. Edmisten, J. Soil studies in the El Verde rain forest. See Ref. 3, H-79–H-87

10. Elton, C. 1942. *Voles, Mice and Lemmings.* Oxford: Clarendon. 496 pp.

11. Greenland, D. J., Kowal, J. M. L. 1960. Nutrient content of the moist tropical forest of Ghana. *Plant Soil* 12:154–73

12. Hewitt, E. J. 1963. The essential nutrient elements: requirements and interactions in plants. *Plant Physiol.* 3:137–360. Academic

13. Jackson, M. L. 1964. Chemical composition of soils. *Chemistry of the Soil. Am. Chem. Soc. Monogr. Ser.,* ed. F. E. Bear, 71–141. New York: Reinhold. 515 pp.

14. Johnson, C. M., Needham, P. R. 1966. Ionic composition of Sagehen Creek, California, following an adjacent fire. *Ecology* 47:636–39

15. Jordan, C. F. 1968. Chemistry of successional vegetation. *Rain Forest Project Ann. Rep. Puerto Rico Nucl. Cent. Publ. No. 119,* 51–52. 147 pp.

16. Jordan, C. F. 1968. A simple, tension-free lysimeter. *Soil Sci.* 105:81–86

17. Jordan, C. F. 1969. Isotope cycles. See Ref. 15, Publ. No. 129, 2–64. 144 pp.

18. Jordan, C. F. 1971. A world pattern in plant energetics. *Am. Sci.* 59:425–33

19. Jordan, C. F., Kline, J. R., Sasscer, D. S. 1972. Relative stability of mineral cycles in forest ecosystems. *Am. Natur.* 106:237–53

20. Jordan, C. F., Kline, J. R., Sasser, D. S. A simple model of strontium and manganese dynamics in a tropical rain forest. *Health Phys.* In press

21. Jordan, C. F. Unpublished data

22. Jordan, P. A., Botkin, D. B., Wolfe, M. L. 1971. Biomass dynamics in a moose population. *Ecology* 52:147–52

23. Keith, L. B. 1963. *Wildlife's Ten-Year Cycle.* Madison: Univ. Wisconsin Press. 201 pp.

24. Kendrew, W. G. 1953. *The Climates of the Continents.* London: Oxford Press. 607 pp.

25. Klinge, H. 1967. Podzol soils: A source of blackwater rivers in Amazonia. *Atas do Simposio sôbre a Biota Amazonica* (Limnologia) 3:117–25

26. Kline, J. R. Radionuclide behavior in tropical soil. See Ref. 15, 57–60

27. Ibid. Neutron activation of tropical soils and plants, 72–78

28. Kline, J. R., Martin, J. R., Jordan, C. F., Koranda, J. J. 1970. Measurement of transpiration in tropical trees with tritiated water. *Ecology* 51:1068–73

29. Kline, J. R. Retention of fallout radionuclides by tropical forest vegetation. See Ref. 3, H-191–H-198

30. Likens, G. E., Bormann, F. H., Johnson, N. M., Pierce, R. S. 1967. The calcium, magnesium, potassium, and sodium budgets for a small forested ecosystem. *Ecology* 48:772–85

31. Likens, G. E., Bormann, F. H., Johnson, N. M., Fisher, D. W., Pierce, R. S. 1970. Effects of forest cutting and herbicide treatment on nutrient budgets in the Hubbard Brook watershed-ecosystem. *Ecol. Monogr.* 40:23–47

32. Lutz, H. J., Chandler, R. F. 1946. *Forest Soils.* New York: Wiley. 514 pp.

33. Lyon, G. L., Peterson, P. J., Brooks, R. R., Butler, G. W. 1971. Calcium, magnesium and trace elements in a New Zealand serpentine flora. *J. Ecol.* 59:421–29

34. Martin, W. E. 1969. Bioenvironmental studies of the radiological-safety feasibility of nuclear excavation. *BioScience* 19:135–37

35. Marshall, G. E. 1964. *The Physical Chemistry and Mineralogy of Soils.* New York: Wiley. 388 pp.
36. McColl, J. G. 1970. Properties of some natural waters in a tropical wet forest of Costa Rica. *BioScience* 20:1096–100
37. Nye, P. H. 1961. Organic matter and nutrient cycles under moist tropical forest. *Plant Soil* 13: 333–45
38. Odum, E. P. 1969. The strategy of ecosystem development. *Science* 164:262–70
39. Odum, E. P. 1971. *Fundamentals of Ecology.* Philadelphia: Saunders. 574 pp.
40. Odum, H. T. Summary: An emerging view of the ecological system at El Verde. See Ref. 3, I-191–I-289
41. Odum, H. T., Pigeon, R. F., Eds. 1970. See Ref. 3
42. Olson, J. S. 1963. Energy storage and the balance of producers and decomposers in ecological systems. *Ecology* 44:322–31
43. Ovington, J. D. 1962. Quantitative ecology and the woodland ecosystem concept. *Advan. Ecol. Res.* 1:103–92
44. Patten, B. C. 1971. A primer for ecological modeling and simulation with analog and digital computers. *Systems Analysis and Simulation in Ecology,* ed. B. C. Patten, 3–121. New York: Academic. 607 pp.
45. Patten, B. C., and Witkamp, M. 1967. Systems analysis of ^{134}cesium kinetics in terrestrial microcosms. *Ecology* 48:813–24
46. Peters, L. N., Olson, J. S., Anderson, R. M. 1970. Trends of foliage and soil data on ^{137}Cs in a tagged Appalachian forest dominated by *Liriodendron tulipifera. Oak Ridge Nat. Lab. Ecol. Sci. Div. Ann. Progr. Rep. 1970. ORNL-4634,* 25–26. 141 pp.
47. Pomeroy, L. R. 1970. The strategy of mineral cycling. *Ann. Rev. Ecol. Syst.* 1:171–90
48. Raines, G. E., Bloom, S. G., Levin, A. A. 1969. Ecological models applied to radionuclide transfer in tropical ecosystems. *BioScience* 19:1086–91
49. Richards, P. W. 1966. *The Tropical Rain Forest.* Cambridge Univ. Press. 450 pp.

50. Rodin, L. E., Bazilevich, N. I. 1967. *Production and Mineral Cycling in Terrestrial Vegetation.* Transl. from Russian. London: Oliver & Boyd. 288 pp.
51. Rosen, R. 1970. *Dynamical System Theory in Biology,* Vol. 1. *Stability Theory and its Applications.* New York: Wiley Interscience. 302 pp.
52. Sasser, D. S., Jordan, C. F., Kline, J. R. 1971. A mathematical model of tritiated and stable water movement in an old-field ecosystem. *Proc. Nat. Symp. Radioecol., 3rd,* ed. D. J. Nelson. In press
53. Schultz, A. M. 1969. A study of an ecosystem: The Arctic tundra. *The Ecosystem Concept in Natural Resource Management,* ed. G. M. Van Dyne, 77–93. New York: Academic. 383 pp.
54. Smith, F. E. 1969. Personal communication cited in Ref. 5
55. Smith, F. E. 1970. Analysis of ecosystems. *Analysis of Temperate Forest Ecosystems,* ed. D. E. Reichle, 7–18. New York: Springer-Verlag. 304 pp.
56. Thomas, W. A. 1969. Accumulation and cycling of calcium by dogwood trees. *Ecol. Monogr.* 39:101–20
57. Trewartha, G. T. 1954. *An Introduction to Climate.* New York: McGraw-Hill. 402 pp.
58. Trippensee, R. E. 1948. *Wildlife Management.* New York: McGraw-Hill. 479 pp.
59. US Dep. Health, Education and Welfare, Public Health Serv. 1961–1968. *Radiol. Health Data Rep.,* Vols. 2–9
60. US Int. Biol. Prog. *Man's Survival in a Changing World.* Nat. Acad. Sci., Nat. Res. Counc., Washington, DC
61. Watt, K. E. F. 1970. Details versus models. *Science* 168:1079
62. Went, F. W., Stark, N. 1968. Mycorrhiza. *BioScience* 18:1035–39
63. Wiklander, L. Cation and anion exchange phenomena. See Ref. 13, 163–205
64. Witkamp, M. Mineral retention by epiphyllic organisms. See Ref. 3, H-177–H-179
65. Woodwell, G. M., Whittaker, R. H. Primary production and the cation budget of the Brookhaven Forest. See Ref. 4, 150–66

19

Reprinted from *BioScience,* **22**(9), 541–543 (1972)

The Metabolism of Some Coral Reef Communities: A Team Study of Nutrient and Energy Flux at Eniwetok

R. E. JOHANNES et al.

Studies of total community metabolism are often relatively uncomplicated in unidirectional currents (Sargent and Austin 1949 and 1954, Odum and Odum. 1955, Kohn and Helfrich 1957, Odum et al. 1959, Milliman and Mahnken 1969, Gordon and Kelly 1962, Kinsey and Kinsey 1967, Qasim and Sankaranarayanan 1970). Measuring changes in chemical and biological characteristics of the water as it impinges upon coral reef communities and is modified by such communities not only provides a variety of indices of community metabolism but also helps identify processes of unanticipated importance in various segments of the community.

In May and June 1971, a team of about 25 scientists formed an expedition called "Project Symbios." We examined the productivity and flux of nutrients in some coral reef communities at Eniwetok Atoll, Marshall Islands. Our research vessel, *Alpha Helix*, with its two well-equipped laboratories, was moored about 400 meters from the primary research site. From here we could carry out a wide variety of analyses and identify quickly any noteworthy phenomena.

To measure reef community metabolism, we established transects parallel to the currents flowing across the shallow windward interisland reefs (Fig. 1). These currents flowed continually from the ocean to the lagoon irrespective of tide or wind conditions. Chemical, physical, and biological characteristics of the seawater were measured before, during, and after the currents flowed through these transects. These data enable us to calculate the net rates of export or import of oxygen, carbon, nitrogen, phosphorus, calcium, various trace metals, detritus, bacteria, and zooplankton.

The most intensively studied transect (Transect II, Fig. 1) is 340 m long and situated just north of Muti (Japtan) Island, about 260 m south of the site of the well-known study of reef ecology made by Odum and Odum (1955). This transect contains the same sequence of biotic zones described by the Odums, beginning upstream with algal pavement, grading into a variety of coral communities, and extending into algae-covered coral rubble.

Transect III (Fig. 1) north of Chinieero Island, is 280 m long and was chosen because unlike Transect II, the transition between algal pavement and coral rubble occurs with no significant intervening coral community.

Supported by NSF Grant GB 24816 to the Scripps Institute of Oceanography for the operation of the Alpha Helix, by grants from the U.S. Atomic Energy Commission to the Eniwetok Marine Laboratory and by a grant from the Janss Foundations.

Some measurements were made on four additional transects.

Flux of Nitrogen and Phosphorus

The ratio of fixed dissolved inorganic nitrogen to dissolved reactive phosphorus in incoming water on Transect II was about 2:1, suggesting initially that of the two elements, nitrogen was more likely to be in short supply. However, we found that there was a net export of NO_3, NH_3, and dissolved and particulate organic nitrogen by the reef community; whereas the concentration of dissolved phosphorus remained very constant across the transect (Table 1). In addition, the C:N (atomic) ratio in suspended particulate material changed from 15 in oceanic waters 3-5 km "upstream" to 7 at the downstream end of Transect II. Our calculations suggested that atmospheric nitrogen fixation must have been occurring at rates of the same high order as those reported for alfalfa fields in order to balance the observed export of fixed nitrogen.

Using the acetylene reduction method, we located a variety of communities within the reef complex in which nitrogen fixation occurred. Communities which reduced the greatest amounts of acetylene were in the spur and groove zone and the heavily grazed *Calothrix* zone seaward of Muti Island. Algal communities growing on corals recently killed by the crown-of-thorns starfish, *Acanthaster planci,* also reduced acetylene. Most of the communities which yielded positive results

were located on the seaward portions of the reef. At this point we do not know the specific organisms responsible for this activity, but additional research at Eniwetok is planned.

The lack of detectable change in dissolved phosphorus concentrations along Transect II (Table 1) is remarkable. There is a reasonably well established relationship between the release of oxygen and the uptake of phosphorus during photosynthesis in complex plankton communities (Steeman Nielsen 1963, Redfield, Ketchum, and Richards 1963). In general, for each phosphorus molecule that disappears from the water, about 138 oxygen molecules will be released to the water. Accordingly, oxygen changes across the reef how how much phosphorus ought to have been consumed in the daytime and produced at night. This amount of phosphorus is about 20 times the minimum concentration change detectable by the analytical method used. Since no phosphorus changes could be detected, recycling of this element within the benthic community on Transect II appears to be unusually efficient. Much of the phosphorus excreted by heterotrophs must be removed within the community by autotrophs with little washout occurring. The efficiency of this system does not seem to diminish measureably at night.

In contrast, there were consistent changes in dissolved phosphorus levels along Transect III. Inorganic phosphorus decreased by about 0.04 μ M l^{-1}, while organic phosphorus

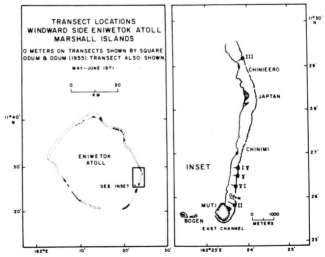

Fig. 1. Map of Eniwetok Atoll, Marshall Islands and transect locations.

increased by 0.03 μ M 1^{-1} irrespective of the time of day. The differing trends on the two transects appear to be related to the presence of a well-developed coral community on one transect and the absence of such a community on the other.

The uptake and release of dissolved organic and inorganic phosphate were examined in a variety of reef organisms. Rates of uptake or release per unit biomass in coral and other invertebrates with symbiotic algae (zooxanthellae) in their tissues, were much lower than in animals without zooxanthellae. These rates suggest that nutrient recycling between coral host and algal symbiont provides an effective mechanism for retaining phosphorus within these associations (Pomeroy and Kuenzler 1969, Johannes et al. 1970).

Rates of loss of ^{32}P from radioactively labelled algal mat communities containing numerous small invertebrates and bacteria were similar to the low loss rates measured in labelled corals. This suggests that these mat communities also possess some mechanism by which leakage of phosphorus into the overlying water is minimized during recycling.

Productivity and Organic Carbon Flux

Separate studies of changes in oxygen and in pH and alkalinity along the transects indicated that gross primary productivity was more than twice as high on Transect III (10 g C/m^{-2}day^{-1}) with negligible coral cover as on Transect II (4 g C/m^{-2}day^{-1}) with a well-developed coral component.[1] These observations along with those of Odum, Burkholder, and River (1959) on *Thalassia* communities suggest that the high gross productivity reported from various coral communities may not be a special characteristic of these communities, but rather of any shallow, well oxygenated tropical benthic community situated in clear water on a stable bottom.

On both transects the data yielded a respiratory quotient (- Δ CO$_2$/ Δ O$_2$) of about 1.0. This observation supports the previously untested assumption of an RQ of 1.0 in the calculations of most earlier workers.

The ratio of photosynthesis to total respiration (P/R ratio) on Transect III was about 2.0, indicating a net export of organic material. Benthic algal fragments appeared to form an important fraction of this export. The P/R ratio of that portion of Transect II dominated visually by corals was less than one, indicating that reduced carbon must be imported to sustain the community. This import occurred in at least two ways. First, net plankton studies showed that there was a net removal of about 1 gm dry weight m^{-2}day^{-1} of macroscopic material from the water, largely in the form of benthic algal

[1] The common assumption has been that daytime respiration rates are identical with nighttime respiration rates. If photorespiration is shown to occur in reef algae, then this assumption is invalid and our productivity estimates, as well as those of many other investigators, will have to be revised upward.

fragments exported by the algal flag "upstream." Secondly, large schools of parrot fish and suggeon fish living in the coral community obtained nutritive energy from beyond its borders by moving periodically to the algal flat to feed.

The contribution by plankton to oxygen changes in water crossing the reef was studied using plankton gently and rapidly concentrated by the method of Pomeroy (Pomeroy and Johannes 1968) from water samples taken at various spots in front of, on, and behind Transect II. Respiration rates were 0.1-1.0 mg-at O$_2$ m^{-3}day^{-1} in water outside the lagoon, and little or no change occurred as the water crossed the reef. These values are typical of open ocean water and verify the dominance of benthic metabolism in the reef environment.

Measurements were made of net photosynthesis and respiration for 15 species of coral which occurred in moderate to large numbers in the vicinity of Transect II. The experiments were done *in situ* and generally lasted about 24 hours. Ambient seawater was flushed through the respiration chambers at hourly intervals to minimize buildup of metabolites and bacteria. Light impinging on the coral heads was measured with an underwater specral radiometer, and downward irradiance was recorded from 380 to 724 mμ wavelength. P/R ratios were generally lower than reported up to this time. Earlier data were based on shorter incubation times. The highest value obtained (for a species of *Acropora*) was less than 1.50 and most were between 1.0 and 1.45.

Similar respirometric studies were performed on three other invertebrates containing symbiotic algae, the tridacnids, *Tridacna squamosa* and *Hippopus hippopus*, and a yet-to-be-identified burrowing sea anemone. A P/R ratio of 1.75 was obtained for *T. squamosa*, indicating that this clam has the potential for obtaining all required energy from its algal symbionts alone. Compensation light intensities were reached at approximately 5 hr 22 min before and after local apparent noon on cloudless days. The P/R ratio for the anemone was approximately 0.5, indicating that the maximum energetic contribution of algae to the association was one-half the total requirement.

Respirometry was also performed on samples of reef substrate and fleshy algal mats from various zones on Transect II. These algal communities supported large numbers and many types of bacteria, protozoa, and microscopic invertebrates. Mean gross production for algal communities was 0.046 mg O$_2$ cm^{-2}hr^{-1}. The highest value found. 0.087 mg O$_2$ cm^{-2}hr^{-1}, was for a sample of algal pavement colonized primarily by the bluegreen alga, *Schizothrix calcicola*. Communities on coral rubble were about one-half as productive as communities on algal pavement.

Particulate and dissolved organic carbon levels did not change significantly across Transect II (Table 1), although particulate organic carbon levels were significantly higher on the transect than in oceanic waters upstream. Viable heterotrophic bacteria, enumerated by plate counts, increased by an average factor of 10 across Transect II.

Since zooplankton have been reported as being too sparse to support coral energy requirements (Johannes, Coles and Kuenzel,

TABLE 1. Results of analysis of seawater samples from the upstream end (Station O) and downstream end (Station 340) of Transect II (Fig. 1) at Eniwetok Atoll[1]

Properties tested	Station 0	Station 340
NH$_3$ (nM/1)	240 (24)	287 (22)
NO$_3$ (nM/1)	109 (9.0)	169 (19)
Dissolved organic N (nM/1)	1790 (131)	2145 (162)
Particulate organic N (nM/1)	157 (13.2)	210 (17)
Reactive phosphorus (nM/1)	174 (6)	169 (6)
Organic phosphorus (nM/1)	152 (8)	155 (7)
Dissolved organic carbon (μ g/1)	1230 (94)	1210 (129)
Particulate organic carbon (μ g/1)	24 (2)	26 (2)
O$_2$ (mg/1)	6.58 (.05)	Day 7.38 (.06)
		Night 6.18 (.05)
Alkalinity (meq/1)	2.295 (.002)	Day 2.285 (.004)
		Night 2.291 (.007)
pH	8.30 (.005)	Day 8.34 (.008)
		Night 8.27 (.005)
Heterotrophic bacteria (viable plate counts) (#/ml)	80	200

[1]Standard errors are in parentheses. Inspection of the means and standard errors in this table will suggest to the reader that some of the differences between upstream and downstream values are not significantly different statistically. This is misleading in some cases. When differences are compared on a paired sample basis, oxygen, alkalinity, pH, and all nitrogen components differ significantly statistically.

1970) and since we found zooplankton densities to be very low on the reef at Eniwetok (also see Odum and Odum 1955), it was of interest to discover that the coral *Fungia* is able to ingest bacterial colonies and can also take up individual bacterial cells. Experiments using marine bacteria stained with acridine orange showed that this uptake is accomplished in the gastrodermal lining of the gastric cavity.

Carbonate Flux

Alkalinity depletion was used to monitor gross $CaCO_3$ production on Transects II and III. The pH was monitored as well in order to provide a more complete record of the CO_2 system and to facilitate estimates of rates of photosynthesis and respiration. Typically, alkalinity was depleted by less than 10μ eq l^{-1} as the water flowed along the transects. These data are convertible to rates of approximately 10 g $CaCO_3$ $m^{-2}day^{-1}$ and constitute the first direct measurement that appears to be published on calcification rates of entire reef communities (D. W. Kinsey has unpublished data on reef calcification and productivity based on short term alkalinity and pH changes at One Tree Island, Great Barrier Reef. His studies began before ours.) No significant diurnal differences in calcification rates were found, nor were differences found between transects II and III. Moderate day to night or transect to transect differences could have been missed, however, due to insufficient resolution of the method.

The pH commonly rose across the transects by about 0.03 by day and fell by 0.01 at night. The pH changes were largely due to photosynthesis/respiration reactions rather than to calcification.

Three prominent calcifying organisms (*Acropora*, *Millepora*, and *Jania*) were incubated in a respirometer to monitor oxygen, alkalinity, and pH changes. For these organisms calcification proved to be more important than photosynthesis and respiration in producing the net pH changes. If these three organisms typify calcifying biota, then the bulk of photosynthesis and respiration on this reef was carried out by non-calcifying organisms.

To estimate calcium carbonate production by reef organisms, several species were incubated in seawater to which Ca^{45} had been added. Species were tested at various times of day and night for periods of one hour. Net calcium carbonate production by *Acropora* sp. averaged 7.6 mg cm^{-2} of coral cover per day. Only slight differences were obtained between living and dead Prolithon suggesting this species is not a significant carbonate producer on the reef.

Different tips of single colonies of *Pocillopora damicornis* varied greatly in Ca^{45} uptake rates. In order to determine if lack of selected nutrients might be the cause of decreased calcification, a single head of *P. damicornis* was divided into branches and exposed to various nutrients for 11 hours prior to testing. The addition of dissolved phosphate, coarse zooplankton, or newly hatched brine shrimp did not favor more uniform calcification rates. Similar experiments were performed on *Acropora* sp. Increased concentration of ammonium ion, phosphate, or glycerol caused no significant change in Ca^{45} uptake rate. Increasing the concentration of nitrate or urea or feeding with newly hatched brine shrimp produced significant decreases in the rate of deposition of calcium carbonate.

Tips of *Acropora* sp. were incubated with Ca^{45} under various light conditions while the temperature was kept constant. Ca^{45} uptake rates in the dark at 3 AM were roughly one-half the rates obtained in full sunlight. The selective removal of 93% of the ultraviolet light from the spectrum had no measureable effect on the rate of calcification. The rate of calcification of this species when darkness was imposed at midday or in the early evening was significantly higher than the calcification rate at night in complete darkness. This suggests that there may be a considerable lag in the effect of darkness on the calcification rate of this species. Results obtained in Hawaii on *Pocillopora damicornis* do not suggest a lag, so the phenomenon is not universal.

Summary

Perhaps the most interesting findings during Project Symbios are those which provide at least a partial explanation of why coral reef communities are so biologically productive though characteristically situated in waters of very low nutrient content; high rates of nitrogen fixation supplement the uptake of fixed nitrogen, and recycling of phosphorus is unusually efficient. Also noteworthy is the observation that the visually unimpressive algal flat community with its comparatively low standing crop of organisms can be extremely productive — more productive than the rich adjacent coral community. Finally, it is worth noting that these phenomena would probably not have been uncovered during this study if we had restricted ourselves to measuring the metabolism of individual reef organisms. Measuring all the major metabolic processes in all species in a complex community is utterly impractical. Measurements of total community metabolism help the investigator identify the processes and organisms within the community that are most likely to reward his scrutiny.

References

Gordon, M. C. and H. M. Kelly. 1962. Primary productivity of a Hawaiian coral reef: a critique of flow respirometry in turbulent waters. *Ecology* 43: 473-480.

Johannes, R. E., S. L. Coles, and N. T. Kuenzel. 1970. The role of zooplankton in the nutrition of some scleractinian corals. *Limnol. Oceanogr.* 15: 579-586.

Kinsey, D. W. and B. E. Kinsey. 1967. Diurnal changes in oxygen content of the water over the coral reef platform at Heron I. *Austr. J. Mar. Freshwat. Res.* 18: 23-34.

Kohn, A. J. and P. Helfrich. 1957. Primary organic productivity of a Hawaiian coral reef. *Limnol. Oceanog.* 2: 241-251.

Milliman, J. D. and C. V. W. Mahnken. 1971. Reef productivity measurements p. 23-26, Appendix. In: J. D. Milliman. Four Southwestern Caribbean atolls: Courtown Cays, Albuquerque Cays, Roncador Bank, and Serrana Bank. *Atoll Res. Bull.* 129.

Odum, H. T., P. R. Burkholder, and J. Rivero. 1959. Measurements of productivity of turtle grass flats, reefs and the Bahia Fosforescente of southern Puerto Rico. *Repr. Instit. Mar. Sci.* 4: 159-170.

Odum, H. T. and E. P. Odum. 1955. Trophic structure and productivity of a windward coral reef community. *Ecol. Monogr.* 25: 291-320.

Pomeroy, L. R. and R. E. Johannes. 1968. Occurrence and respiration of ultraplankton in the upper 500 meters of the ocean. *Deep Sea Res.* 15: 381-391.

Pomeroy, L. R. and E. J. Kuenzler. 1969. Phosphorus turnover by coral reef animals. *Proc. 2nd Conference on Radioecology.* AEC CONF-670503: 474-482.

Qasim, S. Z. and U. N. Sankaranarayanan. 1970. Production of particulate organic matter by the reef on Kavaratti Atoll (Laccadives). *Limnol. Oceanogr.* 15: 574-578.

Redfield, A. C., B. H. Ketchum, and F. A. Richards. 1963. The influence of organisms on the composition of seawater. *In:* The Sea, M. N. Hill (ed.) Wiley, New York. 2: 26-77.

Sargent, M. C. and T. S. Austin. 1949. Organic productivity of an atoll. *Trans. Amer. Geophys. Union* 30: 245-249.

Sargent, M. C. and T. S. Austin. 1954. Biologic economy of coral reefs. *U. S. Geol. Surv. Prof. Paper* 260-E: 293-300.

Steemann Nielsen, E. 1963. Productivity, definition and measurement. *In:* The Sea, M. N. Hill (ed.) Wiley, New York. 2: 129-164.

R.E. JOHANNES, J. ALBERTS, C. D'ELIA, R.A. KINZIE, L.R. POMEROY, W. SOTTILE, and W. WIEBE
Division of Biological Sciences
University of Georgia
Athens, Georgia 30601

J.A. MARSH, JR.
University of Guam Marine Laboratory
P.O. Box EK
Agana, Guam, U.S.

P. HELFRICH, J. MARAGOS, J. MEYER, and S. SMITH
Hawaii Institute of Marine Biology
P.O. Box 1067
Kaneohe, Hawaii 96744

D. CRABTREE and A. ROTH
Loma Linda University
Loma Linda, California 92354

L.R. McCLOSKEY
Marine Biological Laboratory
Woods Hole, Massachusetts 02543

S. BETZER, N. MARSHALL, M.E.Q. PILSON, and G. TELEK
Graduate School of Oceanography
University of Rhode Island
Kingston, Rhode Island 02881

R.I. CLUTTER
South Pacific Islands Fisheries
Development Agency
c/o South Pacific Commission
Noumea, New Caledonia

W.D. DuPAUL and K.L. WEBB
Virginia Institute of Marine Science
Gloucester Point, Virginia 20362

J.M. WELLS, JR.
Wrightsville Marine Bio-Medical Laboratory
7205 Wrightsville Avenue
Wilmington, North Carolina 28401

20

Reprinted with permission from *Ann. Rev. Microbiol.*, **18**, 217–252 (1964)

BIOCHEMICAL ECOLOGY OF
SOIL MICROORGANISMS[1]

By M. Alexander

*Laboratory of Soil Microbiology, Department of Agronomy,
Cornell University, Ithaca, New York*

Contents

Microbiologists interested in biochemical ecology are concerned with the biochemical interrelationships of microorganisms with their environment and the biochemical interactions among populations of microorganisms in their natural habitats. Scant attention has been given in the past to biochemical ecology despite the tremendous increase in scientific literature in many areas of *in vitro* microbial biochemistry and in the descriptive aspects of ecology. However, with the development of considerable detailed knowledge of the soil ecosystem and with the existence of a vast body of descriptive information on the identity and behavior of the subterranean microflora, interest in the interplay between microorganisms and their environment has sharply increased. The present review will attempt to summarize certain of the highlights of recent developments in the biochemical ecology of soil microorganisms although, because of the great array of different soil habitats and the

[1] The survey of the literature pertaining to this review was concluded in December 1963.

217

endless catalogue of microorganisms found therein, only a small proportion of the relevant papers can be cited. Investigations concerned with descriptive ecology or the many agricultural ramifications of microbial activities in soil are outside the scope of the present review, except as they pertain to biochemical ecology. The reader is referred to the several monographs which have appeared in the past few years (1–6) for other aspects of soil microbiology. Of particular relevance to the context of the present discussion is the review by McLaren (7) of the biochemical approach to soil science in general with particular reference to microbial metabolism.

The climax or pioneer community characteristic of a particular ecosystem is a reflection of the interplay of the many physical, chemical, and biological determinants of microbial life and activities, and in seeking to unravel the interrelationships between microorganisms and their environment, the biochemical ecologist endeavors to understand the biochemical influences of the physical environment on the resident microorganisms, the biochemical interrelationships between species in their natural habitat, and the effects of the organisms upon their environment. A clear delineation between these three areas of endeavor is, to be sure, not feasible because of the continuous floral adaptation and fluctuation in response to modifications of the biological and nonbiological determinants and the heterogeneity of the animate and inanimate components of the habitat, but an attempt will be made to keep the three distinct in the interests of brevity and clarity.

Microbial ecologists in general, soil microbiologists included, have too long been obsessed with counting procedures, an enumeratomania (*numerare*, to number; *mania*, madness) coupled with a belief in an almost divine majesty associated with the names of early investigators. The time has long since passed when many studies of this type, which yield mountains of data but few well-documented ecological principles or generalizations, have value or significance. Microbiology is admirably suited for definitive ecological research, for experimental rather than descriptive approaches to ecology, for the establishment of biochemical and physiological principles derived from careful examination of the environment rather than generalizations relying solely upon *in vitro* tests with little regard to natural conditions. Admittedly, it is often not possible to propose basic concepts for the understanding of chemical transformations or microbial interactions in natural habitats because of the complexity of the microflora, the influence of inanimate materials or the occurrence of innumerable competing processes, and investigations with simulated environments or model culture systems and judicious mixtures of type organisms are an essential part of the study of microbial ecology, but these working models are merely *in vitro* means to *in vivo* ends. Descriptive investigations of the microorganisms of terrestrial environments still have great value; far too little information is available on the occurrence, characteristics, and abundance of the algae, protozoa, or higher bacteria and actinomycetes in the soil habitat, but techniques are now at hand—or can be developed from existing procedures—for the creation and delineation of suitable biochemical parameters for microbial ecology.

BIOCHEMICAL INFLUENCE OF THE ENVIRONMENT ON MICROORGANISMS

The microenvironment.—Probably no topic in ecology has elicited as much speculation among soil microbiologists but yet has resulted in as few clear principles as has the microenvironment. Many of the physical and chemical characteristics of the soil are remarkably variable at the specific loci that constitute the micro-sites colonized by the subterranean inhabitants. The supply of available nutrients, O_2, CO_2, and water, the pH, osmotic pressure, and oxidation-reduction potential all vary at the microscopic dimensions that demarcate the habitats of the microflora, and it is not surprising that appreciable differences in population size are observed at micro-sites within the soil or, as shown by Tyagny-Ryadno (8), in bacterial abundance on the surfaces as compared with the interior of aggregates.

Microorganisms apparently in the same habitat are, in fact, often exposed to entirely different environmental influences and population pressures. To understand the forces actually affecting the organisms, a microenvironmental concept rather than the gross macroscopic view of interactions must be adopted. Yet, because of the inherent technical difficulties in biochemical experimentation at the microscopic level, progress in understanding of the microenvironment has been painfully slow. One of the most intriguing ecological problems of nonaquatic habitats, in which the volume occupied by air or water is comparatively small, is related to mechanical interference and the availability of space. Organic nutrients are not uniformly distributed, and it is likely that microscopic sites exist in which readily available nutrients remain unmetabolized because species possessing the requisite enzymes find no portal of entry to the locale well endowed with organic reserves. For example, soluble substances resulting from microbial metabolism or the degradation of plant tissues could be translocated and adsorbed by clay surfaces remote from the mainstream of biological activity and protected by adjacent inanimate particles from microbial penetration. Such a microenvironmental and mechanical interference hypothesis could account for the accumulation of appreciable carbonaceous matter in soil despite the apparent omnipresent shortage of organic carbon. Rovira & Greacen (9), by assessing the influence of aggregate disruption on carbon mineralization by the microflora, presented evidence of an acceleration in decomposition as organic nutrients were made accessible through exposure of substrates present in inaccessible micropores. Viewed also from a microscopic level, it is not inconceivable that, with a physical restriction on cell mass or hyphal extension imposed by particulate materials, space could occasionally limit the size of the population in the microenvironment.

Clays and other colloidal materials dramatically influence the nutrient supply and biochemical activities of the microflora. The inorganic and organic colloids which possess a vast surface in relation to the volume occupied, a gram of clay soil often having a total surface area in excess of 100 square meters, affect proliferation and metabolism by adsorption of cells or filaments, extracellular enzymes, inorganic and organic substrates, and products of metabolism. Using ion exchange resins as model systems, Zvyagintsev (10) noted a relation between pH and the degree of bacterial adsorption. Only certain species were adsorbed, but once bound to the resin, the cells were difficult to remove. When the resin became saturated with bacteria of one species, it would not adsorb additional cells of the same species in significant number although, interestingly, adsorption of cells of other species could occur without removal of the cells derived from the first population. Viruses also adhere to clays and soil particles, the adsorbed virus retaining its infectivity for some time (11).

Clays frequently decrease microbial activities. For example, the respiratory rate of *Bacillus subtilis* fixed to bentonite is less than that of cells free in solution. At certain concentrations, on the other hand, clays have been reported to enhance the respiration of *B. subtilis* and to stimulate nitrogen fixation (12, 13). Commonly, the adsorbed substrate is metabolized more slowly than is the substrate in solution, but an example of the reverse has been revealed by Estermann & McLaren (14), who demonstrated that a monolayer of denatured lysozyme adsorbed to clay surfaces was more rapidly hydrolyzed by extracellular proteinases of strains of *Flavobacterium* and *Pseudomonas* than the same protein in solutions free of colloidal materials. Even a protein adsorbed between the layer lattices of expanding-lattice clay minerals was digested. The enhancement of protein degradation took place only when the substrate was adsorbed, regardless of whether the bacteria were retained or remained in solution. This stimulation, which likely results from the clay functioning as a concentrating surface for substrate and enzyme, may be of significance in microenvironments that are poor in nutrients. Another effect, probably of microenvironmental origin, that appears to involve a clay-microorganism interaction is the relation between the spread of *Fusarium oxysporum* f. *cubense*, causal agent of a wilt of bananas, and the mineralogy of banana soils, the rate of disease spread being related to the type of clay present. It was suggested that this clay mineralogy correlation may result from an effect of the silicate upon the fungal pathogen, the nonpathogenic microflora that serves to check the development of the pathogen, or an altered susceptibility of the host (15).

The explanation of the ability of microorganisms to carry out certain activities in nature at pH values below that at which they function in culture is probably best sought at the microenvironmental level. For example, the formation of nitrate from ammonium proceeds in soils down to a pH value of about 4.0 while nitrification by enrichments in solution culture does not commonly take place below pH 6.0 (16). At specific loci adjacent to the surface of negatively-charged colloidal particles, there is a distinct zone enriched with hydrogen ions so that the acidity immediately adjacent to the particle surface is greater than that in the ambient solution. McLaren (17) has applied this information to explaining the observed change in pH optimum and range for catalytic activity of clay-adsorbed chymotrypsin as compared with the free enzyme, the apparent pH range and optimum of chymotrypsin retained by kaolinite being shifted to higher pH by the clay. These observations with adsorbed enzymes run counter to the report (16) that bacteria in soil nitrify at a lower pH than in solution. Hattori & Furusaka (18) noted that the apparent activity-pH curve for the oxidation of glucose by *Azotobacter agilis* was one pH unit higher for cells adsorbed on an anion exchange resin, an effect which the authors attributed to the fixed bacteria functioning within a hydrogen ion-rich cationic microenvironment surrounding the anionic zone about the resin surface.

Substrates for the population.—The chief microbial producers of energy substrates in pioneer communities deficient in organic matter are algae, which generate fixed carbon at the expense of light energy. At the pioneer stage of microbial successions, the abundance of autotrophic producers relative to heterotrophic consumers is far greater than in the subsequent climax community. As the ecological succession develops and the pioneer species enrich the nutrient reserves of the environment, the initial colonizers are replaced by new communities in which the relative importance of autotrophy has declined until, at the climax phase, the ecosystem is characterized by an autotrophic roof of green plants overlying a subterranean, heterotrophic base utilizing the energy of the organic matter provided from above. The indigenous species constituting the climax community can be classified either as members of the primary population, which metabolize the original energy sources that enter the ecosystem, or as representatives of the secondary population, which utilize products released or synthesized by the primary population or live on the cells of the primary organisms by lysis, parasitism, predation, or by feeding upon dead cell components. The primary microflora includes the algae of the upper zone of light penetration and the heterotrophs that metabolize plant excretions or the organic constituents of the plant and animal remains, the heterotrophic group including both the rapidly growing bacteria and fungi that bring about the initial degradation of readily available carbonaceous nutrients and also those microorganisms utilizing resistant components of the added tissues. The secondary population contains those species growing at the expense of humus and decayed organic remains. In effect, the secondary community includes microorganisms which respond to the substrates generated within the ecosystem; the primary microflora responds to those substrates entering from without.

Despite the wealth of descriptive literature, a clear differentiation between primary and secondary populations is not yet possible. *Pseudomonas* species, probably other gram-negative rods, and certain *Bacillus* species are dominant among the primary bacteria metabolizing amino acids, organic acids, and available constituents of plant remains (19, 20), while fungi exhibiting high growth rates are commonly the initial colonizers of many fresh substrates (21). The actinomycetes, largely *Streptomyces* spp., are with little question late risers in the ecological succession that takes place during plant residue decomposition (20), and hence they are probably members of the secondary population. However, the actinomycetes may be primary organisms when the substrate is a substance to which they are uniquely adapted. Thus, the streptomycetes, despite their slow growth habits, are the chief primary microorganisms when the environment is enriched with chitin (22) or a number of nitrogen-rich organic materials (23).

Much attention has been given to the chemistry of humus, the chief substrate for the secondary population. Humus, the soil organic fraction, includes plant and animal remains in various stages of degradation, cells of the subterranean microflora, compounds formed by the microscopic inhabitants, and products of the reaction of these substances with one another or with inorganic soil constituents. Most evidence indicates that the humic acid fraction is dominated by heteropolycondensates of phenolic constituents, possibly with amino acids participating in the structure of the polymer. Burges et al. (24) demonstrated the presence of a number of phenolic materials in extracts of humic acid fractions, and these authors proposed that the monomers isolated may be bound to one another in the soil organic fraction by ether linkages. The condensation products of the aromatic substances likely make up a large proportion of the carbonaceous substrates of the environment, the persistence of the aromatic polymers in nature reflecting a marked resistance to enzymatic destruction. Polysaccharides are also present in significant amounts to serve as organic substrates; e.g., in several Canadian soils, 6.4 to 7.2 per cent of the carbon in surface soil was in the polysaccharide form while the percentage of the carbon accounted for as polysaccharide was almost twofold higher at great depth (25). The sugar monomers of the polysaccharides include glucose, galactose, mannose, arabinose, xylose, rhamnose, fucose, ribose, and several unidentified sugars (26). The complexity of the substrates for the secondary population is also evident from investigations of the soil nitrogen complexes. In one study, for example, chemical hydrolysis of the organic fraction revealed about 50 ninhydrin-reacting constituents, half of which were identified as specific amino acids or amino sugars (27).

Despite the ubiquity and abundance of organic nutrients, the humus substrates are metabolized slowly by the secondary population, and only a small proportion of the phenolic, sugar, or amino constituents are mineralized in the course of a year. Depolymerization appears to be the rate-limiting step in the decomposition since the concentration of the respective monomers is invariably low; thus, Paul & Schmidt (28), who observed some 15 free amino acids in a silt loam, reported that the level was only about 2 to 4 ppm, an amount probably representing a steady-state concentration between formation and destruction. Only when a readily available form of carbon and nitrogen was added did the concentration of free amino acids become significant.

Further information on the chemistry of bacterial, fungal, actinomycete, algal, and protozoan cells will permit a greater understanding of the substrates maintaining the secondary microorganisms as a significant portion of the carbon must, at one stage or another, move into the cell of one organism and then, in turn, serve as energy source for subsequent species in the food chain. Thus, the large numbers of *Streptomyces* in environments receiving no recent organic materials may result, in part, from the ability of most of these actinomycetes to metabolize chitin, a constituent of the mycelium of a great array of soil fungi (29), although the durability of the streptomycete conidia may likewise be a significant factor in the abundance of representatives of the genus. The finding in humus of bound N-acetylglucosamine (27) is also indicative of a microbial biosynthesis of substrates for other organisms.

The root-inhabiting microflora contains that component of the primary population whose initial substrates are derived from plant excretions or, in certain instances, sloughed off or dying root tissues. The unique ecological niche of these organisms is upon or within the root, and the population includes the rhizosphere bacteria, mycorrhizal fungi, and plant parasites; some of the root zone inhabitants may have an active phase in the nonroot environment as well. Among the root exudation products that may serve as substrates for the primary population of the rhizosphere ecosystem are a large number of amino acids, several monosaccharides and disaccharides, simple organic acids, purines, pyrimidines, glycosides, and several uncharacterized substances (30, 31). The root environment is heterogeneous in substrate composition, and there is a distinct spatial distribution of sites from which the plant-originated nutrients for the microflora are liberated, presumably each point of excretion serving as the locus for the development of the primary microflora most suited to the specific exudates. Thus, ninhydrin-positive compounds are released from the regions of apical meristem, elongation, and of root hair development, and similar spatial patterns of exudation are noted with the sugars. The excretions, in turn, promote microbial development (32). Although many substrates are liberated into the underground habitats with the invasion of the nutrient-rich root into the nutrient-poor soil complex, it is far from clear which of these compounds—amino acids, sugars, organic acids, etc.—are the primary food sources for the dominant species. The abundance of bacteria that require or are stimulated by amino acids is presumptive but far from definitive evidence for a key role for these substances; the bacteria may be responding in the rhizosphere because of their capacity to utilize rapidly nonamino compounds in the excretions, and the amino acids may serve solely as growth factors.

In the rhizosphere habitat, there is a selection, for reasons yet unknown, for bacteria rather than for other broad microbial groupings. Moreover, upon the intrusion of the root into the largely heterotrophic zone, there is a preferential enhancement of the growth of bacteria with shorter generation times and with certain nutritional patterns. The roots likewise favor bacteria whose biochemical activity is, on the average, greater than that of their soil counterparts, this selection for metabolically active bacterial species being particularly pronounced in the rhizoplane (33). In addition to liberating substances that provide the microflora with energy and growth factors, roots produce one and probably more chemotactic substances which establish a unique local microenvironment; for example, the zoospores of *Aphanomyces euteiches* are attracted strongly to the roots of peas and other plants, the greatest site of attraction commonly being the region of elongation just behind the oldest portion of the root cap (34). The identity of the chemotactic compounds concerned in zoospore attraction remains unresolved.

Substrate availability.—A major and often the chief factor limiting the microbial mass in soil is the insufficiency of energy substrates for the microflora, either organic carbon for the heterotrophs or reduced inorganic substances for the chemoautotrophs. Nevertheless, the soil ecosystem is typically rich in organic carbon and often in ammonium ions, the oxidation of which could provide energy for *Nitrosomonas* and related nitrifying bacteria. Much of the carbonaceous nutrient reserve is undoubtedly inherently resistant to biological degradation; a goodly portion, on the other hand, is in the form of easily degraded molecules which apparently exist in an unavailable or inaccessible state. Part of the organic materials in the inaccessible microenvironments seems to be readily utilizable by the microflora, the persistence of the substrates resulting from the physical barrier to microbial penetration (9). Unknown as yet are the age of these inaccessible substrates and the means by which the organic substances reached the micro-sites.

Substrate availability is often determined by adsorption, binding, or complexing of the nutrient with inorganic or organic components of the environment, and the clay-nutrient complexes frequently are protected to an appreciable extent from bacterial or enzymatic degradation. Estermann et al. (35) demonstrated, however, that extracellular proteolytic enzymes can penetrate between the layers of the lattice of clay minerals and hydrolyze proteins contained therein. Not only proteins but polysaccharides, phosphate esters, and other potential nutrients are adsorbed by clays and inorganic soil constituents with a consequent decrease in susceptibility of the sorbed compound to decomposition. However, studies of amino acid metabolism indicate that at least some adsorbed amino acids are rapidly oxidized and disappear in a few days (36, 37). The complexing of proteins with water-soluble constituents of plants also markedly influences the availability of the proteins to microorganisms (38). Of particular importance in governing the rate of substrate and organic matter turnover are the lignins, plant constituents which enter the soil in large quantities and remain there for long periods because of their resistance to enzymatic destruction. The complexing of proteins with lignins or clay diminishes the availability of the protein to the microflora, the lignins offering a greater degree of protection than clay (35). Not only are organic substrates in soil often less readily utilized than the same compounds when free in solution, but inorganic substances that are metabolized with ease in solution frequently are obtained with difficulty in nature. For example, nitrifying autotrophs oxidize very slowly the ammonium fixed by illite and bentonite, and the nitrification of the clay-fixed ammonium is further reduced by potassium (39). Likewise, zinc availability to *Aspergillus niger* may be quite low depending upon local conditions, and the fungus may not be able to obtain the cation despite the presence of considerable amounts of the element in soil (40).

Adaptation to the environment.—There is a great need in the field of microbial ecology for experimental information and suitable parameters to provide a basis to account for the occurrence and dominance of specific microbial groups in individual ecosystems, but, despite the large body of descriptive literature, few basic ecological criteria have been proposed to explain the adaptation of microorganisms to the physical environment and the apparent adaptive value of distinctive physiological and genetic traits. In terrestrial environments endowed with vast microbial populations, there undoubtedly occurs a natural selection for those species most suited to the particular circumstance, and an organism that finds for itself an ecological niche in the soil or in its innumerable microhabitats must be adapted nutritionally and physiologically to the ecosystem. Substrates needed by the organism must be available or be made available by itself or its neighbors, and the fixed physiological limits of the organism must fall within the bounds of the extremes of the particular environment, be they the limits of pH, temperature, aeration, space, or toxins. Little attention, however, has been given to microbial adaptations, biochemical or genetic, as related to colonization of natural habitats or to dominance in specific macro- or microenvironments. Admittedly, microorganisms are potentially everywhere, and it is the character of the habitat which selects and which determines generic and species distribution and dominance, but the critical environmental determinants and the nature of the adaptation to the conditions imposed by the ecosystem remain largely unresolved.

Park (41) has presented evidence that native soil fungi may exhibit greater resistance to bacterial antagonism than do alien fungi. He inoculated into sterile soil a mixed bacterial flora together with fungi derived from several habitats and noted that only the mycelium of the soil isolates remained viable after 16 weeks, such a difference in behavior between native and alien fungi suggesting a natural selection for the more competitive strains under the restrictions imposed by the habitat. Evidence for a morphological selection is found in investigations of the diatom flora, terrestrial diatoms tending to be smaller in size than their aquatic counterparts (42). Adaptation to high osmotic pressures and salt concentrations has been noted among the actinomycetes, *Streptomyces* from saline soil typically tolerating higher osmotic pressures and salt concentrations than their counterparts from non-saline environments (43). On the other hand, Henis & Eren (44) could find no difference in salt tolerance or growth retardation by sodium chloride between *Bacillus* and *Micrococcus* isolates from a highly saline and a nonsaline soil. An adaptation of actinomycetes to pH of the environment is also suggested by the data of Taber (45), who reported that only one of six acid soils contained acid-sensitive actinomycetes while nine of ten alkaline soils had actinomycetes whose growth was retarded by slight acidity. Ul'yanova (46) also suggested a possible adaptation of *Nitrosomonas* to the organic matter level of the habitat, strains less sensitive to inhibition by certain organic compounds being derived from environments rich in organic matter.

Modification of the ability of a microorganism to cope with an altered environment may entail a biochemical adjustment either with or without a change in genotype. A genotypic modification that could be of ecological as well as economic significance is the transformation of noninfective bacteria to strains with an infective capacity. Transformation of this type has been observed between isolates of *Rhizobium* in relation to the transfer of infectivity in the bacterial-legume symbiosis (47, 48). Adaptation is likewise evident in the association of individual genera of fungi or bacteria with soil types or climatic regions. The distribution of *Beijerinckia*, a genus essentially absent from soils in temperate zones but common to tropical regions and especially to laterites, affords the best documentation of a microbial geography. Becking (49) studied the physiology of *Beijerinckia indica* with a view to explaining its association with laterites. He observed that, consistent with the low calcium and phosphate concentrations and the frequently high iron, aluminum, and manganese levels of lateritic soils, *B. indica* required no calcium for nitrogen fixation, and it developed better in phosphate-poor solutions and tolerated higher iron, aluminum, and manganese concentrations than did *Azotobacter*, the aerobic nitrogen-fixing counterpart of temperate climates. This relationship between the mineral nutrition and tolerance of *Beijerinckia* and its distribution is an interesting contribution to the understanding of the environmental determinants of microbial distribution. Recent work has also been concerned with the relationship of algal and fungal groups to soil types and broad environmental and climatic factors (50, 51).

Other ecological determinants.—At any one time, microorganisms are under the influence of many primary and secondary ecological determinants of the physical environment, and it is often not possible to attribute a specific response to an individual environmental factor. Nonetheless, with the wealth of descriptive information dealing with the subterranean inhabitants to serve as a guide, the biochemical ecologist must focus attention upon understanding why a modification in the ecosystem results in a specific microbiological change and why soil types, horizons, or microenvironments having disparate physical and chemical properties possess populations of different composition.

An effect of the environment on microbial activities that is frequently overlooked is the influence of inorganic elements on biochemical changes catalyzed by the microscopic residents. Numerous microbial transformations of geochemical importance or of concern in plant nutrition respond markedly to the level and availability of micro- or macronutrients. Molybdenum, an element often present in large amounts yet biologically unavailable, is essential for nitrogen fixation not only by free-living microorganisms and the *Rhizobium*-legume symbiosis but also, as demonstrated by Bond & Hewitt (52), for the assimilation of free N_2 by the nodulated nonlegume, *Myrica gale*. Similarly, a specific cobalt requirement for nitrogen fixation, previously reported for nodulated legumes, has been observed for the analogous reaction in the nodulated nonlegumes, *Alnus glutinosa* and *Casuarina cunninghamiana* (53). Although the requirement for the element is minute, cobalt-poor soils are not unknown, and the resulting deficiencies are associated with a diminution in the magnitude of nitrogen fixation by the *Rhizobium*-legume symbiosis. Cobalt deficiency has no effect if the symbiotic system is provided with adequate combined nitrogen (54). Calcium supply is also of ecological significance to the *Rhizobium*-plant association, but there appears to be a differentiation between temperate and tropical legumes as only the former are sensitive to a lack of calcium while the tropical species nodulate at particularly low concentrations of the element (55). In this regard, it is of interest that *Rhizobium japonicum* and tropical rhizobia show no response to calcium (56).

The O_2 supply is particularly important to the distribution and metabolism of microorganisms in soil, the level of O_2 in the microenvironment varying considerably with the surrounding pore space, moisture content, and quantity of readily degraded organic carbon. Greenwood (57, 58) examined the effect of O_2 level on the aerobic and anaerobic heterotrophic activity of the population in both carbon transformations and in nitrate dissimilation. In certain locales, particularly those with insufficient aeration, clay minerals may enhance the growth of bacteria, but such a stimulation is less apparent when aeration is adequate (13). The O_2 supply directly affects the oxidation-reduction potential (E_h) of the microenvironment, and the E_h limits in natural environments for algae, sulfur, and iron bacteria and the denitrifiers have recently been established (59). However, it is not yet clear whether E_h per se is a primary determinant of microbial distribution or whether the potential merely reflects the O_2 status and the level of oxidizing substances in the environment. Soils often contain ample quantities of elements in their higher oxidation states, ferric or manganic ions, for example, and O_2-dependent species may sustain themselves in the absence of atmospheric oxygen by using the oxidized forms of other elements as terminal electron acceptors for growth. The demonstration that extracts of *Micrococcus lactilyticus* are capable of catalyzing the reduction of higher oxidation states of a variety of elements (60) makes this a real possibility.

Carbon dioxide is a prominent ecological determinant not only because it affects autotrophic proliferation and alters the micro-site pH but also because of its potential role as a differential inhibitor and as an essential nutrient for heterotrophs. Burges (2) suggested that CO_2 might be of greater consequence in determining population modifications in poorly aerated circumstances than O_2, a slight increase in pCO_2 frequently having a more marked influence upon fungi than the corresponding decrease in pO_2. This suggestion is supported by the observation that subterranean strains of *Rhizoctonia solani* are more tolerant of CO_2 than strains whose chief site of activity is near the soil surface, whereas isolates of this pathogen that attack aerial portions of susceptible plants exhibit the greatest sensitivity (61), the particular habitat seemingly selecting for the predominating *R. solani* types. Pathogenicity of the fungus is diminished by CO_2, moreover, a factor of possible relevance to investigations of the biological control of soil-borne pathogens (62).

Considerable attention has been given to the role of moisture in the ecology of soil fungi, a topic recently reviewed (63). A change in moisture status affects other groups of microorganisms, however, a dramatic example being revealed in the course of a severe drought, during which time the proportion of actinomycetes, initially accounting for less than 30 percent of the count, rose until they made up more than 90 per cent of the viable propagules. It was suggested that the dominance of the actinomycetes resulted from the resistance of their conidia to desiccation (64). Drying followed by remoistening of the soil is a prelude to a flush of microbial activity, measured either in terms of carbon or nitrogen mineralization. This spurt has been attributed to exposure to the microflora of unavailable or inaccessible substrates as a consequence of the drying-wetting sequence (65).

A change in pH alters many physicochemical properties of the ecosystem so that it is frequently difficult to distinguish between those microbiological responses resulting from a direct influence of pH and those which arise from indirect physical, chemical, or often biological changes in the habitat. For example, acidification, by making certain essential nutrients more available, may modify the composition of the community or the transformations catalyzed by the microflora; acidification has been noted to render more available the soil phosphorus, magnesium, iron, calcium, and zinc (40, 66). In plots of a single soil type with pH the sole overt variable, a statistically significant effect of pH on the abundance of total microorganisms and denitrifying bacteria was recorded, the latter organisms being especially sensitive to acidity (67). An indirect influence of acidity on nitrifying bacteria appears from the results of Weber & Gainey (16), who noted that the nitrifying autotrophs oxidized ammonium in solution culture only at pH values greater than about 6.0 although the bacteria were capable of effecting the same reaction in soils to a pH of approximately 4.0.

The vast land area in paddy culture affords a semiaquatic, semiterrestrial habitat which has been the subject of considerable inquiry from agriculturalists and ecologists. The shift to the aquatic type of environment upon flooding or waterlogging is accompanied by a rise in soluble iron, manganese, calcium, magnesium, nickel, and cobalt, abundant fermentable substrates accentuating this change (66, 68). The reducing conditions are also associated with the formation of considerable quantities of soluble sulfide and insoluble ferrous sulfide, reactions which result entirely from microbial metabolism (69). Takijima & Sukuma (70) demonstrated that, in the fermentation of the available carbon in water-saturated environments, several organic acids appeared, with acetic and butyric acids predominating. Upon flooding, the abundance of fungal propagules declines rapidly, but the diminution in bacterial and actinomycete numbers is commonly not as pronounced. Even after the water is removed, the suppression of the mycoflora remains in evidence. The inhibition does not seem to be solely a consequence of O_2 depletion. CO_2, sulfide, or other toxic agents have been proposed as fungicidal principles generated during flooding (71). Menzies (72) observed that, although the microsclerotia of *Verticillium dahliae* survived for long periods when aeration was adequate, the viability of these resistant structures was readily destroyed in flooded or anaerobic soils, apparently by a diffusible fungicidal substance.

Plant roots may modify the microflora by means other than furnishing substrates to support growth. Buxton (73), for example, demonstrated that the development of 15 of 24 rhizosphere isolates was enhanced by root exudates. By contrast, a substance that inhibits nodulation of legumes by *Rhizobium trifolii* and *R. japonicum* is released by certain soybean varieties (74). Many of the plant effects are undoubtedly indirect; a good example was provided by Woldendorp (75), who showed the influence of roots upon denitrification resulted from the diminished O_2 levels or the need for additional electron donors associated with the decomposition of the plant excretions.

INTERACTIONS AMONG MICROORGANISMS IN THE SOIL ENVIRONMENT

Commensalism, protocooperation, and symbiosis.—Because of the vast numbers of physiologically heterogeneous microorganisms restricted to and interacting within a confined space, the soil ecosystem offers a field largely untrod to the ecologist concerned with the physiological and biochemical bases for the associations among microorganisms in nature. In investigations of the interactions between populations of two microbial species, most attention has been centered upon the relationships resulting in the suppression or decline of one of the two groups, but there is developing, albeit slowly, a literature dealing with beneficial associations. In soil are various substances, most as yet uncharacterized but apparently of biological origin, which stimulate some phase of the growth cycle of one or another inhabitant. For example, unknown compounds that favor the germination of *Fusarium solani* chlamydospores are generated during the decomposition of plant remains (76). Similarly, soil leachings stimulate the hatching of *Heterodera rostochiensis* larvae, the data indicating that the active factor is produced by the indigenous microflora (77). Although some of the growth-enhancing factors synthesized microbiologically have a complex structure, many are undoubtedly simple molecules like sugars, amino acids, vitamins, or purine and pyrimidine derivatives. Thus, Vagnerova & Vancura (78) demonstrated that rhizosphere bacteria were capable of releasing amino acids into their surroundings, and these compounds in turn could satisfy the nutrient demands of nonexacting bacteria. The frequency of occurrence in soil of bacteria requiring one or more growth factors is itself evidence that such an interplay between production and utilization of these substances is prominent in microbial ecology.

Several additional instances of commensalism and protocooperation have been recorded recently. Zavarzin (79) found two strains of *Pseudomonas* which, together but not separately, could oxidize manganous ions. Likewise, species of *Nostoc* were noted to be stimulated when grown in two-membered association with isolates of *Caulobacter, Rhizobium, Agrobacterium, Bacillus,* or *Streptomyces.* Although the *Caulobacter* sp. appeared to synthesize indole-3-acetic acid, the auxin did not seem to be the sole stimulant for the alga (80, 81). In more than half of the two-membered associations between algae and other microorganisms examined by Parker & Bold (82), a stimulation of the photoautotroph was noted. The heterotrophic organisms, strains of bacteria, actinomycetes, and fungi, improved growth of the algae by mineralizing nitrogen, altering the availability of O_2 and CO_2, or degrading the algal extracellular polysaccharide. Carbon dioxide, abundantly and continuously evolved in carbon mineralization, also stimulates not only photo- and chemoautotrophs but also the growth and spore germination of many subterranean heterotrophs.

The most thoroughly investigated but still poorly understood symbiosis, that of rhizobia with their leguminous hosts, continues to intrigue microbiologists. Apart from the mechanism of fixation of N_2, probably the outstanding biochemical problems concern the contributions of the individual symbionts to the association and the basis of the specificity of bacterium for plant. There appears to be a symbiotic production of indole-3-acetic acid in the *Trifolium-R. trifolii* association, the plant excreting the tryptophan that is converted to auxin by the bacterium, and it has been proposed that auxin is implicated in development of the infection thread and in the initiation of cell division preceding nodule formation (83). Because the cytological effect of indole-3-acetic acid on *Trifolium repens* root hairs differed from that of the infective bacteria and destruction of the potential auxin-like activity of *R. trifolii* cultures did not prevent the bacterial-induced changes in the root, Sahlman & Fahraeus (84) concluded that the auxin was not solely responsible for root hair curling. They did, however, observe at least one other soluble substance concerned with infection. Ljunggren & Fahraeus (85), in a significant contribution to an understanding of the mechanism of infection that precedes the symbiotic acquisition of N_2, observed that polygalacturonase was always present in legumes inoculated with their homologous bacteria, but there was little or no polygalacturonase in roots exposed to heterologous rhizobia. Moreover, noninfective *R. trifolii*, which was not competent for polygalacturonase induction, became an inducer of the enzyme in the host when the bacterium was transformed to infectiveness. Another major advance in investigations of the symbiosis was the demonstration of the site of N_2 fixation in the nodule. Probably because microbiologists have dominated study of the symbiosis, it has been assumed that nitrogenase was a constituent of the bacteroid rather than the plant tissue component of the nodule; however, by exposing excised soybean root nodules to $^{15}N_2$ for short periods and fractionating the nodule into bacteroid and plant portions, Bergersen (86) demonstrated that the newly fixed nitrogen was not incorporated initially into the bacteroid fraction. Subsequent investigations hopefully will reveal the biochemical basis for the association and the role of the two partners in this geochemically important and ecologically interesting symbiosis.

Competition and limiting elements.—Competition, the rivalry for space or the supply of essential or stimulatory nutrients, appears to be a principal factor in determining the prominence of microbial species in terrestrial environments in which the demand for nutrients or space exceeds the supply. Although investigations of competition in nature have long been neglected in favor of studies of antibiosis and toxin production, a consequence of the emphasis placed upon those microorganisms isolated from soil which produce antimicrobial agents *in vitro*, the significance of competitive phenomena in the self-regulating mechanisms which govern the composition of the climax community has been subjected recently to re-examination. The element which restricts population size does not necessarily seem to be the one present in the lowest concentration but rather the element, the supply of the available forms of which is below the biological demand. In this sense, the chief limiting nutrient for microbial development and abundance appears to be the available carbon level, and in dual-culture model systems, competitive bacteria retard fungal proliferation at low levels of available carbon, a retardation reversed by organic carbon supplementation (87, 88). Once the carbon demand is satisfied, however, other nutrient deficiencies appear; thus, in dual-culture models in carbon-amended soil, the suppression of test fungi by nontoxin-producing bacteria results from the bacterial utilization and immobilization of nitrogen (89). The only elements other than carbon and nitrogen that limit heterotrophic development seem to be phosphorus and sulfur, but the need for these elements is detectable solely upon the addition of considerable exogenous carbon and nitrogen (87, 90). If the development, pathogenicity, or survival of plant pathogens is related to nutrient competition, then modifying the environment to induce the appropriate deficiency should result in disease control, while supplementation with the element in question should lead to a greater incidence or severity of disease. Observations of this type are now available with a number of fungal pathogens; e.g., the *Fusarium solani* f. *phaseoli* root rot was controlled by inducing nitrogen deficiency, whereas nitrogen supplementation increased disease severity (91). Similarly, enrichment with nitrogen favored the activity and survival of *R. solani*, while promoting microbial nitrogen assimilation had the reverse effect (92).

The physiological basis of competitiveness is poorly understood. As a first approximation, it is likely that microorganisms growing readily and capable of exploiting most efficiently the nutrients in limited supply should be the best competitors under conditions of nutritional stress. Evidence to support this contention is found in the demonstration of a correlation between competitive ability of bacteria and growth rate, the faster growing organisms being vigorous competitors in model systems (88). Nevertheless, poor competitors for nutrients are numerous, and these must have compensating mechanisms, such as the capacity to persist for long periods by virtue of resistant structures or the ability to utilize complex or adsorbed organic substrates available to few other species.

Space may be a primary determinant of species dominance or microbial mass in nutrient-rich microenvironments (87), but nutrient excesses are uncommon in terrestrial environments except around decaying tissues and possibly surrounding plant roots, from which there is a prolonged outpouring of a diversity of readily metabolized organic substrates (30). In this, a root-governed habitat, the sand, silt, and clay that serve as the solid backbone of the ecosystem might well offer mechanical impedance to hyphal extension, protozoan movement, or bacterial proliferation.

Amensalism.—Despite the vast literature dealing with antibiosis and antibiotic biosynthesis by soil-derived organisms, the evidence for the existence or formation in soil of characterized antibiotics is equivocal, and the ecological function of these antimicrobial agents remains unresolved. Mirchink & Greshnykh (93), however, presented data indicating that two penicillia were capable of producing antibiotics when introduced into nonsterile soil, the amount increasing upon the addition of sucrose. Bulbiformin, an antibiotic formed by *B. subtilis*, was also stated to be found in soil on the basis of an assay relying upon a peculiar morphological change in *Alternaria tenuis* (94). The existence in soil of diverse unknown substances, presumably of microbial origin, inhibiting spore germination or growth of fungi and bacteria, is quite evident. The inhibitory principle for *Aspergillus fumigatus* in well-aerated soils is heat-labile, water-soluble, diffusible, and migrates to the anode during electrophoresis (95). The substance in anaerobic habitats that is fungicidal to *V. dahliae* is likewise diffusible (72). The fungistatic factors exhibit a wide spectrum of action, inhibiting species representing a variety of genera, though some fungi are little if at all affected. The fungistasis appears to be microbiologically induced since the full inhibitory capacity can be regained by sterile soil after inoculation with bacteria, fungi, or actinomycetes (95). Lingappa & Lockwood (96) noted that products of lignin degradation markedly suppressed the germination of *Glomerella cingulata* conidia although lignin itself was without effect, and these investigators therefore proposed that intermediates in lignin decomposition account in part for soil fungistasis. At the present time, nevertheless, the fungistatic property is little more than an ecological curiosity, and the identification of the active chemicals, the microorganisms responsible for their biosynthesis, and the floral modifications induced as a result of their appearance still require further inquiry.

It is tempting to attribute, particularly by analogy to *in vitro* investigations of antibiosis and the significance in chemotherapy of antibiotics formed by soil microorganisms, the amensalistic relationships in nature to exotic compounds effective as antimicrobial agents at low concentrations. Such an assumption may be misleading inasmuch as many biologically generated, chemically characterized soil constituents known to be present in appreciable amounts are likewise toxic. For example, organic acids commonly liberated during the growth of bacteria and fungi have been detected in soil, and certain of the acids were found to inhibit the growth of fungi, especially in acid conditions (71, 97). Carbon dioxide, the concentration of which may become quite high particularly within the microenvironment, is also concerned in amensalistic interactions, altering not only the biochemical activity of individual species but also the composition of the microbial community at the site of CO_2 accumulation (61, 62). Ammonia formed in nitrogen mineralization is a potent inhibitor for *Nitrobacter* in alkaline habitats, the toxicity being governed by both the pH and the ammonia concentration (98), and the nitrite that remains by virtue of the repression of nitrite-oxidizing autotrophs may, in turn, affect the survival or biochemical activities of sensitive fungi. Thus, Sequeira (99) has observed a correlation between nitrite accumulation and the biologically induced soil toxicity towards *F. oxysporum*.

Parasitism and lysis.—It is simple to detect many types of predatory and parasitic relationships operating within natural habitats. The infection of bacteria and actinomycetes by bacteriophages, the feeding upon bacteria by protozoa, the action of myxobacteria and myxomycetes upon bacteria, and the parasitism of one fungal species by another are but a few of the known interactions. The ecology of the nematode-trapping fungi, in particular, has attracted considerable attention in recent years. In addition, parasites, tentatively considered to be obligate, with lytic activity towards *Pseudomonas* and *Xanthomonas* spp. have been found in soil. These small parasites, which pass through filters of 0.45 μ diameter, are curved rods with a single polar flagellum (100). Teakle (101) presented evidence that *Olpidium brassicae* may be a vector for the soil-borne tobacco necrosis virus which infects the roots of many plant species.

Fungi are quite susceptible to lysis, and many indigenous species are capable of excreting principles effecting the lysis of the hyphae. Chinn & Ledingham (102) noted that, following the germination in soil of *Helminthosporium sativum* spores, the germ tubes that had developed from the spores were rapidly lysed. A large proportion of *Streptomyces* and a lesser percentage of the bacteria are active in mycolysis, and the mycolytic principles digest both living and dead hyphae of *G. cingulata*. Natural but not sterilized soil is likewise capable of effecting the digestion of viable or inactivated mycelium (103). Lysis of *F. oxysporum* by bacteria is associated with the release by the mycolytic isolates of chitinase and a β-1,3-glucanase, enzymes presumably induced by the appropriate polysaccharides of the fungal wall, but the two enzymes together do not bring about the digestion, which suggests the presence of other enzymes in the lytic culture filtrates and the occurrence of other constituents in the cell wall (104).

Because susceptible microorganisms maintain themselves in the presence of their parasites or predators, a balance must exist between the two groups. Cyclic sequences in the composition of the community, with a dominant period for the host or prey followed by the dominance of the parasitic or predatory species, are largely unknown, but cyclic successions undoubtedly are operative within the microenvironment, although existing crude macroenvironmental methodology cannot reveal such changes. In time, a microhabitat depleted of the suscept will be repopulated as the predator or parasite moves on to greater glory or meets its doom upon the enzymic rack of some new predator. In addition to the hypothesis which presumes that the inability of predator and parasite to overwhelm the suscept results in part from the failure of the former to find all cells of the latter, the predator may have alternate species upon which to feed, it may possess an active and even a dominant saprophytic stage, or it may be faced with genetic modifications that result in indigestibility of the formerly delectable species. In the absence of meaningful data, the imaginative ecologist can likely provide other, possibly more suitable hypotheses.

Microbiological control.—The climax communities of soil habitats are in equilibrium with their macro- or microenvironments, a dynamic rather than a static equilibrium, a steady-state balance among the constituent species. The composition of the community is, unless external disturbances or exogenous nutrients disturb the equilibrium, rather stable, and the microflora and fauna adjust to meet the changes induced from without. At the climax stage of ecological succession, self-regulating mechanisms are operative by means of which the community, upon exposure to influences alien to the ecosystem, tends to revert to the initial steady-state condition, and successful colonization by foreign organisms is vigorously resisted by the dominant indigenous species through competition, amensalism, parasitism, etc. This inherent microbiological control assumes economic importance in the biological destruction of the many human, animal, and plant pathogens that enter the environment with diseased tissue or infected wastes, but the nature of the self-regulatory or exclusion mechanisms in the ecosystem is far from clear.

Frequent attempts have been made to modify the self-regulatory or exclusion mechanisms for practical purposes; that is, to induce a directed change in the microbial community by modifying the ecosystem in such a way that pathogenic organisms indigenous to or harbored within soil are either eliminated or brought under control. The traditional and frequently ancient practices of fallowing, crop rotation, or incorporation of plant remains probably owe their success in alleviating disease largely to the changes brought about by the saprophytic species. Differences between soils in the relative efficiency of the control mechanisms leading to the suppression of *Strepto-*

myces scabies have been demonstrated by Menzies (105), who observed that the inhibitory factor could be transferred to soils not active in suppression of the scab actinomycete. The effectiveness of various natural substrates in bringing about the control of *R. solani* and *F. solani* f. *phaseoli* has been examined, and the population of the pathogens and the severity of the diseases they cause were reduced appreciably by certain of the materials used. Often, effectiveness was associated with a modification resulting in a nitrogen deficiency, and the control may be partially attributable to a poor competitiveness by the pathogen for this element (91, 92, 106). Competitive capacity of pathogens is only one of many exploitable physiological traits, however, and attempts to alter the saprophytic community in a directed fashion so as to exploit the physiological deficiencies of intuders or indigenous pathogens must take into account other mechanisms of biological control; e.g., the fungistatic influence of CO_2 (62) or the detrimental effects of inorganic products of microbiol metabolism (71, 99). An approach developed by Chinn & Ledingham (102) makes use of the fact that *H. sativum* spores, but not the germ tubes, were reasonably resistant to lysis so that significant eradication of the pathogen could be obtained by stimulating the fungus to emerge from its resistant stage.

Another approach to the search for means of effecting the directed alteration of the ecosystem to favor a specific segment of the microflora originated from *in vitro* investigations of fungal lysis and of the chemical constitution of the pathogen's hyphal walls. Chitin and a β-1,3-glucan were found in the cell walls of pathogenic fusaria, and the isolates lytic to these fungi were observed to contain chitinase and the appropriate glucanase (104); on the basis of these enzymatic and cell wall relationships, chitin and a β-1,3-glucan were examined as possible selective substrates to enhance the development of an antagonistic population. Addition of either polysaccharide led to control of diseases produced by *Fusarium oxysporum* f. *conglutinans* and *F. solani* f. *phaseoli*, fungi of the genus *Fusarium* containing both chitin and the glucan in their hyphal walls, but the polysaccharides did not alleviate diseases caused by *Pythium debaryanum* or *Agrobacterium tumefaciens*, microorganisms not reported to contain the polysaccharides in their walls (22, 107). Of the carbohydrates examined, these two but no other polysaccharide and also not the monomers of chitin and the glucan were effective (108). Still to be verified, however, is whether the destruction of the pathogen or the disease control results from a selective enhancement of the mycolytic population.

BIOCHEMICAL EFFECTS OF MICROORGANISMS UPON THEIR ENVIRONMENT

Geochemical effects.—The role of microbiological agencies in geochemistry has been scantly explored, yet the sparse literature clearly reflects a key position for the microscopic inhabitants of terrestrial environments in geochemistry and geochemical cycles. The biogeochemical contributions fall into one or more of six general categories: (*a*) increasing the chemical complexity of the ecosystem by biosynthetic reactions of the autotrophs, by humus formation, etc.; (b) decreasing the chemical complexity of the ecosystem by degradation and mineralization of complex molecules; (c) oxidizing the elements in their various inorganic and organic forms; (d) reducing the higher oxidation states of the elements; (e) solubilizing or precipitating geochemicals, often a consequence of oxidation or reduction; and (f) changing the total amount of an element in the ecosystem, as by the fixation or evolution of gaseous forms of carbon, nitrogen, oxygen, hydrogen, or sulfur. The oxidative processes may be linked with energy metabolism, as in heterotrophic carbon oxidation or chemoautotrophic nitrogen oxidation, or the reactions may be incidental to the energy metabolism of the responsible organism, such as in heterotrophic nitrate and sulfate formation. Reductive processes may likewise be either directly linked with energy metabolism, the oxidized forms of the element serving as electron acceptors, or the reactions may be incidental. Examples of the latter are found in the formation of reduced products through O_2 depletion, a fall in oxidation-reduction potential, or a change of pH resulting from microbial activity.

The focal position of microorganisms in the nitrogen cycle and the importance of this element in plant nutrition have been an impetus to inquiries concerned with quantitative and qualitative evaluations of the microbial contribution to the individual transformations that nitrogen undergoes. Estimates of the fixation of N_2 by free-living microorganisms and the consequent return of the element into the food chain of the biosphere have been performed with model systems representing diverse environmental conditions. By use of $^{15}N_2$, Chang & Knowles (109) determined the nonsymbiotic fixation of N_2 in two Canadian soils to range from 0.015 to 0.129 mg nitrogen per kg soil per day, while Parker (110) reported the quantity fixed nonsymbiotically in a fine sandy loam in Australia to be 15 lb per acre per year, although, in a soil under grass where fixation by the rhizosphere and adjoining populations may be coupled with utilization of root excretions, the nitrogen gain was severalfold greater. Probably the principal limiting factor for heterotrophic N_2 fixation, the supply of energy, is not a serious ecological restriction for the blue-green algae, and where the Cyanophyceae flourish, appreciable nitrogen gains have been recorded. Moreover, following inoculation of paddy fields with the N_2-assimilating Tolypothrix tenuis, the alga not only becomes successfully established but also affects favorably the growth of the rice plant (111). The organisms that fix N_2 symbiotically with an autotrophic partner likewise do not suffer from a deficiency of energy sources, and here, too, the nitrogen accretion may be appreciable. Moore (112), for example, observed 250 lb more nitrogen per acre under a Centrosema pubescens-grass mixture in a tropical latosol than under the pure grass stand after a two-year growth period. Despite some four score years of investigation of the Rhizobium-legume symbiosis, only recently has the fixation of N_2 by nodulated nonlegumes come under close scrutiny, and three additional nonlegumes have now to be added to the list of symbiotic N_2 utilizers, Comptonia peregrina (113), Discaria toumatou (114), and Coriaria myrtifolia (115); in no instance has the endophyte been identified or characterized. Quantitative data on the magnitude of nitrogen enrichment by the nonlegume symbiosis are rare, moreover, but the results of a comparative study of the nitrogen content of soil under Douglas fir with that under a fir-*Alnus rubra* plantation indicated that *A. rubra* increased the nitrogen level of its underlying soil by 938 lb per acre in a 26-year period (116). In a sand dune soil, the net nitrogen fixed under a stand of *Casuarina equisetifolia* was reported to be 52 lb per acre per year (117).

Cady & Bartholomew (118) examined the sequence of gaseous products of denitrification in soil. The first volatile product was nitric oxide, but the quantity evolved was small. Subsequently, nitrous oxide and lastly N_2 appeared, the time course of gas evolution suggesting that N_2 was formed by a biological reduction of nitrous oxide. Not all steps in nitrogen volatilization are biological, however, and significant chemical losses may follow the microbial generation of nitrite by oxidative or reductive pathways in acidic environments (119).

Many elements are subject to microbial transformation independent of any relation of the element to cell metabolism or structure. Thus, upon flooding, the levels of soluble iron, manganese, molybdenum, copper, nickel, and cobalt increase (68); part or all of the increase in mobility of certain of these elements undoubtedly arises indirectly from bacterial metabolism at limiting O_2 tensions. Acidification of the environment through anaerobic decomposition of organic materials or through oxidations catalyzed by chemoautotrophic bacteria likewise leads to the solubilization of calcium, magnesium, potassium, and aluminum (66, 120, 121), and thiobacilli may be responsible for the release of copper, iron, zinc or molybdenum from the respective sulfides (122, 123). Another mechanism of solubilization involves 2-ketogluconic acid, bacteria producing this acid releasing a host of elements from minerals, insoluble silicates, and other insoluble inorganic compounds (124). A totally different mechanism of transformation of inorganic substances, one of possible ecological importance, is suggested by the demonstration that *M. lactilyticus* extracts reduce the higher oxidation states of many elements not hitherto known to be acted upon biologically (60). The solubilization of insoluble compounds of phosphorus, a reaction of particular relevance to plant nutrition, has come under close scrutiny, and it appears that a high proportion of the soil and rhizosphere bacteria are capable of dissolving insoluble calcium phosphates (125, 126). Thus, a surprising number of elements undergo microbial transformation, *in vitro* at least, by acidification, chelation, and reduction, but the importance of most of these reactions *in vivo* remains to be established.

Pedogenesis.—The ecological successions, the physiological bases for the successions in microbial communities, and the series of biochemical changes brought about by the sequence of populations that appear and then decline in the transition from the pioneer to the climax communities associated with pedogenesis have been largely ignored in favor of invitroological investigations, but certain principles and concepts have of late been proposed and developed. In the pioneer community, the abundance of producers (microorganisms effecting a net biosynthesis of organic matter through reactions coupled with photo- or chemosynthesis) relative to heterotrophic consumers is far greater than at the climax stage. Concurrently, the rate of generation of organic carbon in the pioneer stages exceeds the rate of carbon mineralization, and heterotrophically useful energy is stored faster than it is consumed; this biologically available energy is bound in humus. At the climax phase, however, the rates of organic carbon accumulation and mineralization are equal, and hence, in the absence of externally induced disturbances of the ecosystem, the energy inflow equals the outflow, the humus concentration reflecting the steady-state between the two opposing transformations. Webley and collaborators (127) have demonstrated that, coincidental with the colonization of rock surfaces by lichens of the pioneer community, there is an increase in abundance of bacteria and fungi. Many of these heterotrophs, which undoubtedly participate in the weathering of rocks and pedogenesis, dissolve calcium, magnesium, and zinc silicates. Roy (128), using samples of rock, weathered rock, and soil materials that resemble the weathering sequence in the conversion of rock to soil, observed that the relative quantity of phosphorus and potassium that becomes biologically available increased with the extent of weathering, despite the quantitative phosphorus and potassium loss from the material as it was subjected to weathering. Chelation appears to be one of the more important mechanisms in the biological destruction of rocks and the dissolution of natural silicates and phosphates, and certain acids produced by lichens as well as 2-ketogluconic, citric, oxalic, and formic acids liberated by heterotrophic bacteria and fungi seem to be among the most effective substances in silicate solubilization and in the liberation of a spectrum of elements from minerals and silicates (121, 127, 129). For the initiation of such microbial weathering, however, organic molecules must first be generated autotrophically, and the destruction of the rocks and minerals will then occur in those microenvironments containing energy sources in the form of cells or excretions of the primary colonizers.

Vertical migration of various elements is characteristic of the genesis of certain soil types, and a number of investigators have examined the formation and translocation of organic iron compounds and the subsequent deposition of the iron in an inorganic state. Plant polyphenols reduce and solubilize inorganic iron, and the resulting organic ferrous complexes participate in the downward migration of the element (130). Organic acids, amino acids, and tannins synthesized or released during the decomposition of litter and plant remains can also combine with and mobilize insoluble iron compounds, and the translocated iron-organic complexes will, upon their degradation at greater depth, liberate the iron in an inorganic, insoluble form (131). This process of extraction, transfer, and deposition of iron as a consequence of microbial metabolism seems to be implicated in the genesis of the podzol profile.

The complexity of biosynthetic reactions in soil is illustrated best by the various steps concerned in the formation of humus, that vast conglomerate of heterogeneous organic substances that makes up the organic fraction. To be sure, humification cannot be attributed solely to conversions catalyzed by microbial enzymes or from precursors generated by the subterranean microflora, but the requisite transformations are entirely dependent upon the metabolism of the microscopic inhabitants and the presence of compounds formed or liberated by them; a part, possibly significant, possibly inconsequential, of humus presumably is derived from plant remains subjected to microbial modification, conjugation, and condensation. Aleksandrova (132) isolated from streptomycete cultures some humic-like materials, the dark products possessing certain similarities to the soil organic fraction, and she proposed that these substances were initial stages in the biogenesis of components of humus. Flaig and collaborators (133) stressed the role of lignin degradation products, aromatic compounds, and polyphenols in the formation of humus. It is well known that the terrestrial microflora actively synthesizes the amino acids and vitamins of the organic fraction, a necessary corollary to the ubiquity of strains requiring the growth factors, but one of the outstanding attributes of soil, its odor, has never been adequately characterized. Recently, however, a considerable degree of purification of the substance responsible for the earthy odor has been achieved with a strain of *Streptomyces griseoluteus* (134). Another novel, albeit incompletely characterized microbial metabolite, an organic nitro compound, was shown to appear in natural conditions during nitrification (135). Humus biochemistry, a field unto itself, cannot be adequately reviewed here, and these selected publications have been cited to serve as further illustration of the prominent biochemical alterations in the ecosystem.

Tracing metabolic pathways.—Despite the great strides in establishing metabolic pathways with individual organisms or enzyme preparations comparable progress has not been made in investigations of biochemical transformations in natural environments; nevertheless, techniques are now available by which the stepwise fate of molecules of ecological importance can be determined. In the sense of biochemical ecology, interest in tracing pathways is concerned largely, if not entirely, with the substances that appear in the environment during the course of a specific transformation; intermediates restricted to the cell's confines are properly within the province of the investigator of cellular rather than environmental metabolism.

Because natural substrates or the products of their dissimilation are often complex, the metabolic pathways commonly unknown, and the concentration of suspected intermediates infinitesimally small, the initial exploratory phase of investigations of pathways in nature must frequently include the establishment of biochemical precedents with simulated environments or model microbial systems. For example, the economically important chlorophenoxy-alkyl carboxylic acids possess exotic structures and are applied to soil at parts per million level so that intermediates arising in their degradation likely appear at significantly lower concentrations. Hence, it was deemed necessary to establish metabolic precedents on the basis of pure culture studies (136, 137); with the monocultural models as guidelines, ascertaining the biochemical pathway in soil ecosystems was not difficult (138). The technique of sequential population induction, an extension of the method of sequential enzyme induction in pure cultures, can likewise be applied effectively to the characterization of the fate of specific molecules. The technique relies upon the fact that populations oxidizing naturally occurring intermediates in a reaction sequence are enriched as the primary population generates the appropriate intermediates so that, upon addition of the intermediate, the soil enriched with the secondary flora by the original substrate immediately effects a rapid oxidation of the second compound without the long delay period usually required for the population to become sufficiently large to bring about a detectable change. On the other hand, a compound not involved in the particular transformation will be oxidized appreciably in the enriched soil only after a considerable time span. This technique has been used by Freney (139) to establish the mechanism by which cysteine-S is converted to sulfate. Similarly, because ethanol and acetate are degraded without the typical delay period that precedes the oxidation of simple organic substrates (140), these two substances appear to be intermediates in the process of carbon mineralization in soil.

Mineralization sequences.—In contrast with monocultures, which rarely are provided with more than a few organic substrates, the mixed natural microflora is repeatedly exposed to a wide array of carbonaceous compounds which, under appropriate conditions, are ultimately mineralized to inorganic products. Some investigators of the carbon cycle have chosen to examine the mineralization of simple molecules in order to establish the patterns and kinetics of the transformation; for example, the rate and extent of oxidation of simple organic compounds and the effect of a readily metabolizable substrate upon the dissimilation of a second have been studied (140, 141, 142). Particular attention has been paid to carbon turnover in conditions of inadequate aeration, the results of Takijima (143) indicating that the chief organic acids appearing during decomposition of the complex organic compounds in soil are acetic and butyric with lesser amounts of formic, oxalic, and other simple mono- and dicarboxylic acids. The organic acids in turn are metabolized in oxygen-poor habitats to yield methane and carbon dioxide, but little hydrogen is evolved (144). As soil receives large amounts of lignin and because certain humus constituents are presumed to have been derived from this resistant polymer, investigations of the intermediates of lignin metabolism have particular relevance to biochemical ecology. In this regard, the finding of several aromatic products in cultures of *Poria*, *Fomes*, and *Trametes* which utilize native lignin is of considerable interest (145), but comparable substances have not yet been detected during the oxidative degradation of polyaromatics in soil.

Coinciding with the increasing environmental contamination by synthetic chemicals, many of them phytotoxic or potentially hazardous to humans or animals, there has been a rise in interest in the microbial detoxication of synthetic and natural compounds possessing antibiological properties. The topic is of immediate concern to the biochemical ecologist because among the synthetic chemicals are compounds which apparently are inherently resistant to degradation, the substance thus often persisting for extended periods in the ecosystem, and other compounds that are converted to uncharacterized toxins. Among the toxic natural products, phlorizin and amygdalin have been studied in order to establish the intermediates liberated into the environment. Börner (146) has demonstrated that the former compound is converted to phloretin, phloroglucinol, p-hydroxycinnamic acid, and p-hydroxybenzoic acid, while cyanide was detected during the mineralization by the microflora of amygdalin (147). Many but far from all phenoxyalkyl carboxylic acids, which are prominent synthetic compounds because of their selective phytotoxicity and widespread use, are metabolized in soil (148), yet in only a few instances have the intermediates or pathways been ascertained. Pure culture studies with isolates of *Nocardia, Pseudomonas, Micrococcus,* and *Flavobacterium* have led to the demonstration of two distinctly different mechanisms for the initial stages in the degradation of these chemicals. One mechanism involves the β-oxidation of the aliphatic moiety of phenoxyalkyl carboxylic acids, a sequence resulting in the formation of new phytotoxic compounds, and the second is initiated by a cleavage of the ether linkage to yield the phenol and the free aliphatic acid. Organisms effecting the latter conversion bring about an immediate detoxication of the substrate (136, 137). By utilizing the sensitivity of gas chromatographic procedures coupled with electron-capture ionization detection and the information obtained with monocultural models, it has been shown that the former mechanism, β-oxidation, applies in soil (138). Milbarrow (149) observed the conversion of α-amino-2,6-dichlorobenzaldoxime to 2,6-dichlorobenzonitrile, a phytotoxin, in *Pseudomonas putrefaciens* cultures and in soil; the reaction in soil was stoichiometric. Munnecke and co-workers (150) reported the conversion of fungicides to products which, in turn, were fungicidal. Intermediates in the degradation of a select few other pesticides have been characterized, but until the results are extended to natural conditions, the observations serve merely as model systems and as indicators of what may have ecological importance. Parenthetically, the ready availability to the microbial community of synthetic compounds, many of which appear to be entirely unrelated structurally to known cellular metabolites, poses the problem of ascertaining the changes that must occur in an organism as it develops the capacity to metabolize substrates with which it had no prior contact.

Recent studies of the biochemistry of nitrogen mineralization have been centered to a large extent upon the amino acids. All of the common amino acids, regardless of the extent of their adsorption, are metabolized rapidly by the subterranean population both aerobically and anaerobically, and no more than a few days are required for the complete disappearance of these compounds. In the presence of O_2, the amino acids seem to be oxidized by way of the corresponding keto acids, although the latter do not accumulate during the transformation (151). Schmidt et al. (36) noted, however, that β-alanine appeared during the aerobic decomposition of amino acids, but the β-alanine did not seem to originate from the decarboxylation of aspartate. Volatile fatty acids but no volatile amines are generated during the anaerobic conversion. About 80 per cent of the α-amino nitrogen in the compounds examined is recovered as ammonia except for tryptophan, from which equimolar quantities of ammonia and indole are formed (37). Durand (152), in an investigation of the degradation of uric acid in soil, reported that the purine was converted to allantoin which, in turn, was oxidized by way of allantoic acid, glyoxylic acid, and urea. Since a large part of the inorganic nitrogen is assimilated by the indigenous population during its growth and converted into relatively immobile cell-nitrogen complexes, the mineralization of cell-bound nitrogen is a key reaction in the ecosystem, the rate often limiting cell mass and even plant growth. Investigation of one or the other of the opposing forces of mineralization and nutrient immobilization, reactions of degradation and of biosynthesis, have been hampered by methodological problems resulting from the inability to differentiate between the two transformations. By use of ^{15}N-nitrate, however, Nömmik (153) obtained a tagged population *in situ*, and he was able thereby to differentiate between the mineralization of microbial and humus nitrogen.

As the supply of inorganic phosphorus in nonaquatic habitats almost invariably exceeds the microbial demand for this element, microorganisms are probably not frequently limited by the rate of phosphorus mineralization. Nevertheless, there is a significant turnover between inorganic and organic forms of phosphate, the data of Birch (154) indicating a rapid incorporation into cells of the orthophosphate liberated during decomposition of readily metabolized substrates and a quick return of the microbial phosphorus to the metabolic pool of this element in the habitat. Sulfur also often appears to undergo ready turnover, the element recycling rapidly between the available inorganic pool and the cell-bound form (90). Barrow (155) has emphasized the relationships between the mineralization of carbon, nitrogen, sulfur, and phosphorus as well as the effect of the abundance of available forms of the elements on the relative rates of mineralization and assimilation.

Oxidative reactions.—Microorganisms are capable of effecting an increase in the oxidation state of a number of elements in addition to carbon, but neither substrates nor products of the oxidations are restricted to inorganic compounds. *Aspergillus flavus*, for example, has been shown to synthesize organic nitro compounds in media containing ammonium salts as the sole nitrogen source (156), while a segment of the soil population produces cystine disulfoxide and cysteine sulfinic acid from the reduced sulfur of

289

cysteine (139). The metabolism of the obligate chemoautotrophs, to be sure, is associated with initial substrate and final product both of which are inorganic, but comparable inorganic oxidations are found in heterotrophs (157), and there is no reason to suspect that the energy-releasing inorganic oxidations in the heterotrophs, in contrast with the comparable reactions in the chemoautotrophs, are coupled with phosphorylation. Consequently, the occurrence in nature of a microbiologically catalyzed, energy-liberating reaction involving inorganic reactants or products does not constitute *a priori* evidence that the biological agents are autotrophic. Indeed, in many instances, no proliferation of autotrophs is observed during the oxidation.

Nitrification, the biological conversion of the nitrogen in organic or inorganic compounds from a reduced to a more oxidized state, is brought about by heterotrophic and autotrophic microorganisms. Proliferation of *Nitrosomonas* and *Nitrobacter* is dependent upon nitrogen oxidation, and it is therefore possible to estimate population size from the quantity of nitrate formed, assuming the autotrophs to be solely responsible for the conversion. The calculated population size often far exceeds that observed in soil, suggesting either the participation of heterotrophic nitrifiers or the inadequacy of the counting method developed for solution culture investigations (158). The availability of a selective inhibitor, 2-chloro-6-(trichloromethyl)pyridine, that suppresses only one of the two groups of microorganisms (159) should facilitate the determination of the relative significance of autotrophic and heterotrophic communities to nitrogen oxidation. Intermediates in autotrophic nitrification are not known to exist outside the confines of the bacterial cell. Hydroxylamine, the likeliest candidate for the first product of ammonium oxidation (160, 161), has not been detected extracellularly during active nitrification by *Nitrosomonas* spp., and the postulate that nitrohydroxylamine is an intermediate appears untenable (161). With *Nitrobacter*, on the other hand, all the nitrite-nitrogen oxidized is recovered as nitrate in intact cells and cell extracts (162). During nitrification in soil, organic compounds containing nitrogen in an oxidized state are formed (135), and the appearance of these metabolites may reflect a heterotrophic contribution to the process since such substances are excreted by fungi capable of nitrate biosynthesis (156). However, the identity of the compounds in soil remains unclear, and the possibility that they follow rather than precede nitrate has yet to be ruled out.

The heterotrophic oxidation of other elements is not uncommon, and the activity likewise is widespread in the microbial realm. Jensen (163) observed the conversion of about one-sixth of the thiourea-sulfur to sulfate by strains of *Aspergillus* and *Penicillium*, but certain *Aspergillus niger* isolates oxidized about 80 percent of the cysteine-sulfur to sulfate. Casida (164) demonstrated the accumulation of free orthophosphate in *Pseudomonas fluorescens* cultures supplied with orthophosphite, an anion that is a phosphorus source for many bacteria. The heterotrophic oxidation of manganese, too, has been the subject of recent inquiry (79). Hence, although the ecological significance of heterotrophs to oxidations of these and other elements has yet to be established and despite the fact that in only one instance, sulfate formation from amino acids (139), has an organic pathway been clearly demonstrated to be functional in natural environments, emphasis upon autotrophy as the sole mechanism in nature for the oxidations appears to be premature.

Reductive reactions.—In a number of investigations of microorganisms or of enzyme preparations, a host of elements were demonstrated to undergo biological reduction. Woolfolk & Whitely (60) have shown the coupling of the hydrogenase of *M. lactilyticus* with the reduction of one or more of the higher oxidation states of nitrogen, sulfur, iron, manganese, arsenic, molybdenum, selenium, tellurium, as well as of other elements, but inorganic phosphorus salts were not reduced. Tsubota (165), however, had earlier found that *Clostridium butyricum* and *Escherichia coli* produced phosphite and hypophosphite from orthosphosphate; soil under anaerobiosis also effected the conversion of orthophosphate to phosphite, hypophosphite, and a phosphorus-containing gas. The reduction of inorganic phosphorus compounds is of particular interest in view of the low E_h of the several systems. Ceric ions can be reduced to the cerous state by *F. oxysporum* (Gunner and Alexander, unpublished data), and chlorate and inorganic carbon are converted to more reduced forms in both culture and in soil. Many of the transformations, to be sure, are still laboratory curiosities, but the biological potential for such reactions and the diversity of microbial types in terrestrial environments suggest that biogeochemical reductions of many of the elements hitherto not investigated may occur in suitable habitats.

Reduction of the inorganic forms of many elements is of considerable ecological significance since the conversion from higher to lower oxidation states may alter biological availability, toxicity to plants or microorganisms, and mobility of the element in the ecosystem. Toxicity of reduction products liberated at low O_2 tensions may be one mechanism of fungal suppression (71), and inhibition of seed germination has likewise been attributed to sulfide generated biologically in soil (69). Manganese and iron availability is increased upon reduction, but nitrate reduction commonly leads to a diminished quantity of nitrogen in the environment because molecular nitrogen, nitrous oxide, or nitric oxide are evolved by many of the bacteria adapted to utilize nitrate as an alternative to O_2 as electron acceptor. In habitats containing or receiving oxidized nitrogen, losses of the nitrogen are pronounced when the demand for O_2 exceeds the supply, a condition that is particularly apparent in water-saturated soils endowed with sufficient available carbon (58) or in the rhizosphere when the plant is providing the microflora with readily metabolized root exudates (75). There are also indications that other products and reactions in addition to those already well established may be involved in nitrogen volatilization (166).

Microorganisms are undoubtedly directly responsible for a number of reductions in inadequately aerated habitats, but many of the biologically induced reactions are probably nonphysiological or nonenzymatic. A variety of fermentation products, in the absence of viable organisms or enzymic intervention, have a reducing effect, and the consumption of O_2 and the fall in E_h associated with growth are linked with similar changes. Reductions of inorganic compounds which proceed in part by nonenzymatic means have recently been demonstrated in cell-free preparations of bacteria (60). It is difficult, with mixed populations proliferating in complex environments, to interpret the sparing effect of nitrate on ferrous formation from ferric iron (69) and on sulfide formation from sulfate (167), but nonbiological factors such as poising of the E_h at a high potential conceivably could play a part. The inhibition by one element of the reduction of another may likewise be physiological, the former functioning as a second acceptor for the electrons released in substrate oxidation, effectively funneling electrons into another channel, or possibly, as shown recently for the nitrous oxide reductase of *M. denitrificans* (168), by serving as a repressor for the synthesis of a specific reductase. However, extension from the test tube to the field is inevitably hazardous since, for example, the sparing compound might act in a mixed microbial community not as a repressor of a specific enzyme of one organism capable of using two electron acceptors but rather as a selective inhibitor of the species effecting the reduction of the element spared. In this regard, studies of the sparing of sulfate reduction in natural habitats should prove valuable inasmuch as the responsible bacteria are not known to have either a nitrate or an O_2 respiration although nitrate and O_2 retard sulfide appearance in soil.

From the ecological viewpoint, it would be interesting to know which of the elements reduced enzymatically function as electron acceptors to sustain growth; i.e., are there equivalents among the other elements of the O_2 respiration of aerobes, the nitrate respiration of denitrifying bacteria, the sulfate-linked growth of the anaerobic vibrios, or the CO_2-dependent proliferation of the methane bacteria? Moreover, of the elements—rarely or abundantly distributed—that are reduced microbiologically, whether the reductions are coupled directly with growth or not, which of them are acted upon by identical enzymes or microorganisms and which are associated with unique enzymes or individual microbial types? The finding that in at least one instance, the sulfite and nitrite reductases of *E. coli* (169), two reductive activities are properties of the same enzyme suggests that a single species may be capable of the transformation of several elements with no requirement for adaptation to the new oxidant.

Microbiological effects on plants.—A consideration of the biochemical effects of soil microorganisms upon the environment would not be complete without some comment on their influence upon the plant. Parasitic interrelationships, however, will not be reviewed as this topic is properly within the province of the pathologist rather than the ecologist. Abundant evidence of both detrimental and stimulatory effects is now at hand, stimulatory reactions herein being considered separately from those mineralization and solubilization processes which provide plants with essential nutrient elements in a readily assimilable form. For example, culture filtrates of isolates of *Arthrobacter*, *Bacillus*, *Streptomyces*, and *Fusarium*, all widely distributed genera, were found to contain auxin- and gibberellin-like substances promoting growth of suitable test plants (170, 171). The synthesis of growth-promoting substances is not the sole means by which the subterranean residents may enhance plant development and vigor inasmuch as antibiotics like chloramphenicol are absorbed by plants cultivated in artificial conditions (172); should these antimicrobial agents be indeed produced in the rhizosphere at sites and under circumstances favorable to root uptake, the antibiotics may function as systemic fungicides or bactericides to ward off susceptible pathogens.

Phytotoxic compounds are frequently synthesized or released during the decay of plant remains. Water-soluble phytotoxic substances inhibitory to germination and seedling growth are elaborated in the early stages of decomposition, but their concentration or activity declines with time, and often aqueous extracts of older residues enhance plant growth (173). Microbial suppression of widely different plant genera has been reported, and a variety of morphological changes in primary and secondary roots and in root hairs have been observed (174). As yet, no complete picture of the chemical identity of the toxins has developed, but detrimental influences have been ascribed to sulfide, acetate, butyrate, oxalate, phenylacetate, *o*- and *m*-hydroxyphenylacetate, patulin, and a multitude of other characterized and uncharacterized substances (69, 70, 175, 176).

Microbial fallibility and molecular recalcitrance.—It is usually regarded as axiomatic in microbial ecology that all organic molecules, regardless of complexity, are degraded by the soil inhabitants, and the technique of elective culture is predicated upon the assumption that some organism in nature, soil or mud typically being the source material of choice, will utilize as carbon and energy source any organic compound. This concept, commonly taken as ecological dogma, may be termed the "Principle of Microbial Infallibility." To the regret of modern agricultural and industrial technologists, microorganisms are unfortunately quite fallible, and the dogma must be modified in the light of the contributions to ecology made by investigations of the microbiology of synthetic chemicals.

Clearly, organic compounds synthesized by enzymatic means must be biodegradable, at least in certain circumstances, and the absence of vast accumulations of resistant organic molecules is mute evidence of the reliability of the microflora. The infallibility principle may then be applied with some degree of confidence to natural products, and it may be proposed that, with suitable and selective culture methods, strains possessing the capacity to initiate the degradation of every naturally occurring organic compound can be isolated from soil; nevertheless, natural organic substances destroyed in one environment may persist in another set of conditions, the relative susceptibilities of petroleum hydrocarbons or peats to aerobic and anaerobic decay being two of many possible illustrations. Some organic molecules formed biologically exhibit a remarkable recalcitrance to microbial destruction even in near optimal circumstances, radiocarbon dating of humus, for example, indicating an average age for the soil organic fraction of hundreds of years. Since a significant portion of the organic fraction is undoubtedly juvenile, a large part must resist microbial destruction for very long periods, but the identity of the recalcitrant constituents or linkages remains obscure.

Microbial fallibility has, without a doubt, been revealed most dramatically by the prolonged duration of toxicity in soils receiving pesticides. Many of these molecules or their inhibitory derivatives resist enzymatic detoxication and destruction, and they persist for months or years even when exposed to a large, heterogeneous population operating in favorable circumstances. There is no evidence suggesting a genotypic or phenotypic change in the indigeneous flora in response to the presence of the recalcitrant pesticidal substrate, and the toxicity of the parent molecule or its derivatives is not relieved at a significant rate. Among the compounds exhibiting resistance to biological inactivation in soil and persisting without major structural change are: a range of chloro-substituted phenoxyalkyl carboxylic acids and chlorophenols (148, 177); s-triazine herbicides which retain their activity in excess of 15 months (178); chloro-substituted benzyl diisopropyldithiocarbamate, a phytotoxic molecule not completely destroyed in soil 6 years after application (179); the insecticides chlordane, benzene hexachloride, heptachlor, aldrin, and dieldrin, these compounds or their derivatives still suppressing insects in periods ranging up to 12 years after the initial soil treatment (180); certain quaternary ammonium compounds (181); and others. Ludzack & Ettinger (182) reported that many ethers, glycols, surfactants, and detergents resist degradation in aquatic habitats or sewage. Not infrequently, the property of molecular recalcitrance may reside in a substance formed from the original chemical; thus, epoxidation products of aldrin and heptachlor are still recoverable several years after application to the environment of even minute amounts of the original insecticides (180).

The specific structural features determining molecular recalcitrance are frequently obscure or, more often, unknown. In many instances of prolonged persistence of organic compounds, the resistance is undoubtedly inherent in the molecular structure itself, although some substances may owe their durabilities to poor solubility characteristics, adsorption, complexing, or microenvironmental barriers that make the materials unavailable or inaccessible. A number of investigators have considered the influence of chemical structure and substitution on degradation of individual substrates by pure cultures and by enzymes obtained from them (183–187), but it is essential for the biochemical ecologist to differentiate between the effect of structure on decomposition by a specific organism and the influence of molecular architecture on degradability by a heterogeneous population such as that in soil. To be sure, the metabolic capabilities and inadequacies of the microflora reflect the summation of individual capacities of the indigenous species, but studies of monocultural systems are limited to the properties and idiosyncrasies of the test organisms, whereas the existence of a compound that is resistant to degradation by a vast and diverse community presumably indicates a true resistance of the molecule itself to biodegradation, an interaction of chemical structure and enzymatic composition of microorganisms in general. As yet, few of the configurations, linkages, or functional groups that confer resistance upon molecules are known, but it appears that principles can be established to account for the resistance to microbial destruction of at least some molecules. Thus, the presence in phenols or phenoxy compounds of a halogen in a position *meta* to the phenolic hydroxyl and the linkage of the aliphatic moiety of phenoxyalkyl carboxylic acids to the aromatic ring are characteristics associated with resistance (148, 177). However, far more information is required with respect to the effect of molecular structure on biodegradation and the underlying physiological mechanisms before a reasonable picture of fallibility and recalcitrance can emerge.

SUMMARY

Considerable progress has been made in the last few years in the application of biochemical concepts and techniques to ecological problems. Nevertheless, no more than the merest outer film surrounding the entire discipline of microbial ecology has been removed, a surface scratching which still has demonstrated that the field of microbial ecology is now ripe for biochemical exploration and interpretation. Only a few of the fruitful areas have been examined, even in the most cursory fashion, but the preliminary steps reveal a broad vista for future investigation. The physiological basis of microbial distribution, the geochemical functions of the population, the biochemical relationships between species and communities, the phenotypic and genotypic modifications of microorganisms in response to environmental change and population pressure, the ecological and biochemical significance of the primary ecological determinants, and the metabolic basis for dominance of individual genera in specific ecosystems are but a few of the problems awaiting careful and perceptive inquiry by experimental rather than descriptive approaches and by means of an *in vivo* rather than an *in vitro* view of microbial biochemistry and ecology.

LITERATURE CITED

1. Pochon, J., and De Barjac, H., *Traité de Microbiologie des Sols.* (Dunod, Paris, 685 pp., 1958)

2. Burges, A., *Micro-organisms in the Soil.* (Hutchinson, London, 188 pp., 1958)

3. Parkinson, D., and Waid, J. S., Eds., *The Ecology of Soil Fungi.* (Liverpool Univ. Press, Liverpool, 324 pp., 1960)

4. Alexander, M., *Introduction to Soil Microbiology.* (Wiley, New York, 472 pp., 1961)

5. Krassilnikov, N. A., *Soil Microorganisms and Higher Plants.* (Transl. publ. by Office of Technical Services, U. S. Dept. of Commerce, Washington, D. C., 474 pp., 1961)

6. Garrett, S. D., *Soil Fungi and Soil Fertility.* (Pergamon Press, Oxford, 165 pp., 1963)

7. McLaren, A. D. *Science*, 141, 1141–47 (1963)

8. Tyagny-Ryadno, M. G., *Soviet Soil Sci.*, (Engl. Trans.), 1958, 1378–87 (1958)

9. Rovira, A. D., and Greacen, E. L., *Australian J. Agr. Res.*, 8, 659–73 (1957)

10. Zvyagintsev, D. G., *Mikrobiologiya*, 31, 339–43 (1962)

11. Miyamoto, Y., *Virology*, 9, 290–91 (1959)

12. Lahav, N., and Keynan, A., *Can. J. Microbiol.*, 8, 565–72 (1962)

13. Macura, J., and Pavel, L., *Folia Microbiol.* (*Prague*), 4, 82–90 (1959)

14. Estermann, E. F., and McLaren, A. D., *J. Soil Sci.*, 10, 64–78 (1959)

15. Stotzky, G., and Martin, R. T., *Plant Soil*, 18, 317–37 (1963)

16. Weber, D. F., and Gainey, P. L., *Soil Sci.*, 94, 138–45 (1962)

17. McLaren, A. D., *Science*, 125, 697 (1957)

18. Hattori, T., and Furusaka, C., *J. Biochem.* (*Tokyo*), 48, 831–37 (1960); 50, 312–15 (1961)

19. Holding, A. J., *J. Appl. Bacteriol.*, 23, 515–25 (1960)

20. Rybalkina, A. V., and Kononenko, Ye. V., *Soviet Soil Sci.*, (Engl. Trans.), 1959, 537–49 (1959)

21. Domsch, K. H., *Arch. Mikrobiol.*, 35, 229–47 (1960)

22. Mitchell, R., and Alexander, M., *Soil Sci. Soc. Am. Proc.*, 26, 556–58 (1962)

23. Porter, J. N., and Wilhelm, J. J., in *Developments in Industrial Microbiology*, 2, 253–59 (Rich, S., Ed., Plenum Press, New York, 306 pp., 1960)

24. Burges, A., Hurst, H. M., Walkden, S. B., Dean, F. M., and Hirst, M., *Nature*, 199, 696–97 (1963)

25. Graveland, D. N., and Lynch, D. L., *Soil Sci.*, 91, 162–65 (1961)

26. Gupta, U. C., Sowden, F. J., and Stobbe, P. C., *Soil Sci. Soc. Am. Proc.*, 27, 380–82 (1963)

27. Waldron, A. C., and Mortenson, J. L., *Soil Sci.*, 93, 286–93 (1962)

28. Paul, E. A., and Schmidt, E. L., *Soil Sci. Soc. Am. Proc.*, 25, 359–62 (1961)

29. Blumenthal, H. J., and Roseman, S., *J. Bacteriol.*, 74, 222–24 (1957)

30. Rovira, A. D., *Soils Fertilizers*, 25, 167–72 (1962)

31. Vrany, J., Vancura, V., and Macura, J., *Folia Microbiol.* (*Prague*), 7, 61–70 (1962)

32. Schroth, M. N., and Snyder, W. C., *Phytopathology*, 51, 389–93 (1961)

33. Katznelson, H., and Bose, B., *Can. J. Microbiol.*, 5, 79–85 (1959)

34. Cunningham, J. L., and Hagedorn, D. J., *Phytopathology*, 52, 616–18 (1962)

35. Estermann, E. F., Peterson, G. H., and McLaren, A. D., *Soil Sci. Soc. Am. Proc.*, 23, 31–36 (1959)

36. Schmidt, E. L., Putnam, H. D., and Paul, E. A., *Soil Sci. Soc. Am. Proc.*, 24, 107–9 (1960)

37. Greenwood, D. J., and Lees, H., *Plant Soil*, 12, 69–80 (1960)

38. Handley, W. R. C., *Plant Soil*, 15, 37–73 (1961)

39. Welch, L. F., and Scott, A. D., *Soil Sci.*, 90, 79–85 (1960)

40. Schneider, K. H., and Siegel, O., *Landwirtsch. Forsch.*, 11, 270–74 (1958)

41. Park, D., *Trans. Brit. Mycol. Soc.*, 40, 283–91 (1957)

42. Hayek, J. M. W., and Hulbary, R. L., *Proc. Iowa Acad. Sci.*, 63, 327–38 (1956)

43. Klevenskaya, I. L., *Mikrobiologiya*, 29, 215–19 (1960)

44. Henis, Y., and Eren, J., *Can. J. Microbiol.*, 9, 902–4 (1963)

45. Taber, W. A., *Can. J. Microbiol.*, 6, 503–14 (1960)

46. Ul'yanova, O. M., *Mikrobiologiya*, 31, 77–84 (1962)

47. Balassa, R., *Nature*, 188, 246–47 (1960)

48. Lange, R. T., and Alexander, M., *Can. J. Microbiol.*, **7**, 959–61 (1961)
49. Becking, J. H., *Plant Soil*, **14**, 297–322 (1961)
50. Shtina, E. A., *Trans. Intern. Congr. Soil Sci., 7th, Madison, 1960*, **2**, 630–34 (1960)
51. Mishustin, E. N., and Pushkinskaya, O. I., *Izv. Akad. Nauk SSSR, Ser. Biol.*, No. 5, 641–60 (1960)
52. Bond, G., and Hewitt, E. J., *Nature*, **190**, 1033–34 (1961)
53. Bond, G., and Hewitt, E. J., *Nature*, **195**, 94–95 (1962)
54. Ozanne, P. G., Greenwood, E. A. N., and Shaw, T. C., *Australian J. Agr. Res.*, **14**, 39–50 (1963)
55. Andrew, C. S., and Norris, D. O., *Australian J. Agr. Res.*, **12**, 40–55 (1961)
56. Harris, K. N., and Woodbine, M., *Intern. Congr. Microbiol., 8th, Montreal, 1962*, p. 58 (Abstr.)
57. Greenwood, D. J., *Plant Soil*, **14**, 360–76 (1961)
58. Greenwood, D. J., *Plant Soil*, **17**, 365–91 (1962)
59. Baas-Becking, L. G. M., Kaplan, I. R., and Moore, I., *J. Geol.*, **68**, 243–84 (1960)
60. Woolfolk, C. A., and Whitely, H. R., *J. Bacteriol.*, **84**, 647–58 (1962)
61. Durbin, R. D., *Am. J. Botany*, **46**, 22–25 (1959)
62. Papavizas, G. C., and Davey, C. B., *Phytopathology*, **52**, 759–66 (1962)
63. Griffin, D. M., *Biol. Rev. Cambridge Phil. Soc.*, **38**, 141–66 (1963)
64. Meiklejohn, J., *J. Soil Sci.*, **8**, 240–47 (1957)
65. Birch, H. F., *Plant Soil*, **12**, 81–96 (1960)
66. Siuta, J., *Soviet Soil Sci.*, (Engl. Trans.), **1962**, 500–8 (1962)
67. Valera, C. L., and Alexander, M., *Plant Soil*, **15**, 268–80 (1961)
68. Ng, S. K., and Bloomfield, C., *Plant Soil*, **16**, 108–35 (1962)
69. Yezhov, Yu. I., *Soviet Soil Sci.*, (Engl. Trans.), **1962**, 165–70 (1962)
70. Takijima, Y., and Sukuma, H., *Nippon Dojo-Hiryogaku Zasshi*, **32**, 560–64 (1961)
71. Mitchell, R., and Alexander, M., *Soil Sci.*, **93**, 413–19 (1962)
72. Menzies, J. D., *Phytopathology*, **52**, 743 (1962)
73. Buxton, E. W., *J. Gen. Microbiol.*, **22**, 678–89 (1960)
74. Elkan, G. H., *Can. J. Microbiol.*, **7**, 851–56 (1961)
75. Woldendorp, J. W., *Plant Soil*, **17**, 267–70 (1962)
76. Tousson, T. A., Patrick, Z. A., and Snyder, W. C., *Nature*, **197**, 1314–16 (1963)
77. Ellenby, C., *Nature*, **198**, 1110 (1963)
78. Vagnerova, K., and Vancura, V., *Folia Microbiol. (Prague)*, **7**, 55–60 (1962)
79. Zavarzin, G. A., *Mikrobiologiya*, **31**, 586–88 (1962)
80. Bunt, J. S., *Nature*, **192**, 1274–76 (1961)
81. Bjälfve, G., *Physiol. Plantarum*, **15**, 122–29 (1962)
82. Parker, B. C., and Bold, H. C., *Am. J. Botany*, **48**, 185–97 (1961)
83. Kefford, N. P., Brockwell, J., and Zwar, J. A., *Australian J. Biol. Sci.*, **13**, 456–67 (1960)
84. Sahlman, K., and Fahraeus, G., *Kgl. Lantbruks-Hogskol. Ann.*, **28**, 261–68 (1962)
85. Ljunggren, H., and Fahraeus, G., *J. Gen. Microbiol.*, **26**, 521–28 (1961)
86. Bergersen, F. J., *J. Gen. Microbiol.*, **22**, 671–77 (1960)
87. Stotzky, G., and Norman, A. G., *Arch. Mikrobiol.*, **40**, 341–69 (1961)
88. Finstein, M. S., and Alexander, M., *Soil Sci.*, **94**, 334–39 (1962)
89. Marshall, K. C., and Alexander, M., *Plant Soil*, **12**, 143–53 (1960)
90. Stotzky, G., and Norman, A. G., *Arch. Mikrobiol.*, **40**, 370–82 (1961)
91. Snyder, W. C., Schroth, M. N., and Christou, T., *Phytopathology*, **49**, 755–56 (1959)
92. Papavizas, G. C., and Davey, C. B., *Phytopathology*, **51**, 693–99 (1961)
93. Mirchink, T. G., and Greshnykh, K. P., *Mikrobiologiya*, **30**, 1045–49 (1961)
94. Vasudeva, R. S., Singh, P., Sen Gupta, P. K., Mahmood, M., and Bajaj, B. S., *Ann. Appl. Biol.*, **51**, 415–23 (1963)
95. Weltzien, H. C., *Zentr. Bakteriol. Parasitenk., Abt. II, Orig.*, **116**, 131–70 (1963)
96. Lingappa, B. T., and Lockwood, J. L., *Phytopathology*, **52**, 295–99 (1962)
97. Strider, D. L., and Winstead, N. N., *Phytopathology*, **50**, 781–84 (1960)
98. Aleem, M. I. H., and Alexander, M., *Appl. Microbiol.*, **8**, 80–84 (1960)
99. Sequeira, L., *Phytopathology*, **53**, 332–36 (1963)
100. Stolp, H., and Starr, M. P., *Bacteriol. Proc.*, p. 47 (1963)
101. Teakle, D. S., *Nature*, **188**, 431–32 (1960)

102. Chinn, S. H. F., and Ledingham, R. J., *Can. J. Botany*, **39**, 739–48 (1961)

103. Lockwood, J. L., and Lingappa, B. T., *Phytopathology*, **53**, 917–20 (1963)

104. Mitchell, R., and Alexander, M., *Can. J. Microbiol.*, **9**, 169–77 (1963)

105. Menzies, J. D., *Phytopathology*, **49**, 648–52 (1959)

106. Papavizas, G. C., Davey, C. B., and Woodard, R. S., *Can. J. Microbiol.*, **8**, 915–22 (1962)

107. Mitchell, R., *Phytopathology*, **53**, 1068–71 (1963)

108. Mitchell, R., and Alexander, M., *Plant Disease Reptr.*, **45**, 487–90 (1961)

109. Chang, P., and Knowles, R., *Bacteriol. Proc.*, p. 19 (1963)

110. Parker, C. A., *J. Australian Inst. Agr. Sci.*, **26**, 288 (1960)

111. Watanabe, A., *J. Gen. Appl. Microbiol.*, **8**, 85–91 (1962)

112. Moore, A. W., *Empire J. Exptl. Agr.*, **30**, 239–48 (1962)

113. Ziegler, H., and Hüser, R., *Nature*, **199**, 508 (1963)

114. Morrison, T. M., *Nature*, **189**, 945 (1961)

115. Bond, G., *Nature*, **193**, 1103–4 (1962)

116. Tarrant, R. F., and Miller, R. E., *Soil Sci. Soc. Am. Proc.*, **27**, 231–34 (1963)

117. Dommergues, Y., *Agrochimica*, **7**, 335–40 (1963)

118. Cady, F. B., and Bartholomew, W. V., *Soil Sci. Soc. Am. Proc.*, **24**, 477–82 (1960)

119. Clark, F. E., Beard, W. E., and Smith, D. H., *Soil. Sci. Soc. Am. Proc.*, **24**, 50–54 (1960)

120. Simon-Sylvestre, G., and Boischot, P., *Ann. Agron.*, **13**, 549–74 (1962)

121. Bromfield, S. M., and Williams, E. G., *J. Soil Sci.*, **14**, 346–59 (1963)

122. Ivanov, V. I., *Mikrobiologiya*, **31**, 795–99 (1962)

123. Kramarenko, L. E., *Mikrobiologiya*, **31**, 694–701 (1962)

124. Duff, R. B., Webley, D. M., and Scott, R. O., *Soil Sci.*, **95**, 105–14 (1963)

125. Katznelson, H., Peterson, E. A., and Rouatt, J. W., *Can. J. Botany*, **40**, 1181–86 (1962)

126. Kobus, J., *Acta Microbiol. Polon.*, **11**, 255–63 (1962)

127. Webley, D. M., Henderson, M. E. K., and Taylor, I. F., *J. Soil Sci.*, **14**, 102–12 (1963)

128. Roy, B. N., *Nature*, **195**, 472 (1962)

129. Schatz, A., *J. Agr. Food Chem.*, **11**, 112–18 (1963)

130. Coulson, S. B., Davies, R. I., and Lewis, D. A., *J. Soil Sci.*, **11**, 30–44 (1960)

131. Kaurichev, I. S., and Nozdrunova, Ye. M., *Soviet Soil Sci.*, (Engl. Trans.), **1961**, 1057–64 (1961)

132. Aleksandrova, I. V., *Soviet Soil Sci.*, **1962**, 1330–34 (1962) (Engl. trans.)

133. Flaig, W., Salfeld, J. C., and Haider, K., *Landwirtsch. Forsch.*, **16**, 85–96 (1963)

134. Romano, A. H., and Safferman, R. S., *J. Am. Water Works Assoc.*, **55**, 169–76 (1963)

135. Turtschin, F. B., Bersenjewa, S. N., Kortizkaja, I. A., Shidkick, G. G., and Lobowikowa, G. A., *Trans. Intl. Cong. Soil Sci.*, *7th, Madison, 1960*, **2**, 236–45 (1960)

136. Taylor, H. F., and Wain, R. L., *Proc. Roy. Soc.* (London), **156B**, 172–86 (1962)

137. MacRae, I. C., and Alexander, M., *J. Bacteriol.*, **86**, 1231–35 (1963)

138. Gutenmann, W., Loos, M. A., Alexander, M., and Lisk, D. J., *Soil Sci. Soc. Am. Proc.*, **28**, 205–7 (1964)

139. Freney, J. R., *Australian J. Biol. Sci.*, **13**, 387–92 (1960)

140. Stevenson, I. L., and Katznelson, H., *Can. J. Microbiol.*, **4**, 73–79 (1958)

141. Drobnik, J., *Can. J. Microbiol.*, **7**, 769–75 (1961)

142. Drobnik, J., *Folia Microbiol.* (*Prague*), **7**, 126–31 (1962)

143. Takijima, Y., *Nippon Dojo-Hiryogaku Zasshi*, **31**, 435–40 (1960)

144. Yamane, I., and Sato, K., *Soil Sci. Plant Nutr.* (*Tokyo*), **9**, 32–36 (1963)

145. Ishikawa, H., Schubert, W. J., and Nord, F. F., *Arch. Biochem. Biophys.*, **100**, 131–39 (1963)

146. Börner, H., *Beitr. Biol. Pflanz.*, **36**, 97–137 (1961)

147. Hine, R. B., *Phytopathology*, **51**, 10–13 (1961)

148. Burger, K., MacRae, I. C., and Alexander, M., *Soil Sci. Soc. Am. Proc.*, **26**, 243–46 (1962)

149. Milbarrow, B. V., *Biochem. J.*, **87**, 255–58 (1963)

150. Munnecke, D. E., Domsch, K. H., and Eckert, J. W., *Phytopathology*, **52**, 1298–1308 (1962)

151. Van Driel, W., *Acta Botan. Neerl.*, **10**, 209–47 (1961)

152. Durand, G., *Compt. Rend.*, **252**, 1687–89 (1961)

153. Nömmik, H., *Acta Agr. Scand.*, 11, 211–26 (1961)
154. Birch, H. F., *Plant Soil*, 15, 347–66 (1961)
155. Barrow, N. J., *Australian J. Agr. Res.*, 11, 317–30 (1960)
156. Marshall, K. C., and Alexander, M., *J. Bacteriol.*, 83, 572–78 (1962)
157. Hora, T. S., and Iyengar, M. R. S., *Arch. Mikrobiol.*, 35, 252–57 (1960)
158. Alexander, M., Marshall, K. C., and Hirsch, P., *Trans. Intern. Congr. Soil Sci.*, Madison, *1960*, 2, 586–91 (1960)
159. Shattuck, G. E., Jr., and Alexander, M., *Soil Sci. Soc. Am. Proc.*, 27, 600–1 (1963)
160. Burge, W. D., Malavolta, E., and Delwiche, C. C., *J. Bacteriol.*, 85, 106–10 (1963)
161. Falcone, A. B., Shug, A. L., and Nicholas, D. J. D., *Biochem. Biophys. Res. Commun.*, 9, 126–31 (1962)
162. Aleem, M. I. H., and Alexander, M., *J. Bacteriol.*, 76, 510–14 (1958)
163. Jensen, H. L., *Arch. Mikrobiol.*, 28, 145–52 (1957)
164. Casida, L. E., Jr., *J. Bacteriol.*, 80, 237–41 (1960)
165. Tsubota, G., *Soil Plant Food*, 5, 10–15 (1959)
166. Cady, F. B., and Bartholomew, W. V., *Soil Sci. Soc. Am. Proc.*, 27, 546–49 (1963)
167. Hesse, P. R., *Plant Soil*, 14, 249–63 (1961)
168. Pichinoty, F., and D'Ornano, L., *Ann. Inst. Pasteur*, 101, 418–26 (1961)
169. Kemp, J. D., Atkinson, D. E., Ehret, A., and Lazzarini, R. A., *J. Biol. Chem.*, 238, 3466–71 (1963)
170. Panossian, A. K., Arutunian, R. S., and Marshavina, Z. V., *Z. Allgem. Mikrobiol.*, 3, 42–46 (1963)
171. Katznelson, H., Sirois, J. C., and Cole, S. E., *Nature*, 196, 1012–13 (1962)
172. Pramer, D., *Exptl. Cell Res.*, 16, 70–74 (1959)
173. Patrick, Z. A., Tousson, T. A., and Snyder, W. C., *Phytopathology*, 53, 152–61 (1963)
174. Bowen, G. D., and Rovira, A. D., *Plant Soil*, 15, 166–88 (1961)
175. McCalla, T. M., Guenzi, W. D., and Norstadt, F. A., *Z. Allgem. Mikrobiol.*, 3, 202–10 (1963)
176. Aoki, H., Sassa, T., and Tamura, T., *Nature* 200, 575 (1963)
177. Alexander, M., and Aleem, M. I. H., *J. Agr. Food Chem.*, 9, 44–47 (1961)
178. Sheets, T. J., Crafts, A. S., and Drever, H. R., *J. Agr. Food Chem.*, 10, 458–62 (1962)
179. Klingman, G. C., *Weeds*, 10, 336–37 (1962)
180. Lichtenstein, E. P., and Polivka, J. B., *J. Econ. Entomol.*, 52, 289–93 (1959)
181. Marth, P. C., and Mitchell, J. W., *Proc. Am. Soc. Hort. Sci.*, 76, 673–78 (1960)
182. Ludzack, F. J., and Ettinger, M. B., *J. Water Pollution Control Federation*, 32, 1173–1200 (1960)
183. Webley, D. M., Duff, R. B., and Farmer, V. C., *Nature*, 183, 748–49 (1959)
184. Jensen, H. L., *Acta Agr. Scand.*, 10, 83–103 (1960)
185. Kearney, P. C., Kaufman, D. D., and Beall, M. L., *Biochem. Biophys. Res. Commun.*, 14, 29–37 (1963)
186. Hirsch, P., and Alexander, M., *Can. J. Microbiol.*, 6, 241–49 (1960)
187. Davies, J. I., and Evans, W. C., *Biochem. J.*, 82, 50P–51P (1962)

IV
Systems Analysis

Editor's Comments on Papers 21 Through 25

Nowhere in ecology has systems analysis been a more natural and important development than in studies of the cycles of essential elements. The origin and development of systems analysis in ecology will be the subject of another volume in this series, by B. C. Patten. Such terms as "system" and "model" were common in the literature on cycles of elements well before the formal discipline of systems analysis was incorporated into ecology. Most of the papers in Part III and some in Part II look upon their subject as a system, and well they might at any time after Tansley's (1935) coining of the term "ecosystem" and certainly after Lindemann's (1942) benchmark paper on the energetics of Cedar Bog Lake. What sets apart the recent systems approach is its formal analysis of the system, its total condensation of the system to equations, and its use of computer simulation.

Mathematical analyses similar to the simpler modeling of the present systems analysts have been the stock-in-trade of meteorologists, physical oceanographers, and geochemists for nearly fifty years (Sverdrup et al., 1942). Equations are used to express the continuity of movement of air and water through a defined three-dimensional space, accompanied by changes in physical and chemical properties. Biological oceanographers, a species of marine ecologist, began using this kind of analysis to examine the movement of oxygen, phosphorus, nitrogen, and carbon in the ocean (Riley, 1951). Riley extended this method to models (and he called them that) of primary and secondary production in the ocean (Riley, Stommel, and Bumpus, 1949). Unfortunately for present purposes, most of Riley's early papers are monographic in length and cannot be reproduced here, although several of them are benchmarks, both as early examinations of the cycle of essential elements in an ocean basin (Riley, 1951) and as early examples of modeling of the flux of materials through a community or ecosystem (Riley, 1947). The following quotation from Riley (1947) expresses some of the problems besetting not only biological oceanography but all of ecology.

> A complex field such as oceanography tends to be subject to two opposite approaches. The first is the descriptive, in which several quantities are measured simultaneously and their inter-relationships derived by some sort of statistical method. The other approach is the synthetic one, in which a few reasonable although perhaps oversimplified assumptions are laid down, these serving as a basis for mathematical derivation of relationships.
>
> Each approach has obvious virtues and faults. Neither is very profitable by itself; each requires the assistance of the other. Statistical analyses check

the accuracy of the assumptions of the theorists, and the latter lend meaning to the empirical method. Unfortunately, however, in many cases there is no chance for mutual profit because the two approaches have no common ground. Until such contact has been established no branch of oceanography can quite be said to have come of age. In this respect physical oceanography, one of the youngest branches in actual years, is more mature than the much older study of marine biology. This is perhaps partly due to the complexities of the material. More important, however, is the fact that physical oceanography has aroused the interest of a number of men of considerable mathematical ability, while on the other hand marine biologists have been largely unaware of the growing field of bio-mathematics, or at least they have felt that the synthetic approach will be unprofitable until it is more firmly backed by experimental data.

Although Riley wrote these words more than twenty-five years ago, many eminent ecologists remain aloof to what he called the synthetic approach and we now call modeling. What those unconvinced ecologists may have overlooked is the fact that synthetic models of the type Riley describes do not require precise values for their success in early stages of development. What they do need is a clear and correct compartmental model of the system under consideration. If it is a food web, then all significant populations and pathways must be known if the model is to have utility for understanding the real world. As Smith (1970) has stated it: "Hypothetical simple systems are far more sensitive to changes in the organizational structure of relations between components than to changes in the values of the components themselves."

It is in later interactions between the statistical and modeling approaches that the need for statistically meaningful values for stocks and fluxes are necessary. Not until that stage in the analysis of ecosystems can we decide which values deserve the kind of expensive attention that yields really precise results.

Analysis of ecosystems by alternate modeling and analytical measurement of standing stocks and fluxes of one or more elements is such a large-scale undertaking that shortcuts are highly desirable. One means by which some useful information can be obtained is through the use of microcosms. A "microcosm" is simply a small system, like an aquarium or terrarium. Microcosms may be carefully defined and constructed from a set of known species, or they can be simply small samples of the real world, such as soil or water, confined in such a way that samples and rate measurements are easily obtained. Microcosms do not always behave like the real world in all respects, but they provide us a first approximation of how the world works at great savings of both time and funds. Patten and Witkamp developed a powerful tool by combining the microcosm approach with a tracer experiment and a model of the system. There have been many tracer experiments of microcosm scale, but the opportunities presented by interaction with a model have rarely been exploited.

A more common route of entry into modeling is to use an existing set of data to develop an ecosystem model. One of the first to explore the potential utility of ecosystem models was H. T. Odum, who began experimenting with analog circuits in the late 1950s. His first paper on modeling is reprinted here. Although much of the emphasis was on energy flux, the use of modeling in studying the cycles of chemical elements was also explored. Odum stayed with analog simulation for a decade, developing a biological analog computer language of his own which contained such typical biological functions as storage, delayed response, and feedback. Jerry S. Olson at the Oak Ridge National Laboratory was another pioneer in analog simulation of the flux of elements through ecosystems. Many of the first generation of ecological modelers got their introduction from Olson, using the excellent analog facilities at Oak Ridge. The paper reproduced here may be the first simulation of the flux of a specific nutrient element in a particular ecosystem, using data from field observations as a basis.

For the simulation of simple, highly condensed models, analog modeling may have distinct advantages in speed, simplicity, and sometimes accuracy. However, there are few analog facilities large enough or sensitive enough to permit one to wire up a complex, nonlinear model of an ecosystem. Digital computers have improved rapidly, and they are now the logical choice for most ecological modeling. The IBP models are taking that route. One of the few of them to have appeared in the open literature at this time is the ocean upwelling model of Walsh and Dugdale. In addition to using computers for simulation of their biome, Walsh and Dugdale have combined automated analysis with computerized data reduction on board their oceanographic vessel. The vessel spends eight to twelve hours, through the middle of each night, steaming along a grid over a suspected upwelling area, while automatic equipment continuously measures and stores data on salinity, temperature, phosphate, ammonia, nitrate and chlorophyll. The computer is programmed to map each of these, interpolating and also extrapolating from the data points stored in its memory. A suspected region of high or low values is predicted by extrapolation, and the persons who are monitoring the program can change the ship's course to intercept the area of interest. At the end of the track the scientists have a set of distribution maps with which to select a study area for the day. Plumes of upwelling water can be followed in this way from their first appearance through their successional stages. This is getting close to what modelers call a study in real time, in which the analysis of data keeps pace with the generation of new data. This is an important departure from conventional biological oceanography in which analysis of the data comes too late to change the course of the ship or its program of observations.

With the rise in the use of systems analysis and the closing of the gap between statistical and synthetic studies of communities and ecosystems, studies of the cycles of the essential elements can be said to be on a truly scientific basis. It must be emphasized, however, that the present state of the art is still rudimentary. Systems analysis is placing new demands on our methods of data collection. The interactions between synthesis and statistics are only beginning, and we can expect to see rapid changes in the way that we study ecosystems in the next decade.

21

Reprinted from *Radioecology*, 121–125 (1963)

ANALOG COMPUTER MODELS FOR MOVEMENT OF NUCLIDES THROUGH ECOSYSTEMS

JERRY S. OLSON

Health Physics Division, Oak Ridge National Laboratory, Oak Ridge, Tennessee

INTRODUCTION

Analog computers should fulfill an increasing need in the interpretation of data in radioecology Ecological tracer experiments and case studies of radioactive contamination of the environment often require us to interpret the net changes in radioactivity in some part or parts of our system in terms of simultaneous gains and losses. The basic operation of the analog computer is to keep a running balance of such gains and losses for all the major parts or "compartments" of a system, be it a reactor or an ecosystem.

Such computers are coming into frequent use for kinetic interpretation of physiological tracer experiments, concerned with the movement of nuclides between cells or organs of a single individual. They should eventually become even more helpful in more complicated ecological systems where other methods of unscrambling simultaneous transfers are not satisfactory. Ecological applications of electronic analog computers have been made to population problems (Wangersky and Cunningham, 1957a,b) and to the world-wide biogeochemical cycle of carbon (DeVries, 1958). H.T. Odum (1960) has used an electrical network analog for simulating certain features of an ecosystem. Neel and Olson (1962) have used the Oak Ridge National Laboratory Analog Computer Facility for similar problems which can be viewed in relation to the transfer of energy, carbon, biomass, nutrients, and radionuclides through ecosystems.

However, the analog computer is more than a handmaiden for deriving numbers from empirical research. It provides a physical simulation of the processes to which the model pertains. It is already serving as a valuable stimulus and aid in developing a body of quantitative ecological theory concerning the operation of ecosystems.

The present paper diagrammatically illustrates how the running balance, or integration, of gains and losses is accomplished. The Methods section points out the basic idea for a single compartment, namely photosynthetic vegetation. The main section, on Transfer of Carbon and Carbon-14, extends this example to a simple ecological model covering four compartments (vegetation tops, roots, litter soil organic matter). Finally, a discussion of transfer of fission products and mineral nutrients through the ecosystem suggests how this basic methodology can eventually be extended further to models with more compartments and with other refinements.

METHODS

The basic unit of the analog computer is the integrator, which continually adds or subtracts voltages to the voltage already stored on a condenser contained within the integrator. If the photosynthetic rate P of producing vegetation were defined as a function of time and fed into the integrator (Fig. 1), then the output of the integrator would show the production which had occurred in the system since time zero. Schematic diagrams represent an integrator by a rectangular box next to a triangle which points in the "downstream" direction of transfer of material in the model. (Actually

INTEGRATOR

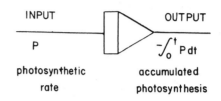

INPUT $\qquad$ OUTPUT

P

photosynthetic rate

$-\int_0^t P\,dt$

accumulated photosynthesis

Figure 1. Simple integration of a single input (photosynthetic rate) to give output of cumulative photosynthesis on an integrator of an analog computer.

NEGATIVE FEEDBACK

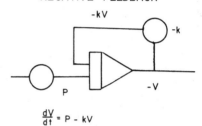

$$\frac{dV}{dt} = P - kV$$

Figure 2. Integration of a positive input (photosynthesis) minus a negative feedback controlled by the feedback potentiometer setting k.

the sign of this voltage is inverted, so the negative voltage is shown on the output.) For example, if photosynthetic rate is constant, then accumulated photosynthesis would increase uniformly through time, just as would the level of water in a cylindrical bucket that was being filled by a steady inflow of water.

Now if we imagine instead a bucket with a hole that leaks at a rate proportional to the amount of water accumulated in it, the rate of rise of the water level increases rapidly at first, and then gradually more slowly until the rate of leakage at last nearly balances the rate of steady input. Similarly, in an ideal case in which the rate of loss of organic carbon was a constant fraction of the organic carbon accumulated by photosynthetic vegetation, the standing crop of vegetation would increase until losses equalled input and a steady state was attained. Either of these idealized situations can be simulated by the analog computer diagram of Figure 2. If we let the amount of accumulation (either vegetation biomass, or volume of water in the bucket) be V, then the rate of change of V is the sum of P and −kV, i.e.,

$$\frac{dV}{dt} = P - kV \qquad (1)$$

Since −V is simply the output of the integrator, a potentiometer is used to tap off a fraction k of this voltage to give the product −kV. This negative product, or negative feedback, is added to the positive part of the input represented by P, so the integrator actually works on the sum of these two terms. (A positive feedback has also been used to represent the buildup in photosynthetic rate as the amount of photosynthetic tissue increase, but this was omitted from Figure 2.)

The resulting integral is the familiar exponential equation which gradually approaches an upper asymptote in the way we had intuitively surmised. The level which is approached as a steady state is equal to the ratio of the income P to the "negative feedback" parameter k. For the initial condition V = 0 at t = 0, the model formula for the accumulation V in the vegetation is obtained by integrating with respect to the time up to the value t:

$$V = \int_0^t (P - kV)dt = \frac{P}{K}(1 - e^{-kt}) \quad (2)$$

Chart records actually drawn by the analog computer are shown on Figure 3. The uppermost graph represents the constant input just assumed, which is here assumed to be 400 grams of carbon photosynthesized per square meter per year. Below this is the expected exponential equation for vegetation. The remaining graphs were obtained by using fractions of the output of the vegetation compartments as inputs for additional compartments as explained below.

A recent Oak Ridge National Laboratory report (Neel and Olson, 1962), based on a thesis by Robert Neel, provides further discussion and examples of the principles and techniques involved in this kind of electronic analog computer simulation. Earlier uses of electrical network analogs by De-Vries (1958) and H.T. Odum (1960) are cited in it, along with some technical limitations of such networks and of the electronic computers discussed here.

TRANSFER OF CARBON AND CARBON-14

The first step in extending this basic methodology to models with several compartments is the preparation of a block design like that of Figure 4 showing the pathways of transfer. For carbon, this shows how one fraction of carbon in the above-ground portion of vegetation is delivered to the plant roots (R), and a small fraction is considered to be consumed by animals (which are not here shown as separate boxes, for the sake of simplicity of illustration). A third fraction contributes to dead organic litter (D). Both roots and litter in turn serve as inputs for the organic carbon or humus in the mineral soil (S). The k parameters here correspond with total loss rates, and the ϕ's with partial transfer coefficients, indicating the redistribution just mentioned. The numbers cited here were somewhat arbitrary, for illustrative purposes.

The second step which can follow directly from this block diagram is the construction of an analog computer circuit diagram (Figure 5). Corresponding with each compartment is a little sub-circuit like that of Figure 2. Additional potentiometers are used to indicate what fraction of the output from one compartment serves as input for the next compartment. An initial condition other than zero volts could be provided for on any compartment; as if, for example, we wished to start counting t = 0 at a time when there was already a cover of vegetation V_0 present on the ground.

Instead of the rather unrealistic assumptions, such as of constant rate of photosynthesis, various elaborations have been introduced to this model one at a time. For example, in Figure 6, we have an

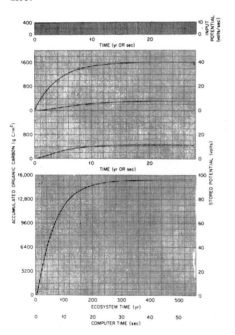

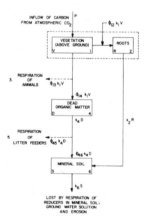

PARTIAL TRANSFER COEFFICIENTS		DECAY PARAMETERS	
ϕ_{12} — 0.20		k_1 — 0.25	
ϕ_{13} — 0.00		k_2 — 0.25	
ϕ_{14} — 0.80		k_4 — 0.693	
ϕ_{45} — 0.50		k_6 — 0.0156	
ϕ_{46} — 0.50			

Figure 3. Hypothetical accumulation of carbon in above-ground vegetation, roots, organic litter, and inorganic matter in mineral soil, from steady photosynthetic input into model system described in Figure 4.

Figure 4. Box diagram indicating pathways of movement of carbon in the larger compartments of a model ecosystem, and arbitrary selection of parameters for illustrations in Figures 3 and 6.

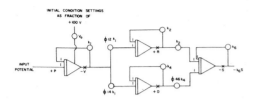

X - POINTS OF ANALOG COMPUTER CIRCUIT CONTINUOUSLY MONITORED BY
VOLTAGE CHART RECORDERS

Figure 5. Simple analog computer diagram showing transmission of changes in voltage from input on left, through integrators corresponding to above-ground vegetation (V), roots (R), dead organic litter (D), and soil organic matter (S).

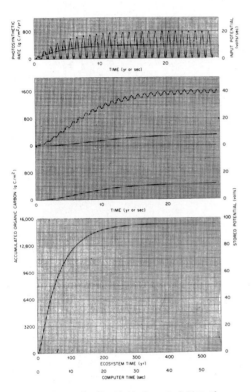

Figure 6. Hypothetical accumulation of carbon in a system with an annual cycle of high and low photosynthetic rate, initially averaging 0 grams of carbon per square meter per year, gradually rising to average of 400 grams of carbon per square meter per year. Sinusoidal oscillations attenuated in vegetation and almost eliminated in roots and litter under the assumed conditions. Note scale shift for soil organic matter.

input which oscillates to simulate the change in production between summer and winter. Furthermore, the mean annual rate of photosynthesis is allowed to vary instead of being kept as a constant over a period of years. Sinusoidal terms are evident in the

graph for the vegetation compartment, though somewhat damped in amplitude compared with the oscillations in photosynthetic rate. Each additional stage of transfer, such as that to roots or litter, further attenuates the amplitude of the oscillatory term, so that graphs for these compartments are virtually identical with those that would be expected for a model with an input having no annual oscillation of photosynthesis. Because of the very slow turnover assumed for soil organic carbon, the use of a compressed scale for both axes was necessary for showing this compartment on the same size graph paper.

While elementary differential equations can be used to solve these equations analytically, the solutions become awkward even for the four-compartment model considered here. They become more difficult to handle when additional compartments are added, and quite intractable when certain "nonlinear" features are introduced into the ecological model.

Fortunately, the electronic analog computer circuit has a variety of special devices such as function generators, servomultipliers and diodelimiters, which make the computer about as satisfactory for working with nonlinear models as with the highly simplified linear model considered previously.

One of the special devices, the function generator, was used by Neel and Olson (1962) as a first approach to extending the present model to see how a sudden increase of carbon-14 in the atmosphere as a result of thermonuclear tests would modify the carbon-14 content of the other parts of the same model ecosystem considered above (Figure 7). It was assumed that the "normal" amount of carbon-14

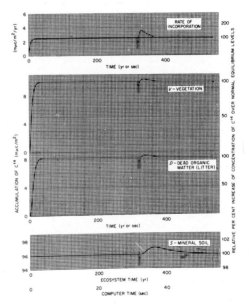

Figure 7. Response of model ecosystem described in Figure 4 to initial accumulation of "normal" carbon-14 in proportion to total carbon, and subsequent increase in carbon-14 fixation due to thermonuclear tests. Note lag in response of soil compartment.

produced in cosmic rays had previously been incorporated in the ecosystem in direct proportion to the content of total carbon, during the development of vegetation and litter. The soil humus was assumed to have had its equilibirum content of organic carbon and carbon-14 already. At the time indicated by the arrow, the analog computer was switched to the "hold" position and the function generator was switched into the circuit. The function generator then fed in a somewhat higher rate of incorporation of carbon-14 (Figure 7, top) which had previsuly been set up to correspond to the increase projected by Broecker and Olson (1960).

Because of the relatively fast time constants assumed for vegetation and litter, these compartments responded fairly promptly in showing an increase in carbon-14. The slow time constant which was assumed for the turnover of humus incorporated in the mineral soil resulted in a slower response to the "pulse" of carbon-14, and a markedly slower return to the "pre-nuclear test" levels (assuming that atmospheric nuclear testing and nuclear warfare were not resumed for centuries following the 1958 moratorium on testing). The absolute quantity of carbon-14 projected for each square meter of ground surface was somewhat higher for soil humus than for vegetation and litter, but because of the larger storage of carbon, the percentage increase was much less.

Examples of the kinds of ecological conclusions which might follow from more extended analyses of this type include the following: (1) The substantial increase in specific activity of atmospheric carbon-14 from weapons tests will have a diluted influence on the specific activity of carbon-14 with each transfer through the ecosystem, but the amounts of dilution and promptness of response will vary markedly depending on the parameters of different kinds of ecosystems. (2) If the increase in radiation hazard relative to existing background can be considered small (Totter, Zelle and Hollister,1958) for a comparatively radiosensitive organism like man, who derives food fairly directly from photosynthetic producers, then the radiation influence on other parts of the ecosystem from this source is even less likely to be important in comparison with such sources of radioactivity as potassium-40 and fission products. (3) Much more widespread atmospheric contamination which might arise in the event of nuclear war would similarly be less spectacular at first than that from fission products, but the long half-life and the capacity for storage of carbon-14 in humus as well as in the oceans would mean that any environmental problem involving carbon-14 might persist for many generations to follow. (4) The increase in specific activity of carbon already produced from nuclear tests provides an unprecedented opportunity for gathering information on the probable holdup of carbon-14 in different compartments of the biosphere and for evaluating the consequences of further contamination of this type. (5) Information on the turnover of carbon is valuable for basic research on the time parameters of carbon transfer in ecosystems, since these parameters influence the rates of many other ecological processes.

The interpretation of such measurements of response to changes in atmospheric carbon-14 would have to go hand in hand with definite hypotheses concerning the several pathways of income and loss of carbon. Such interpretations would presumably be expedited by analog computer models appropriate for the systems where measurements were being taken.

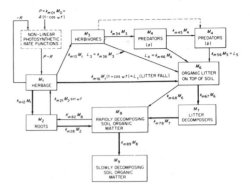

Figure 8. Flow of biomass and organic matter (M) through an herbaceous terrestrial ecosystem. Main units of M are grams per square meter. Heavy lines show main pathway of flow. Losses to atmosphere as carbon dioxide occur at many stages. The k_m's express instantaneous rates of transfer, as a fraction of source compartment, moving between the compartments numbered.

TRANSFER OF FISSION PRODUCTS THROUGH ECOSYSTEMS

In 1961, several obvious limitations of the preceding small-scale models were overcome. (1) Additional special computer features such as servomultipliers and diode limiters were used to make the behavior of the models correspond more closely to natural processes. (2) The large capacity of the National Laboratory Analog Computer Facility at Oak Ridge was more fully utilized for a larger number of ecological compartments (Figure 8), including the various trophic levels of the animal food chain, litter decomposers, and separate compartments for rapidly decomposing organic matter and slowly decomposing organic matter. (3) Finally, a second console of the computer was operated in tandem with the first in order to represent the transfer of mineral nutrient elements (or their radioactive nuclides) which follow some pathways of movement different from carbon.

With slight changes in definitions of terms, the first console can be used for models of transfer of energy, carbon, or of biomass. The transfer of free energy of course involves loss of availability at every stage of transfer, so there is no recycling. Each transfer of carbon involves a release of carbon dioxide which (for terrestrial systems) mostly becomes mixed with the general atmospheric pool before being reincorporated into the biomass as a result of new photosynthesis. Most of the biomass of organisms and soil organic matter follows the same pathway as carbon, from photosynthetic parts of vegetation, down through roots, or through litter and microorganisms, to soil.

By contrast, on the second console whose block diagram is shown in Figure 9, there is a major pathway of uptake from some exchangeable pool of nutrient material in the soil, to roots, up into plant tops, and only then along the above-mentioned pathway to litter and back to soil. For either mineral nutrient elements, or the majority of fission products or other radioactive nuclides, this kind of

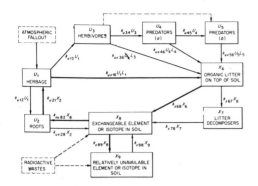

Figure 9. Flow of inorganic nutrient elements or radioactive nuclides through the system whose biomass transfers were shown in Figure 8. In U, units are expressed per gram of biomass or organic matter, while in X, units are per square meter. (The k_{ux} and k_{xu} are factors for appropriate conversion between these units; L represents rates of litter fall in grams per square meter.) Pathways of movement of nuclides from radioactive fallout or waste disposal can be traced.

pathway would be followed, and would involve the possibility of repeated cycling of materials from soil to organisms and back to soil. For some elements like cesium, there is a strong tendency toward fixation into a relatively unavailable form within the soil On the other hand, even such a readily fixable element can be recycled by root uptake before it happens to come into contact with sites of fixation, or may even be taken up directly by roots from decomposing litter (Witherspoon, 1963).

Further explanation of the terminology involved in Figures 8 and 9 can be consulted in the forest ecology section of the 1961 Oak Ridge National Laboratory Health Physics Progress Report (Olson et al., 1961). Time does not permit coverage of the analog computer circuit diagrams which have been developed on the basis of these block diagrams. The preceding discussion may indicate how electronic analog computers can be used for a variety of models for the movement of radionuclides through ecosystems.

SUMMARY

The interpretation of information on the rates of income and loss of radionuclides (or of nutrients, carbon, biomass, or free energy) between compartments of an ecosytem involves assumptions about the

system which can be formalized in mathematical models. Electronic analog computer techniques can be viewed either as a means of solving the differential equations describing the income and losses for each compartment of the system, or as a device for stimulating the natural processes operating in the system.

An example of a small analog computer model for the transfer of organic carbon or carbon-14 from photosynthetic vegetation, to roots or organic litter, to humus in mineral soil was discussed and further details are given in an Oak Ridge National Laboratory Report (Neel and Olson, 1962). Newer work in 1961 involved the use of two consoles, one to simulate the transfer of energy, of carbon, or of biomass, and a second one to simulate the cycling movement of inorganic nutrient elements or of radioactive nuclides.

ACKNOWLEDGMENTS

This work was performed under contract number W-7405-eng-26 between the U. S. Atomic Energy Commission and the Union Carbide Corporation.

REFERENCES

Broecker, W.S., and E.A Olson. 1960. Radiocarbon from nuclear tests, II. Science 162(3429): 712-721.

DeVries, H. 1958. Variation of concentration of radiocarbon with time and location on earth. Proc. Koninkl. Ned. Akad. Vetenschap. B-61: 94-102.

Neel, R.B., and J.S. Olson. 1962. Use of analog computer models for simulating the movement of isotopes in ecosystems. Oak Ridge Natl. Lab., U. S. AEC report ORNL-3172. xii, 108 pp.

Odum, H.T. 1960. Ecological potential and analogue circuits for the ecosystem. Am. Scientist 48(1): 1-8.

Olson, J.S., et al. 1961. Forest Ecology. In Health Physics Div. Ann. Prog. rep. for period ending July 31, 1961. Oak Ridge Natl. Lab., U. S. AEC report ORNL-3189. pp. 105-128.

Totter, J.R., M.R. Zelle, and H. Hollister. 1958. Hazard to man of carbon-14. Science 128(3337): 1490-1495. (Also: U. S. AEC report WASH-1008).

Wangersky, P.J., and W.J. Cunningham. 1957a. Timelag in prey-predator population models. Ecology 38 (1): 136-139.

- - - 1957b. Timelag in population models. Cold Cold Spring Harbor Symposia on Quantitative Biology 22: 329-338.

Witherspoon, J.P. 1963. Cycling of cesium-134 in white oak trees on sites of contrasting soil type and moisture. I. 1960 Growing Season. In this volume, pp. 127-132.

Reprinted from *Ecology*, **48**(5), 813–824 (1967) with permission of the publisher, Duke University Press, Durham, North Carolina

SYSTEMS ANALYSIS OF [134]CESIUM KINETICS IN TERRESTRIAL MICROCOSMS[1]

BERNARD C. PATTEN AND MARTIN WITKAMP

Radiation Ecology Section, Health Physics Division,
Oak Ridge National Laboratory, Oak Ridge, Tennessee

(Accepted for publication September 12, 1966)

Abstract. Laboratory experiments were conducted to determine patterns and rates of [134]Cs exchange in microecosystems composed of different combinations of radioactive leaf litter, soil, microflora, millipedes, and aqueous leachate. Rate constants were determined by fitting models to data with an analog computer. Simulations with the models permitted comparisons of different microcosms in terms of time to radiocesium equilibrium, steady state concentrations, concentration factors, input and output fluxes, turnover rates, and stability.

Rate constants varied with different compartment combinations, indicating both qualitative and quantitative differences in the cesium exchange patterns within different systems. This result is generalized: transfers of energy and matter in ecosystems are functions of networks which define intercompartmental interactions; internal coupling should be considered a significant variable in investigations of ecosystem processes.

Points of particular interest derived from the computer simulations are (i) organism concentration factors for a material may vary in different systems, depending upon how the organism is coupled to other compartments; (ii) total flux of a material in a steady state system may vary considerably from that in another system which receives identical input; (iii) material turnover within compartments and in the system as a whole tends to increase as more compartments are added; and (iv) stability of material concentration does not appear to increase with system complexity.

[1] Research sponsored by the U. S. Atomic Energy Commission under contract with the Union Carbide Corporation.

INTRODUCTION

Investigations of mineral cycling in terrestrial ecosystems are hindered by difficulties of separating soil organisms, plant roots, dead organic matter, and the mineral soil. Consequently, most studies consider only two compartments, such as soil-litter or soil-plant, and usually only at some experimental endpoint (harvest). Investigations of mineral flows between multiple compartments have been few (Jansson 1958, Remezov 1959, Witherspoon 1964, Witkamp and Frank 1964, Neel and Olson 1962, Olson 1965), and only the last two of these have given explicit attention to exchange kinetics. Whittaker (1961) has attempted to determine the kinetics of ^{32}P in aquatic microcosms. The present study employs two innovations to develop fuller understanding of mineral cycling in a series of experimental microecosystems, one involving laboratory syntheses of progressively more complex systems, and the other involving use of an analog computer to determine the kinetics of intercompartmental transfers.

Microcosms of increasing complexity were synthesized by adding one compartment at a time. Five compartments were used: ^{134}Cs-labeled oak leaves, mineral soil, microflora, millipedes and an aqueous leachate. Radioactivity in each compartment was determined as a function of time, with ^{134}Cs in the microflora computed as the difference between activity leached from sterile and nonsterile leaves. This stepwise approach to construction of multicompartmental systems permits evaluation of the effects of each added compartment on the mineral's dynamics.

Empirical data take the form of curves of radiocesium activity in each compartment graphed against time. Such information by itself does not specify transfer pathways, rates and fluxes along these pathways, or transfer mechanisms. It is possible, however, to determine these system characteristics by adjusting an analog computer model to fit the experimental data. This procedure results in a description of radiocesium kinetics for each microcosm in terms of a system of differential equations. The equations specify and quantify transfer pathways, and permit additional information to be developed by computer simulations of experiments which would be difficult or impossible to perform in the laboratory.

Miss Bonnie McGurn assisted in conducting the laboratory experiments, and Drs. D. A. Crossley and D. E. Reichle provided helpful criticism and discussion.

MATERIALS AND METHODS

Experimental

Microcosms were established in 60 ml glass funnels, inside diameter 4 cm, with fritted, medium porosity filters. The funnels were seated on 125 ml suction flasks each containing a counting tube for collection of leachate. Triplicates of the following systems were established: (i) sterile white oak (*Quercus alba* L.) leaves, (ii) leaves and microflora, (iii) leaves, microflora and millipedes, and (iv–vi) same as i–iii but placed on the surface of a 1 cm layer of silt-loam (pH 6.4, 12.6% organic matter, 27% moisture). The microcosms were maintained under identical conditions at room temperature (about 25°C).

The radioactive leaves were collected in July 1961 from small trees trunk-labeled (Witherspoon 1964) with ^{134}Cs a year before. Each funnel received 1 g of air-dried leaf material cut slightly smaller than 4 cm diameter. Leaves and soil were sterilized with 12% ethylene oxide (16 hr, 8 psi, 40°C) after which microflora and millipedes were added. Five millipedes (*Dixidesmus erasus* Loomis), averaging 392 mg total fresh weight and newly-collected from neighboring forests, were added to each funnel requiring millipedes. Dead millipedes were replaced daily if necessary and assayed for ^{134}Cs activity. Microflora was introduced in 1 ml of supernatant from 10 g wet weight of oak leaves shaken for 15 min with 100 ml of sterile water. This inoculum was administered to funnels requiring microflora as part of the first leaching at the start of the experiments. For each leaching, 8 ml of sterile water was dripped from a pipette as uniformly as possible over the contents of each funnel. This procedure was repeated every 2–3 days to approximate mean precipitation in Oak Ridge (2.5 cm/wk). During leaching, 10 cm Hg vacuum was applied. Leachate volumes were recorded and radiocesium activity of 5 ml aliquots was determined by NaI(Tl) scintillation counting. Between leachings, positive pressure of about 1 cm Hg was maintained in the flasks to prolong millipede survival by providing aeration. After 18 days and 9 leachings, the experiments were terminated and ^{134}Cs activity of remaining litter, soil and millipedes measured.

Computational

A block diagram of the litter-soil-microflora-millipede-leachate microecosystem, indicating all possible transfer routes, is shown in Figure 1. This system is reducible to fewer compartments by letting ^{134}Cs transfers be zero along appropriate pathways. All the microcosms thus implied in Figure 1 can be regarded as closed with respect

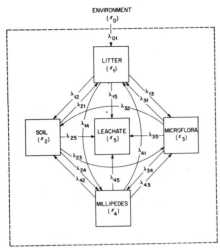

ENVIRONMENT
(x_0)

FIG. 1. Block diagram of litter-soil-microflora-millipede-leachate microecosystem showing all possible routes of cesium transfer and associated rate constants, $\lambda_{ij}(i = 0,1,\ldots,4;\ j = 1,2,\ldots,5)$.

to the radioisotope, even though radioactivity was removed in successive leachings, by considering the cumulative leachate as a sink (no back transfers to other compartments) within the system of definition. Thus, the only loss of activity for computational purposes is by radioactive decay, and this is negligible in an 18-day experiment (¹³⁴Cs half-life = 2.07 y).

Let X (a constant) be the initial radioactivity introduced in oak litter, and hence the total activity of the microcosm throughout an experiment. This activity will move from one compartment to another with time. If $X_j(t)$ is the total amount of radioactivity in compartment j at time t, then, since no activity is lost in an experiment, $\Sigma_j X_j = X =$ constant. If the mass of compartment j at time t is $m_j(t)$, then the concentration or activity-density of radiocesium in the compartment at time t is $x_j = X_j/m_j$. Letting $\lambda_{ij}(t)$ be a function expressing the rate of ¹³⁴Cs transfer from compartment i to compartment j, the rate of change of radioisotope concentration in the j'th compartment is the balance between incomes and losses:

$$\dot{x}_j = \Sigma_i \lambda_{ij} x_i - \Sigma_i \lambda_{ji} x_j, \qquad (1)$$

$$(i,j = 1,2,\ldots,5;\ i \neq j;\ 0 \leq \lambda_{ij}, \lambda_{ji} < \infty;$$
$$0 \leq x_i, x_j \leq 1)$$

where $\dot{x}_j$ is the first derivative of x_j with respect to time. Since the experimental systems were all

non-stationary (at least one $\dot{x}_j \neq 0$ during the period of observation) and since a transient solution of a differential equation corresponds to a unique transient behavior, equations (1) can be made to describe uniquely ¹³⁴Cs exchanges in the microcosms by finding appropriate functions, $\lambda_{ij}(t)$. To do this a simplest possible model was programmed for study on an Electronic Associates TR–10 analog computer. This proved adequate for description of the experimental results, and further refinements were unnecessary.

The model assumed first order kinetics: that transfer of ¹³⁴Cs from compartment i to compartment j is directly proportional to the radioisotope concentration in compartment i. With this assumption, the λ_{ij} in equations (1) become rate *constants*, having units $\{t^{-1}\}$.

To normalize the system equations so that different microcosms with different values of X could be compared, the fraction of radioactivity in each compartment relative to total system activity, $y_j = x_j/X$, was used as a variable for work on the computer. This has no effect on the rate constants to be determined, as the normalized system equations show:

$$\dot{y}_j = \frac{\dot{x}_j}{X} = \Sigma_i \lambda_{ij} \frac{x_i}{X} - \Sigma_i \lambda_{ji} \frac{x_j}{X}$$
$$= \Sigma_i \lambda_{ij} y_i - \Sigma_i \lambda_{ji} y_j. \qquad (2)$$
$$(0 \leq y_i,\ y_j \leq 1)$$

The computing procedure was to fit program-generated curves of $y_j(t)$ to empirical curves by adjusting potentiometers representing the λ_{ij} and λ_{ji}, thereby obtaining numerical values for these parameters and simultaneous solutions of the system equations (1 and 2) for each experimental microcosm.

RESULTS

The five compartments can be labeled as follows: $i,\ j = 1 \equiv$ oak litter, $2 \equiv$ soil, $3 \equiv$ microflora, $4 \equiv$ millipedes and $5 \equiv$ leachate (Fig. 1). The environment is compartment 0. Possible combinations of litter-leachate and other compartments are 1–5, 1–2–5, 1–3–5, 1–4–5, 1–2–3–5, 1–2–4–5, 1–3–4–5, and 1–2–3–4–5. Of these, 1–4–5 and 1–2–4–5 are not experimentally feasible because of difficulties of sterilizing millipedes. The remaining combinations, labeled I-VI, respectively, were investigated in six experiments whose results are summarized in Figure 2 and Table 1.

Combination I (litter-leachate).—As shown in Figure 2a, ¹³⁴Cs was transferred from litter to leachate at a rate of $\lambda_{15} = 0.037$ day⁻¹, 3.7% per

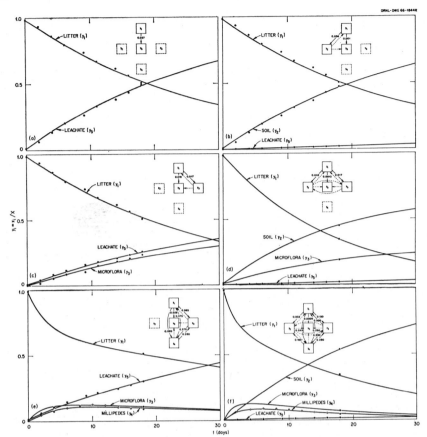

Fig. 2. Analog computer curves of changes in relative distribution (y_j) of radiocesium in the microcosms of Combs. I–VI (Figs. a–f, respectively). Means of three replicates are shown as solid circles. Open circles represent estimates of relative cesium concentration in millipedes based on radioassays of animals which died. Insets show compartment combinations and rate constants (day^{-1}) for each indicated transfer route. Pathways over which no cesium was transferred are illustrated by broken arrows.

day. System equations will not be written explicitly hereafter, but for the purpose of illustrating how the microcosm behaviors depicted in Figure 2 are described by the model (equations 1 and 2), the system equations for the present experiment are

$$\begin{cases} \dot{x}_1 = -0.037\,x_1 \\ \dot{x}_5 = 0.037\,x_1 \end{cases} \quad \text{or} \quad \begin{cases} \dot{y}_1 = -0.037\,y_1 \\ \dot{y}_5 = 0.037\,y_1 \end{cases}$$

Combination II (litter-soil-leachate).—Addition of soil sharply reduced transfer of radiocesium from litter to leachate (Fig. 2b). The total loss rate from litter was the same as in Combination I

$(\lambda_{12} + \lambda_{15} = 0.037 \text{ day}^{-1}$, Table 1), but now 0.036 day^{-1} went to soil and only 0.001 day^{-1} to the leachate. Once in the soil radiocesium was held tightly, as indicated by non-leachability $(\lambda_{25} = 0)$ and absence of back transfer to litter $(\lambda_{21} = 0)$.

Combination III (litter-microflora-leachate).—Loss of ^{134}Cs from litter was essentially the same as in Combination I, 0.036 day^{-1} (Table 1). Of the total litter loss, 0.017 day^{-1} was transferred to microflora and 0.019 day^{-1} to the leachate (Fig. 2c, Table 1). In agreement with *in vitro* leaching experiments with microflora, radiocesium in

TABLE 1. Rate constants, day^{-1}

Parameters	Combinations					
	I (1-5)	II (1-2-5)	III (1-3-5)	IV (1-2-3-5)	V (1-3-4-5)	VI (1-2-3-4-5)
λ_{12}	—	0.036	—	0.042	—	0.033
λ_{13}	—	—	0.017	0.017	0.065	0.120
λ_{14}	—	—	—	—	0.020	0.045
λ_{15}	0.037	0.001	0.019	0.0013	0.025	0.0009
$\Sigma\lambda_{1j}$	0.037	0.037	0.036	0.060	0.110	0.199
λ_{21}	—	0	—	0	—	0
λ_{23}	—	—	—	0	—	0
λ_{24}	—	—	—	—	—	0
λ_{25}	—	0	—	0	—	0
$\Sigma\lambda_{2j}$	—	0	—	0	—	0
λ_{31}	—	—	0	0	0	0.500
λ_{33}	—	—	—	0	—	0
λ_{34}	—	—	—	—	0.375	0.250
λ_{35}	—	—	0	0	0	0
$\Sigma\lambda_{3j}$	—	—	0	0	0.375	0.750
λ_{41}	—	—	—	—	0.475	0.250
λ_{42}	—	—	—	—	—	0.160
λ_{43}	—	—	—	—	0.050	0.250
λ_{45}	—	—	—	—	0	0
$\Sigma\lambda_{4j}$	—	—	—	—	0.525	0.660

the microflora was non-leachable ($\lambda_{35} = 0$). No reverse transfer from microflora to litter was indicated ($\lambda_{31} = 0$). The rate of leachate gain was reduced compared to when microflora were absent (Comb. III vs. I), but not as much as when soil was present (Comb. III vs. II).

Combination IV (litter-soil-microflora-leachate)—Soil and microflora together had a synergistic effect which markedly increased the rate of cesium loss from litter ($\lambda_{12} + \lambda_{13} + \lambda_{15} = 0.060$ day^{-1}, Table 1). As shown in Figure 2d and Table 1, transfer from litter to soil was increased to $\lambda_{12} = 0.042$ day^{-1} compared to 0.036 in Combination II. Accumulation by the microflora, however, was unaffected by soil ($\lambda_{13} = 0.017$ as in Comb. III). Leachability of the litter ($\lambda_{15} = 0.0013$) was essentially as in Combination II. Hence, the synergism appears to be an effect mainly on the ability of soil to acquire cesium from the litter, which is difficult to understand in view of previous results indicating that microflora do not alter the rate of loss from the litter (Comb. III vs. I). Furthermore, no direct exchanges between soil and microflora are indicated ($\lambda_{23} = \lambda_{32} = 0$), and consequently the character of the microflora effect on litter-soil exchange is not ap-

parent. As in Combinations II and III, soil and microflora radioactivity was non-leachable ($\lambda_{25} = \lambda_{35} = 0$), and also there was no cesium feedback from these compartments to the litter ($\lambda_{21} = \lambda_{31} = 0$).

Combination V (litter-microflora-millipedes-leachate).—The open circles in Figure 2e represent estimates of y_4 for total millipedes based on radioassays of animals which died and were removed. The fit of the model to these data is not too satisfactory: the millipede curve does not rise high enough between 6 to 14 days and does not fall rapidly enough toward the end of the observation period. Whether this difficulty is primarily with the model or with the extrapolated data is uncertain, but the fact that a better fit was obtained in the next, more complex combination which included soil (Fig. 2f) is good reason for confidence in the model.

The Combination V results (Fig. 2e, Table 1) indicate that when both microflora and millipedes are present the total rate of cesium loss from litter ($\lambda_{13} + \lambda_{14} + \lambda_{15} = 0.110$, Table 1) was tripled compared to Combinations I–III and almost doubled compared to Combination IV. The rate of transfer from litter to microflora was four times greater than in Combination III ($\lambda_{13} = 0.065$ vs. 0.017), presumably due to action of millipedes upon the litter. This action is also reflected in increased leachability of litter cesium ($\lambda_{15} = 0.025$ vs. 0.019 in Comb. III).

Despite these indications of millipede effects on litter, the rate of radiocesium transfer from litter to millipedes was fairly low, $\lambda_{15} = 0.020$ day^{-1}. In contrast, uptake rate from the microflora was about 19 times greater ($\lambda_{34} = 0.375$). Based on these results, and prompted by laboratory observations of no visible fungal mycelia in microcosms containing *Dixidesmus* but abundant growths otherwise, it was originally thought that the model denoted a 19-fold perference of *Dixidesmus* for microflora over litter. This is not necessarily true, as illustrated by the following examples of two-compartment exchanges.

Let compartment i have biomass m_i and cesium concentration x_i, compartment j have biomass m_j and concentration x_j, and suppose $x_i > x_j$. (i) If biomass is transferred from i to j ($\lambda_{ij} > \lambda_{ji} = 0$), then compartment i loses mass but its cesium concentration does not change (homogeneous distribution of the mineral within the mass is assumed for purposes of the argument). Compartment j gains biomass which is of higher radioisotope concentration than that within the compartment, and therefore x_j increases. As the process continues, both activity-densities approach equality, $x_j \rightarrow x_i$. (ii) If biomass is transferred from j to i ($0 = \lambda_{ij} < \lambda_{ji}$), compartment i gains biomass with cesium concentration x_j, and therefore x_i decreases. Compartment j loses mass but its concentration of radiocesium is unchanged. Continued exchange leads in the direction $x_i \rightarrow x_j$. (iii) If both forward and reverse exchanges occur simultaneously ($\lambda_{ij}, \lambda_{ji} > 0$), the cesium concentrations will move toward an equilibrium defined by $x_{j(eq)} = \dfrac{\lambda_{ij}}{\lambda_{ji}} x_{i(eq)}$. The second example, where compartment i gains biomass and consequently has its cesium concentration reduced, illustrates why biomass and tracer transfers cannot both be represented by the same rate constants. A 19-fold difference in rates of cesium input from two compartments cannot be interpreted as a 19-fold preference.

As with the microflora in this and preceding combinations, millipede radioactivity was non-leachable ($\lambda_{35} = \lambda_{45} = 0$). However, unlike the microflora which did not transfer ^{134}Cs back to the litter ($\lambda_{31} = 0$), the millipedes did so at a high rate ($\lambda_{41} = 0.475$). In addition, the animals also transferred radiocesium to the microflora at a rate of $\lambda_{43} = 0.050$ day^{-1}.

Combination VI (litter-soil-microflora-millipedes-leachate).—Results with this most advanced microcosm are given in Figure 2f and Table 1. Numerous points of comparison with previous combinations are possible, and the most systematic way to proceed is down the last column of Table 1.

The rate of cesium transfer from litter to soil ($\lambda_{12} = 0.033$) was less than in other combinations in which soil was present. In Combination IV, compared to II, microflora increased the rate of movement from litter to soil. With millipedes present (Comb. VI), this effect is apparently nullified. One reason is that the rate of movement from litter to microflora was about 7 times greater ($\lambda_{13} = 0.120$ in Comb. VI vs. 0.017 in Comb. IV). Another is that the rate of transfer from litter to millipedes was more than doubled in the presence of soil ($\lambda_{14} = 0.045$ vs. 0.020 in Comb. V). The basis for this soil-millipede interaction is not apparent except, perhaps, that the normal habitat of these animals was better approximated here than in Combination V, resulting in more normal rates of food consumption. Leachability of litter cesium, $\lambda_{15} = 0.0009$ day^{-1}, was low as in Combinations II and IV, which also had soil present. That microflora and millipedes substantially increased the mobility of litter cesium is clear from the total litter loss rate ($\lambda_{12} + \lambda_{13} + \lambda_{14} + \lambda_{15} = 0.199$), which is almost 6 times greater than in Combinations I–III, more than 3 times greater than in Combination IV, and almost twice as great as in Combination V.

Soil is clearly a cesium sink at the low concentrations in these experiments, since none of the microcosms studied contained a compartment capable of acquiring the mineral from soil ($\lambda_{2j} = 0$ for all $j \neq 2$; Table 1).

Unlike previous combinations with microflora, radiocesium in this combination was transferred at a high rate from microflora to the litter ($\lambda_{31} = 0.500$). This may be due to millipede feeding since Combination V indicated that these animals obtain cesium from the microflora ($\lambda_{34} = 0.375$ in Comb. V vs. zero in Combs. III and IV without millipedes). The transfer to litter rather than soil ($\lambda_{32} = 0$) verifies the experimentally observable fact that mycelia and other microbial components are associated with litter more than soil. Lack of transfer from microflora to litter in Combination V is unexplained.

The transfer rate from microflora to millipedes, while still high ($\lambda_{34} = 0.250$), is nevertheless reduced compared to Combination V. Soil, the only variable between these two combinations, must be responsible. A physiological difference between microflora in this combination and Combination V is indicated by doubling of the radiocesium loss rate in the presence of soil $(\lambda_{31} + \lambda_{32} + \lambda_{34} + \lambda_{35} = 0.750$ day$^{-1})$. Nothing is known of the microflora species composition, but it must have been different from Combination V, and this could account for some difference in transfer to millipedes.

Cesium transfer from millipedes to litter was reduced almost half compared to Combination IV $(\lambda_{41} = 0.250)$. This is due partly to soil affinity for excreted cesium $(\lambda_{42} = 0.160)$, but the combined loss to soil and litter $(\lambda_{41} + \lambda_{42} = 0.410)$ is still somewhat less than the lost to litter in Combination V $(\lambda_{41} = 0.475)$. The high transfer from millipedes to microflora $(\lambda_{43} = 0.250$ vs. 0.050 in Comb. IV) may be due to improved nutritional and moisture conditions with soil present. This contributed to a higher observed elimination rate for *Dixidesmus* than in Combination V $(\lambda_{41} + \lambda_{42} + \lambda_{43} + \lambda_{45} = 0.660$ vs 0.525), although this difference appears to be due largely to the method of obtaining millipede data: after each leaching, dead millipedes were removed for radioassay and replaced by new animals. Data from the dead specimens were used to represent live animals. No difficulties apparently were encountered through cesium leaching from the dead material $(\lambda_{45} = 0)$, but microbial decomposition $(\lambda_{43} = 0.250)$ had begun. Assuming no direct transfer to microflora from live millipedes, corrected elimination rates for *Dixidesmus* in Combs. V and VI, respectively, would be 0.475 and 0.410 day^{-1}.[2] These values compare favorably with 0.29 day^{-1} (biological half life $= 2.4$ days at 20°C), reported for assimilated cesium by Reichle and Crossley (1965), considering the higher temperature of the present experiments and the fact that both assimilated and unassimilated components are taken into account.

In general, the Table 1 results show that, under otherwise similar conditions, rate constants change in microcosms composed of different combinations of the same compartments. From this it can be concluded that patterns of intercompartmental coupling are prime variables to consider in studies of energy and material flows in ecosystems.

DISCUSSION

Pathways of mineral transfer in coupled microecosystems are both identifiable and quantifiable by the procedures employed, namely the simul-

taneous solution of systems of differential equations with an analog computer. (i) Qualitatively, actual transfer routes of those which are possible were determined. These are illustrated in the insets of Figure 2, the pathways over which no transfers occurred being shown as broken arrows. The numbers of actual routes in the systems of Combinations I–VI were, respectively, 1, 2, 2, 3, 6 and 9. The ratios of actual pathways to possible ones were 1.00, 0.50, 0.50, 0.33, 0.67 and 0.56 (ii) Quantitatively, rates of cesium transfer along the actual routes are given by the values of the parameters, λ_{ij}.

In general, two systems are identical if their parts are the same and the connections between the parts are the same. In Figure 3, systems 1

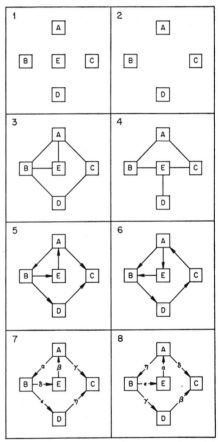

FIG. 3. Examples of non-identical systems (discussion in text).

[2] The values 0.475 and 0.410 may not be significantly different because of subjectivity in determining best fit in the curve-fitting procedure. Numerical approaches for use with digital computers are now being explored. A preliminary value of λ_{41} is 0.414 day^{-1} (G. M. Van Dyne, personal communication), and that by another 0.470 day^{-1}.

and 2 are different because their components are different; systems 3 and 4 differ because the connections are not the same in the "qualitative" sense given above; in systems 5 and 6 the connections are different in a "relational" or "directional" sense; and systems 7 and 8 differ in the "quantitative" sense that their rate functions are different. None of the systems of Figure 3 are identical, and consequently none will behave precisely the same as another whether it be in reaction to some environmental stimulus, in utilization of energy, or in the cycling of minerals, if these be appropriate system activities. The "structure and function" of these systems is said to be different. Similarly, the six microecosystems of this study have been shown to differ in radiocesium kinetics, with examples (Fig. 2, insets) of each of the reasons for system differences, illustrated in Figure 3, being represented.

Analog Computer Simulations

The microcosm models which have been obtained are operational, and it now becomes possible to study comparative transient and stationary behavior of these systems on the computer, in effect performing experiments which would be difficult or impossible to conduct in the experi-

mental laboratory. The number of experiments possible is infinite, and the ones actually performed were selected to bring out aspects of microecosystem behavior of general ecological interest.

To generate system behavior, a driving or forcing function, $\lambda_{01} x_0$, was introduced via the litter compartment (Fig. 1). In the six experiments preceding, there was only an initial cesium concentration in the litter, but no subsequent input from the environment ($\lambda_{01} = 0$). Driving the systems through the litter is somewhat artificial, just as establishing the microcosms with tagged litter is, because in nature each compartment is exposed directly to cesium input as environmental fallout. The experiments might have been more realistic, therefore, if an initial cesium dose had been rained on as part of the first leaching rather than introduced as tagged litter. This would not have changed any system parameters (Table 1), but it would have yielded additional parameters of interest (λ_{0j}; $j = 1, 2, \ldots, 5$). The simulations could then have been driven by multiple inputs. The present model is still very useful for relative comparisons of cesium behavior in the six microcosms. Relationships between radioactivity and voltage, and real time and computer time are clarified in the appendix.

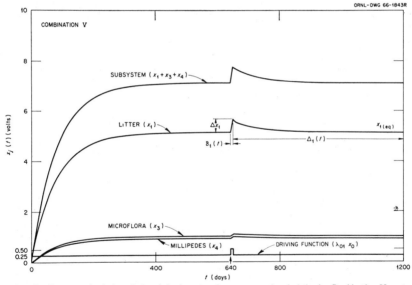

Fig. 4. Computer simulation of time behavior of cesium concentration (x_j) in the Combination V system, with constant input and perturbation after equilibrium. Soil and leachate did not equilibrate and are omitted. Shown is the driving function of 0.25 v, increased to 0.50 v between 640 and 648 days; also litter, microflora and millipede compartments, and the subsystem of these three compartments. The variables $\delta_j(t)$, $\Delta_j(t)$, $x_{j(eq)}$ and Δx_j are illustrated for the litter compartment.

Experiments of 1200 days (3.3 years) duration were run (Fig. 4). The driving function arbitrarily selected was a constant $\beta^{-1}\lambda_{01} x_0 = 0.25$ v sec^{-1}, obtained by letting $x_0 = 10$ v, $\lambda_{01} = 0.0125$ day^{-1}, and β (time-scale factor, see appendix) $= \frac{1}{2}$ sec day^{-1}. All compartments were radiocesium-free initially $(x_j(0) = 0$, all $j > 0)$. Figure 4 shows results for the Combination V system. A similar graph was obtained for each of the other microcosms from which the data of Table 2 were then read off and computed.

Figure 4 illustrates general aspects of system behavior in these simulations under the influence of the constant "radiocesium" input. Three com-

TABLE 2. Equilibrium characteristics of compartments, and subsystems composed of equilibrated compartments.

Combinations	Litter	Soil	Microflora	Millipedes	Leachate	Equilibrium Subsystems
			Compartments			
Time to Equilibrium (days)						
I	264	—	—	—	∞	264
II	328	∞	—	—	∞	328
III	264	—	∞	—	∞	264
IV	168	∞	∞	—	∞	168
V	616	—	448	432	∞	616
VI	232	∞	186	168	∞	232
^{134}Cs Concentration (v)						
I	3.54	—	—	—	∞	3.54
II	3.52	∞	—	—	∞	3.52
III	3.52	—	∞	—	∞	3.52
IV	2.20	∞	∞	—	∞	2.20
V	5.10	—	1.01	0.93	∞	7.04
VI	2.28	∞	0.47	0.33	∞	3.08
Concentration Factors						
I	14.16	—	—	—	∞	14.16
II	14.08	∞	—	—	∞	14.08
III	14.08	—	∞	—	∞	14.08
IV	8.80	∞	∞	—	∞	8.80
V	20.40	—	4.04	3.72	∞	12.27
VI	9.12	∞	1.88	1.32	∞	5.76
Input (v day^{-1})						
I	0.13	—	—	—	0.13	0.26
II	0.13	0.13	—	—	0.003	0.26
III	0.13	—	0.06	—	0.07	0.26
IV	0.13	0.09	0.04	—	0.002	0.26
V	0.57	—	0.38	0.48	0.13	1.56
VI	0.44	0.13	0.36	0.22	0.02	1.17
Output (v day^{-1})						
I	0.13	—	—	—	0.00	0.13
II	0.13	0.00	—	—	0.00	0.13
III	0.13	—	0.00	—	0.00	0.13
IV	0.13	0.00	0.00	—	0.00	0.13
V	0.56	—	0.38	0.49	0.00	1.43
VI	0.45	0.00	0.35	0.22	0.00	1.02
Turnover (% day^{-1})						
I	3.70	—	—	—	0.00	3.70
II	3.70	0.00	—	—	0.00	3.70
III	3.60	—	0.00	—	0.00	3.60
IV	6.03	0.00	0.00	—	0.00	6.03
V	11.00	—	37.50	52.50	0.00	101.00
VI	19.89	0.00	75.00	66.00	0.00	160.89
Stability (%)						
I	21.60	—	—	—	—	21.60
II	18.08	—	—	—	—	18.08
III	23.44	—	—	—	—	23.44
IV	18.08	—	—	—	—	18.08
V	20.40	—	24.96	29.04	—	21.61
VI	10.43	—	12.23	10.64	—	10.02

partments—litter, microflora and millipedes—went through a transient response followed by a steady state. Soil and leachate never reequilibrated because they are cesium sinks at the concentrations considered. Equilibrial characteristics of interest include (i) *time to equilibrium*, {days}; (ii) radiocesium *concentration*, {v}; (iii) *concentration factor*, a dimensionless property defined for present purposes as the ratio of compartment concentration to the voltage of the forcing function, 0.25 v; (iv) cesium *input flux* (input to compartment $j = \Sigma_i \lambda_{ij} x_i$), {v day^{-1}}; *output flux* (output of $j = \Sigma_i \lambda_{ji} x_j$), {v day^{-1}}; (vi) *turnover rate*, calculated for compartment j either as $\Sigma_i \dfrac{\lambda_{ij} x_i}{x_j}$ or as $\Sigma_i \dfrac{\lambda_{ji} x_j}{x_j} = \Sigma_i \lambda_{ji}$, {day^{-1}}; and (vii) *stability*.

Stability is the capacity of a system, subsystem, or compartment to minimize effects of a disturbance. To assess this for the six microcosms, the systems were disturbed after compartments which would equilibrate had reached equilibrium. This was accomplished by doubling the input to $\beta^{-1}\lambda_{01} x_0 = 0.5$ v sec^{-1} for 8 days and then returning it to its former level (Fig. 4: the driving function is doubled between 640 and 648 days). The disturbance caused a perturbation in equilibrated compartments (the nonsteady state ones were ignored; as cesium sinks they do not feed back to the remaining system and hence cannot affect it). Removal of the disturbance was followed by return to former equilibrium levels (Fig. 4: the graph shows deflections between 640 and 648 days, followed by recovery of equilibrated compartments—litter, microflora and millipedes—and of the subsystem comprised of these compartments). If $x_{j(eq)}$ is the equilibrium cesium concentration of compartment or subsystem j, Δx_j the perturbation of the same unit, $\delta_j t$ the duration of the perturbation, and $\Delta_j t$ the time for j to return to the original equilibrium (see Fig. 4), then a good measure of the stability of j is provided by $(x_{j(eq)}) (\delta_j t)/(\Delta x_j) (\Delta_j t)$. This is because the less the disturbance, (Δx_j), relative to the equilibrium concentration, $(x_{j(eq)})$, and the more rapid the recovery time, $(\Delta_j t)$, relative to the duration of the disturbance, $\delta_j t$, the greater is the stability of j. The measure is dimensionless.

The following discussion will be limited to a general comparison of principal system differences which appear in the table.

Equilibration time.—Given a nonzero input, in order for a system, subsystem, or compartment to achieve a steady state it must also have an output which is nonzero. Thus, while soil and leachate

in all of the microcosms studied had cesium inputs, they lacked outputs (Fig. 2, insets), and consequently never equilibrated (Table 2). Similarly, the microflora of Combinations III and IV accumulated radiocesium, but did not lose it. All other compartments eventually equilibrated at cesium levels defined by the equilibrium solutions of equations (1).

As shown in Table 2, the time to achieve this varied almost 4-fold, from 168 days for the litter of Combination IV and the millipedes of Combination VI to 616 days for the litter of Combination V (values to the nearest 8 days). Two general points illustrated by the tabulated data are that different compartments of the same microcosm and the same compartments in different microcosms take different periods of time to reach equilibrium.

Equilibrium Concentration of ^{134}Cs.—Compartments which do not equilibrate are indicated in Table 2 by "∞." Equilibrium concentrations, $x_{j(eq)}$, varied from 0.33 to 5.10 v in individual compartments, and from 2.20 to 7.04 v in subsystems composed of the steady state compartments. The litter value of 5.10 for Combination V is unusually high and more than double the equilibrium level of this compartment in Combination VI. The only difference between these two microcosms is that one contained soil. Since the environmental input to all of these systems was identical ($\lambda_{01} x_0 = 0.125$ v day^{-1}), the data of Table 2 indicate considerable variability in steady state cesium levels. This is a function of the components present and how they are coupled together.

Concentration Factors.—The observed range was 1.32 for millipedes in Combination VI to 20.40 for litter in Combination V. Since the driving function was constant, these figures are multiples (4×) of the equilibrium cesium levels. Expressed as concentration factors, though, they give some additional insights. The change in relative cesium-concentrating ability of *Dixidesmus* in Combinations V and VI can be interpreted in terms of the defining equations. The equilibrium solutions of equations (1) are

$$0 = \Sigma_i \lambda_{ij} x_{i(eq)} - \Sigma_i \lambda_{ji} x_{j(eq)} ,$$

signifying input = output,

$$\Sigma_i \lambda_{ij} x_{i(eq)} = \Sigma_i \lambda_{ji} x_{j(eq)} ,$$

and giving for the equilibrium concentration of compartment j

$$x_{j(eq)} = \left(\Sigma_i \lambda_{ij} x_{i(eq)} \right) / \Sigma_i \lambda_{ji} . \qquad (3)$$

Note (definitions p. 822) that the numerator

is the compartment's input flux and the denominator its turnover rate.

$$\text{(V)} \quad x_{4(eq)} = \frac{\lambda_{14}x_{1(eq)} + \lambda_{34}x_{3(eq)}}{\lambda_{41} + \lambda_{43}} = \frac{(0.020)\ (5.10) + (0.375)\ (1.01)}{0.475 + 0.050} = 0.92 \text{ v}$$

$$\text{(VI)} \quad x_{4(eq)} = \frac{\lambda_{14}x_{1(eq)} + \lambda_{34}x_{3(eq)}}{\lambda_{41} + \lambda_{42} + \lambda_{43}} = \frac{(0.045)\ (2.28) + (0.250)\ (0.47)}{0.250 + 0.160 + 0.250} = 0.33 \text{ v}$$

Note that these values are almost identical to those given in Table 2, which were read off the computer output graphs. These calculations indicate, subject to the limitations of counting dead millipedes for radioactivity, as discussed earlier, that the equilibrium cesium concentrations—and hence the concentration factors for *Dixidesmus*—differ between Combinations V and VI. Correcting for loss to microflora through decomposition of the dead animals (i.e., assuming $\lambda_{43} = 0$), the concentration factors would still be different because $x_{4(eq)} = 0.29$ v in Combination V and 0.54 v in VI. In general, concentration factors for given compartments, j, would be expected to change in communities with different compartment compositions (different i's) as the input flux (numerator of equation 3), and possibly also the turnover (denominator), changed due to differences in the i's. This is borne out in Table 2 for all equilibrated compartments and subsystems.

Input and Output Fluxes.—Compartments, subsystems or systems in steady state with respect to a material have input = output. This is illustrated in Table 2. For equilibrated compartments in the six microcosms, inputs and outputs ranged from 0.13 to 0.56 v day⁻¹. For subsystems of equilibrated compartments, the input range was 0.26 to 1.56 v day⁻¹, and the output 0.13 to 1.43 v day⁻¹. These data illustrate the following points about material equilibria in coupled systems, all a consequence of network organization. Within a system, different compartments will tend to have different stationary input/output fluxes. Between systems driven by identical forcing functions, the same compartments may have quite different inputs and outputs. Therefore, the total flux of a material in a steady state system may vary considerably from that in another system which receives identical input from the environment.

Turnover.—The turnover rate of a compartment, j, is most easily computed as the sum of output rates, $\sum_i \lambda_{ji}$, rather than as the ratio of input flux to concentration, $\sum_i \frac{\lambda_{ij}x_i}{x_j}$. Turnover is thus equivalent to elimination rate. Calculations of turnover as $\sum_i \lambda_{ji}$ can be made directly from Table 1, and consequently are independent of the hypothetical simulations. The turnovers listed in

The equilibrium equations for millipedes in Combinations V and VI are

Table 2, therefore, are real values for the experimental microcosms.

Table 2 indicates that each compartment tends to have higher turnover or elimination rates in systems with more compartments. This is because there are more compartments, i, from which to receive inputs in $\sum_i (\lambda_{ij}x_i)/x_j$, and to which to transfer outputs in $\sum_i \lambda_{ji}$. The result does not apply necessarily to compartments whose elimination is controlled by internal physiology, and the millipede data have already been qualified (p. 819, and footnote 2). The corrected daily millipede turnover rates are 47.5% in Combination V, and 41.0% in Combination VI. As more compartments are accrued in a system, total system turnover tends to increase. This is illustrated in the right-hand column of Table 2, and is consistent with general notions about energy and material flows in relation to ecosystem complexity.

Stability.—Data tabulated in Table 2 range from 10.43×10^{-2} to 29.04×10^{-2} for individual compartments, and 10.02×10^{-2} to 23.44×10^{-2} for the subsystems of equilibrated compartments. It is indicated that different compartments of a system can have different, although here generally similar, stabilities, and that the stability of cesium concentration in a given compartment varies from system to system. There appears to be no correlation between system complexity and ability to minimize and damp perturbations of radiocesium concentrations, which is contrary to inference from the stability theory of MacArthur (1955).

CONCLUSIONS

This investigation demonstrates that mineral kinetics in small laboratory microecosystems of limited complexity can be successfully modeled, and then their transient and stationary behavior studied comparatively with an analog computer. Environmental conditions and the compartments employed were the same for all systems, but the compartment combinations were different, and this was the principal experimental variable. It resulted in different patterns and degrees of intercompartmental coupling, producing radiocesium kinetics which were unique for each microcosm. The variable kinetics defined in turn variable system behavior. System and compartment attributes which changed with the nature of the coupling networks were duration of transient response, equilibrial cesium concentrations, concentration factors, input and output fluxes, turnover rates, and stability.

These results focus attention on the exceeding importance in natural, complex ecosystems of the organizational networks which define compartment interactions. Only one of numerous substances simultaneously transferred in the present microcosms, probably each with unique kinetics, was actually studied. Changes in the concentration or dynamics of any of these, especially major nutrients, would probably have altered the cesium kinetics. The multiplicity of material transfers and interactions conceivable in macroecosystems, together with the effects of intrasystem coupling as revealed in this investigation, make it apparent that to understand ecosystems ultimately will be to understand networks.

APPENDIX

Relationships between the microcosms and their models are as follows. Let

$x_j \equiv$ radionuclide activity-density of compartment j, in appropriate units, $\{a\}$.

$t \equiv$ real time, $\{days\}$.

$\lambda_{ij} \equiv$ rate constant for transfer of radioisotope from compartment i to j, $\{t^{-1}\}$.

$\lambda_p \equiv$ rate constant for physical radioactive decay $\{t^{-1}\}$; $\lambda_p = -0.0009$ day^{-1} for ^{134}Cs.

$v \equiv \{volts\}$.

$\alpha_j \equiv$ scale factor relating radioactivity to computer voltage for compartment j, $\{va^{-1}\}$.

$\tau \equiv$ computer time, $\{sec\}$.

$\beta = \tau t^{-1} \equiv$ scale factor relating real time to computer time, $\{sec\ day^{-1}\}$.

The system equations (1), prior to being corrected for radioactive decay, would be

$$\dot{x}_j = \Sigma_i \lambda_{ij} x_i - \Sigma_i \lambda_{ji} x_j - \lambda_p x_j \,.$$

Correction of the primary data before curve-fitting eliminates physical decay from further consideration, both in determination of rate constants, and in simulations with the determined coefficients. In (1), both sides of each equation have the units $\{at^{-1}\}$. With voltage scaling, (1) becomes

$$[\alpha_j \dot{x}_j] = \Sigma_i \frac{\alpha_j (\lambda_{ij}) [\alpha_i x_i]}{\alpha_i} - \Sigma_i \frac{\alpha_j (\lambda_{ji}) [\alpha_j x_j]}{\alpha_j}$$

$$= \Sigma_i \left(\frac{\alpha_j}{\alpha_i}\right) (\lambda_{ij}) [\alpha_i x_i] - \Sigma_i (\lambda_{ji}) [\alpha_j x_j] \,,$$

and the new units are $\{vt^{-1}\}$. The ratios (α_j/α_i) are input gains on integrators, and terms in brackets are scaled computer variables. Substituting $\beta\tau^{-1}$ for t^{-1} to achieve time scaling,

$$\beta[\alpha_j \dot{x}_j] = \Sigma_i \left(\frac{\alpha_j}{\alpha_i}\right) (\lambda_{ij}) [\alpha_i x_i] - \Sigma_i (\lambda_{ji}) [\alpha_j x_j] \,,$$

or

$$[\alpha_j x_j] = \Sigma_i \left(\frac{\alpha_j}{\alpha_i}\right) \left(\frac{\lambda_{ij}}{\beta}\right) [\alpha_i x_i] - \Sigma_i \left(\frac{\lambda_{ji}}{\beta}\right) [\alpha_j x_j] \,,$$

with units $\{v\tau^{-1}\}$. Thus, the real systems, with units $\{at^{-1}\}$, are converted to computer systems with units $\{v\tau^{-1}\}$.

When determining values of λ_{ij} in the Results section, voltage scale factors were $\alpha_i = \alpha_j = 10v/100\%$ radioactivity $= 0.1\{va^{-1}\}$. The timescale factor was $\beta = 15$ sec/30 days $= \frac{1}{2}\{\tau t^{-1}\}$. The equations actually used instead of equation (2) were then

$$[0.1\dot{y}_j] = \Sigma_i 2(\lambda_{ij}) [0.1y_i] - \Sigma_i 2(\lambda_{ji}) [0.1y_j] \,,$$

where the y's are dimensionless.

In the simulations (Discussion), x_j was in arbitrary units, requiring no explicit scale factors, α_j. Computing time was $\tau = 600$ sec, representing a real time of 1200 days ($\beta = \frac{1}{2}\{\tau t^{-1}\}$). Scaled equations for all compartments except litter were

$$[\dot{x}_{j \neq 1}] = \overset{5}{\underset{i=1}{\Sigma}} 2(\lambda_{ij}) [x_i] - \overset{5}{\underset{i=1}{\Sigma}} 2(\lambda_{ji}) [x_j] \,.$$

For the litter, driven by a forcing function as input,

$$[\dot{x}_1] = \overset{5}{\underset{i=0}{\Sigma}} 2(\lambda_{i1}) [x_i] - \overset{5}{\underset{i=1}{\Sigma}} 2(\lambda_{1i}) [x_1] \,.$$

LITERATURE CITED

Jansson, S. L. 1958. Tracer studies on nitrogen transformations in soil with special attention to mineralisation-immobilization relationships. Ann. Royal Agric. Coll. Sweden 24: 101–361.

MacArthur, R. 1955. Fluctuations of animal populations, and a measure of community stability. Ecology 36: 533–536.

Neel, R. B. and J. S. Olson. 1962. Use of analog computers for simulating the movement of isotopes in ecological systems. Oak Ridge Nat. Lab. Rep. 3172: 1–111.

Olson, J. S. 1965. Equations for cesium transfer in a Liriodendron forest. Health Phys. 11: 1385–1392.

Reichle, D. E. and D. A. Crossley. 1965. Radiocesium dispersion in a cryptozoan food web. Health Phys. 11: 1375–1384.

Remezov, N. P. 1959. Methods of studying the biological cycle of elements in forests. Sov. Soil Sci.: 59–67.

Whittaker, R. H. 1961. Experiments with radiophosphorus tracer in aquarium microcosms. Ecol. Monogr. 31: 157–188.

Witherspoon, J. P. 1964. Cycling of Cs^{134} in white oak trees. Ecol. Monogr. 34: 403–420.

Witkamp, M. and M. L. Frank. 1964. First year of movement, distribution and availability of Cs^{137} in the forest floor under tagged tulip poplars. Rad. Bot. 4: 485–495.

Reprinted from *Amer. Scientist*, **48**, 1–8 (Mar. 1960)

ECOLOGICAL POTENTIAL AND ANALOGUE CIRCUITS FOR THE ECOSYSTEM *

By HOWARD T. ODUM

The Ecosystem and the Ecomix

A PATCH of forest is a mysterious thing, growing, repairing, competing, holding itself against dispersion, oscillating in low entropy state, getting its daily quota of free energy from the sun. It is an ecosystem.

Understanding the basic nature of the ecosystem is a principal objective of ecology. Ecosystems are phenomena such as forests, deserts, lakes, reefs, lagoons, microcosm cultures, and polluted streams. They usually contain three kinds of processes, (a) photosynthesis, (b) respiration, and (c) circulation. Although extremely diverse, ecosystems have some basic structures and functional processes in common. To permit quantitative comparison on similar bases, a growing number of measures of general structure and function have been devised such as photosynthetic production, community metabolism, biomass, species variety, efficiencies, storage ratios, chlorophyll per area, assimilation ratio, turnover, etc. The time is now ripe for the further synthesis of the new data into generalized theorems of the ecosystem.

In ecosystems as in many other kinds of open systems, energy is supplied in concentrated form from the outside driving a sequence of branching energy flows, maintaining complex structure, recycling materials, and finally passing out from the system in a dispersed state of high entropy. The rate adjustments are set by natural selection which constitutes a fourth law of thermodynamics applicable to those open steady states which have self-reproduction and maintenance (Odum and Pinkerton, 1955).

The flow of energy in an ecosystem is represented by energy flow diagrams like that in Figure 1 (Odum, 1956). Organisms in an ecosystem in their food and energy roles participate at one of five relative positions in the flow circuit. These levels are usually called trophic levels as follows: primary photosynthetic producing plants, P; herbivore animals, H; carnivore animals, C; second order top carnivores, TC; and decomposer microorganisms and other components whose position is uncertain and for practical purposes must be lumped in a miscellaneous category pending elucidation of their exact role, D. These trophic levels are indi-

---
* These studies were aided by a grant from the National Science Foundation NSF G3978 on Ecological Microcosms.

1

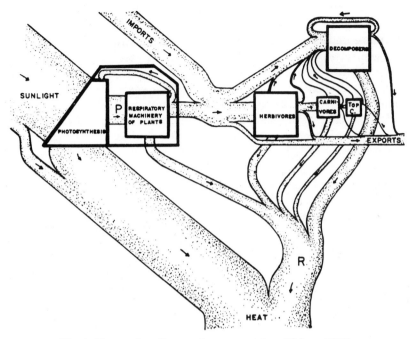

Fig. 1. Energy flow diagram for an ecosystem (Odum, 1956).

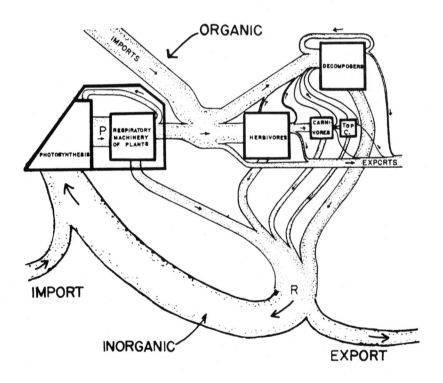

ECOMIX CYCLE

Fig. 2. Ecomix cycle diagram for an ecosystem.

cated in Figures 1, 2, and 3 by the boxes as labeled. Whereas the flow of energy is unidirectional towards dispersed, unavailable form according to the second law of thermodynamics, the materials such as carbon, nitrogen, phosphorus, and trace elements circulate in a cyclic manner being elevated into high energy combinations by plants, passing subsequently through a sequence of diminishing energy levels in the consumers. The cycle of the materials of an ecosystem can be represented by a diagram like that in Figure 2. Since the elemental ratios in the primary photosynthetic production tend to be similar to those in the respiratory-regenerator aspect of the ecosystem according to Redfield's principle (Redfield, 1934), one may consider the cycle of the raw materials as a group. The word ecomix is used in Figure 2 to represent the particular ratio of elemental substances being synthesized into biomass and subsequently released and recirculated. For example, the ratios of some of the elements in the ecomix of a planktonic system are indicated in the overall equation for the primary producers process:

1,300,000 Cal. radiant energy + 106 CO_2 + 90 H_2O + 16 NO_3 + 1 PO_4 plus mineral elements⟶ 13,000 Cal. potential energy in 3258 gm protoplasm (106 C, 180 H, 45 O, 16 N, 1 P, 815 gm mineral ash) + 154 O_2 and 1,287,000 Cal. heat energy dispersed.

This and other explanations and examples of ecosystems may be found in an ecological text (Odum and Odum, 1959).

The Ecological Analogue of Ohm's Law

The familiar Ohm's law states that the flow of electrical current, A, is proportional to the driving voltage, V, with R, the resistance, a property of the circuit.

$$A = \frac{1}{R} V \quad \text{(Ohm's Law)} \tag{1}$$

or, in an alternative form,

$$A = CV \tag{2}$$

where $C = 1/R$ is the conductivity.

In terms of steady state thermodynamics, Ohm's law is a special case of the more generalized theorem that the flux, J, is proportional to the driving thermodynamic force, X, with C the conductivity (Denbigh, 1951).

$$J = CX \tag{3}$$

Just as the product of voltage, V, and amperage, A, is power (wattage), so, in the general case, the product of thermodynamic force, X, and flux, J, is power, JX.

The ecosystem also has a flow of material under the driving influence of a thermodynamic force. The flux is the flow of food through a food chain circuit (Fig. 2) as expressed in units such as carbon per square meter of ecosystem area per unit of time. The force is some function of the concentration gradient of organic matter and biomass above and below the food circuit. A number of authors have related consumption to concentration. Jenny, Gessel, and Bingham (1949) have shown that the rate of organic decomposition by microorganisms in soil is proportional to the concentration of organic matter. A similar relationship is involved in the equation for the oxygen sag downstream from pollution outfall. Sinkoff, Geilker, and Rennerfelt (unpublished report obtainable from the Taft Environmental Health Center) use an analogue circuit for simulation of oxygen sag curves based on the principle of decomposition rate being dependent on the amount of organic matter. In comparing the ecosystem to the Ohm's law analogue, the consumption of living food as well as dead organic matter is considered to be dependent on the concentration of the food.

The validity of this application may be recognized when one breaks away from the habit of thinking that a fish or a bear catches food and thinks instead that accumulated food by its concentration practically

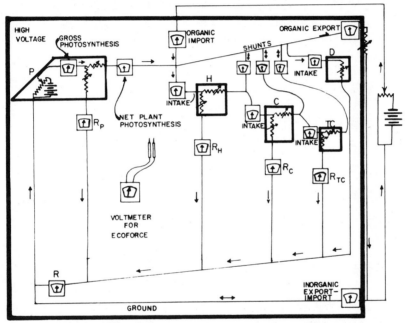

Fig. 3. Electrical analogue circuit for a steady state ecosystem like the one in Fig. 2. The flow of electrons corresponds to the flow of carbon.

forces food through the consumers. Any aggression by the fish is paid for by the food. When there are no consumers there is a state of high resistance. Usually, ecosystems rapidly develop circuits to drain reservoirs of organic free energy, often being self-organized to maintain suitable biomass structure for the purpose.

Thus we may write the equation for the force and flux in the ecosystem in the form of equations 1, 2 and 3 as follows:

$$J_e = C_e X_e \qquad (4)$$

where X_e is the thermodynamic force (ecoforce); J_e the ecoflux; and C_e the ecological conductivity of the food circuit. The application of equation (4) and the elucidation of the nature of the ecoforce follow in a subsequent section.

An Analogue Circuit for the Biogeochemical Cycle of the Ecosystem

Since the form of equation (4) relating ecoflux and ecoforce is the same as the form for Ohm's law, an electrical circuit can be constructed analogous to the flows of the ecomix in Figure 2. This has been done as diagrammed in Figure 3. Like the biological system in Figure 2, the electrical system in Figure 3 is an open steady state. Application of more complex analogue circuits with feedbacks, oscillations, and transient phenomena remain for the future.

In the electrical circuit of Figure 3 resistances are grouped at the locations of the producing and consuming populations. Batteries supply the concentrated energy representing the sun and the energy imported as organic matter from the outside. The various branching flows of food energy to consumers are presented with branching electrical wires. Variable resistances and switches permit the observer to set up various special situations and combinations. Milliammeters are placed in each circuit to permit rapid visual examination of the electrical flow which represents the flow of carbon and associated ecomix. As in the real ecosystem the energy is in the state of the flowing matter and is radiated from the computer as heat during passage from the high energy state of the battery to ground level. The amount of energy dispersed in any flow is readily measured by the product of the amperage and the voltage. A voltmeter with leads is available for measuring the voltage drop in any circuit adjustment.

Determination of the Ecoforce from Ecosystem Data Using the Electrical Analogue Circuit

Although approximate, fairly complete data now exist on rates of flux in circuits of real ecosystems. Data from one stable ecosystem, a fresh water stream, Silver Springs, Florida (Odum, 1957) available in the form of Figure 2, were put into the electrical analogue circuit (Figure 3) on a scale of 13.9 milliamperes per gm/M²/year of carbon flow. The variable resistances were adjusted so that the rates of current flow were in scale with the average rates of flow of carbon estimated by various ecological means in the field. The voltage drops between the various parts of the ecosystem and the ground were then measured with the voltmeter. The results are included in Figure 4. The ecoforce, defined as a linear function of flux (equation 4), was thus measured directly from real data for an ecosystem using the electrical analogue circuit as a computation device.

Biomass Concentration, Ecoforce, and Ecopotential

As indicated previously, the flow of energy in a food chain circuit may be intuitively related to the concentration of food, just as the rates of reaction in simple chemical systems are related to the concentrations of reactants. However, the flow of energy between complex, self-reproducing entities organized within the ecosystem need not have, *a priori*, similarities to the chemical systems even though organic chemicals are involved in both.

The organic matter accumulated in the biomass of part of an ecosystem may be defined as ecopotential, E, equal to the free energy per unit carbon. This free energy, F, is the chemical free energy of the packages of biomass, prorated over the area of the ecosystem. Thus, ecopotential is a function of the concentration of biomass and organic matter.

The product of ecopotential and ecoflux has the dimensions of power.

$$EJ_e = \frac{\Delta F}{C} \frac{dC}{dt} \tag{5}$$

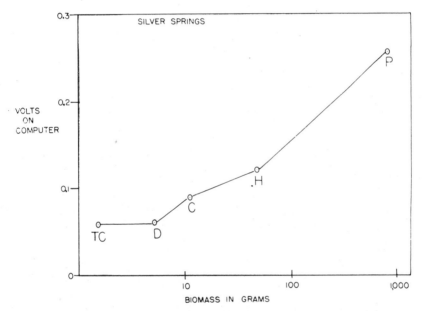

Fig. 4. Measured rates of metabolism of the organisms in the trophic levels of Silver Springs, Florida, were set in the circuit of Fig. 3 by adjusting the resistances. Then the voltages at this steady state were measured between the trophic level and ground. Since Ohm's law is a linear relationship, the voltages on the circuit are the ecoforces as defined in equation (4). The linearly defined ecoforce is some function of the organic food matter upstream from the trophic level. That is, the driving impetus to the metabolic circuits is some function of the concentration of food supply. At any point in the circuits the food supply is the standing concentration of biomass and other edible organic matter in the trophic level just above.

Measurements of the biomass concentration of organic matter in Silver Springs are available. In the graph of Fig. 4, the ecoforce is plotted as a function of the logarithm of biomass available to drive the flow. As one might predict from the basic similarity of ecological systems to chemical systems, the ecoforce is not the biomass concentration but may be a logarithmic function of it.

where E is ecopotential; J_e, ecoflux; ΔF, free energy change; C, carbon; and t, time.

Ecopotential is defined in energy units by equation (5); ecoforce was defined without specifying its dimensions except that it was linearly related to flux (equation 4). What is the relationship of ecopotential E and ecoforce X_e? In the case of Ohm's law, the potential and the force are identical. In the chemical flow systems, however, away from equilibrium, the potential is the logarithm of the force. The question arises as to the relationship of ecopotential and ecoforce in the ecosystem. Stating the question in another way, how is the flux related to the potential? Linearly? Logarithmically?

On the ordinate in Figure 4 are plotted the biomass concentrations from Silver Springs. On the abscissa are plotted the voltages from the electrical computer set for the average flux in Silver Springs. It is apparent that the relationship of biomass and voltage is not linear, but may be logarithmic. If the voltage of the computer represents the ecoforce defined linearly (equation 4) and the biomass concentration is a potential (equation 5), then ecopotential and ecoforce are not equal or linearly related. Many more such data need to be tested.

To avoid confusion, it should be stated here that the concepts discussed here have nothing to do with the misnomer, biotic potential, which is not a potential in the energy sense but is a specific growth rate. Chapman's efforts (1928) to draw an analogy between Ohm's law and biotic potential are fallacious as can be recognized by dimensional analysis.

Hints About Ecosystems Derived from the Analogue Circuit

The construction and manipulation of the analogue is a powerful stimulant to the imagination concerning the behavior of ecosystems. The following are some suggestions from the analogue for experimental testing in the real ecosystems. The tests employed were made with the circuit in steady states resembling natural systems such as Silver Springs.

1. Competition exists when two circuits are in parallel.

2. Consumer animals compete with plant respiratory systems.

3. When unusual biomass and ecoforce distributions (potentials) are postulated, circuits reverse direction with food passing in unusual direction. For example, with large rates of import of organic matter, energy flows into the plants heterotrophically increasing plant respiration over its photosynthesis.

4. As sources of power, the primary producers and the import system compete.

5. If shunts exist with bacteria in important roles, a steep pyramid of metabolism develops.

6. If consumer respiration is increased, gross photosynthesis is also increased due to the lowered resistance.

7. Doubling the power supply doubles the metabolism at all levels.

8. Cutting off export increases metabolism of consumers.

9. Cutting off top carnivores does very little to the remainder of the energy flows.

10. Increasing import increases respiratory metabolism and diminishes gross photosynthesis.

11. Cutting out herbivores reduces photosynthesis and increases bacterial and plant respiration.

12. Higher trophic levels compete in part with the trophic level which it consumes.

13. A change in plant respiration has a major compensatory effect on the consumers.

14. A decrease in respiration increases the voltage (biomass concentration) upstream.

15. A short circuit is comparable to a forest fire.

ACKNOWLEDGMENTS

The author acknowledges the stimulation of theoretical discussions with Mr. Robert Beyers and Mr. Ronald Wilson of the Institute of Marine Science and with Dr. E. P. Odum, Dr. J. Olson, Dr. F. Golley, Dr. A. Smalley, and Dr. E. Kuenzler, participants in the 1959 Ecological Society Symposium on Energy Flow.

REFERENCES

CHAPMAN, R. N. 1928. The quantitative analysis of environmental factors. *Ecology, 9*, 111–122.

DENBIGH, K. G. 1951. Thermodynamics of the steady state. Methuen.

JENNY, H., S. P. GESSEL, and F. T. BINGHAM. 1959. Comparative study of decomposition rates of organic matter in temperate and tropical regions. *Soil Science, 68*, 419–432.

ODUM, E. P. with collaboration of H. T. ODUM. 1959. Fundamentals of Ecology. Saunders, Philadelphia.

ODUM, H. T. 1956. Primary production in flowing waters. *Limnolog. and Oceanogr., 1*, 102–117.

——— 1957. Trophic structure and productivity of Silver Springs, Florida. *Ecol. Monogr., 27*, 55–112.

ODUM, H. T., and R. C. PINKERTON. 1955. Time's speed regulator: the optimum efficiency for maximum power output in physical and biological systems. *Amer. Sci., 43*, 331–343.

REDFIELD, A. C. 1934. On the proportions of organic derivations in sea water and their relation to the composition of plankton, pp. 176–192 in James Johnstone Memorial Volume, Liverpool Univ. Press, 348 pp.

24

Reprinted from *Bull. Bingham Oceanog. Collection*, **19**, 72–80 (1967)

Mathematical Model of Nutrient Conditions in Coastal Waters

By

Gordon A. Riley[1]

Bingham Oceanographic Laboratory
Yale University

TABLE OF CONTENTS

ABSTRACT

A mathematical model is developed to illustrate the distribution of nitrate and phosphate in coastal waters. The model depends on the existence of a deep water source of nutrients at the edge of the continental shelf and determines nutrient distribution in relation to horizontal and vertical mixing and biological rates of regeneration and utilization. It is shown to be applicable to the coastal region off southern New England with respect to nutrient concentrations, N:P ratios, and productivity levels. General conclusions are that the usual pattern of exchange betweeen inshore and offshore waters tends to enrich the coastal zone irrespective of enrichment by freshwater drainage, and that nitrate is more likely to be a limiting factor than phosphate, because of its inherently slower rate of regeneration.

INTRODUCTION

Coastal waters generally are more productive than the open sea. Two factors are believed to be responsible, in varying degrees according to local circumstances. The first is shoreward transport, from the edge of the continental shelf, of deep and nutrient rich water, which then becomes available to surface phytoplankton populations in the inshore waters as a result of tidal vertical mixing. The other is enrichment by freshwater drainage.

[1] Present address: Institute of Oceanography, Dalhousie University, Halifax, Nova Scotia, Canada.

Ketchum and Keen (1955) have analyzed salt balance in the coastal waters from Cape Cod to Chesapeake Bay. The annual river flow into this area is less than 1% of the volume of sea water within the area considered. This drainage obviously must be mixed with a much larger volume of offshore water in order to produce the observed salinity of the coastal region. Analyses that are available (for example, Riley, 1959) indicate that the nutrient content of river water is not markedly higher and at times is much lower than that of deep, offshore water. Freshwater drainage, therefore, is believed to be a minor and almost insignificant source of enrichment in this region except in local and semi-enclosed areas such as Narragansett Bay and Long Island Sound. In the Sound, despite its limited exchange with outside waters and abundant freshwater supply, enrichment from oceanic waters appears to be about equal to that derived from drainage (Harris, 1959).

Ryther and Yentsch (1958) have shown that the New England coastal waters of 25 to 50m depth have an annual phytoplankton production of about 160 g C/m². Production declines in a seaward direction, the average estimates being 135 g C/m² in the depth range of 50 to 1,000m and 100g between 1,000 and 2,000m. Although these differences are slight, they support a concept to be developed here, namely that inshore waters can support a moderately high level of productivity even though enrichment by drainage is relatively insignificant.

The deep water at the edge of the continental shelf commonly contains 15 to 24 μg-at $NO_3 - N/l$ and 1.0 to 1.5 μg-at P/l, with an N:P ratio of about 15:1. In much of the coastal area the ratio is lower than this. As an extreme example it is about 8:1 in Long Island Sound at the time of the midwinter nutrient maximum (Riley and Conover, 1956), and in summer it is likely to be as little as 2:1, even when nitrate, nitrite, and ammonia are all included in the ratio (Harris, 1959). Maximum phosphate levels inshore are equal to or greater than those in deep water offshore; the alteration in the ratio is due to a decrease in nitrogen. In Long Island Sound and probably most of the coastal area, nitrogen is a more limiting factor than phosphate, as might be expected when the N:P ratio in the water is so much lower than that of normal phytoplankton.

Riley (1959) found that the phosphate content of river drainage generally was equal to or less than that of Sound waters, whereas nitrate usually was higher. Thus the anomaly in the N:P ratio is not associated with drainage but must be inherent in the oceanic system. Anomalously low N:P ratios have been noted in certain other situations. Harvey (1945) and Riley (1951) have postulated that the nitrogen cycle operates more slowly than that of the phosphate, requiring a longer time for nitrogen to be returned to the water in soluble form after utilization by phytoplankton. In the present case this would mean that as nutrients move shoreward and are recycled enroute, more nitrogen than phosphorus will remain behind in a bound condition. The situation in

Long Island Sound would then be merely an exaggerated expression of a phenomenon that is common to the whole coastal region, and the latter should be treated as a unit in any quantitative analysis of nutrient problems.

The present paper will examine these hypotheses from a theoretical point of view. A simple mathematical model of coastal circulation will be postulated, and this will be used to examine the nutrient situation and to determine whether simple hypotheses stated in quantitative terms will lead to a realistic distribution of nutrients.

THE PHYSICAL MODEL

Imagine a hypothetical series of stations at 25 km intervals, crossing the shelf at right angles to the coast. Station 1 is in shallow water near shore, and Station 6, 125 km distant, is at the edge of the shelf. These stations are reference points which will be used to compute nutrient concentrations along a transcoastal profile by means of equations written in finite difference form.

The profile will be simplified to a two-layered system of surface and bottom water in which mixing between layers is limited, but mixing within layers is rapid enough to maintain a vertically uniform concentration of nutrients within each layer. The amount of mixing between layers is postulated to decrease in a seaward direction, amounting to 10% interchange of water per day between the two layers at Station 1, 5% at Station 2, 2% at Station 3, and 1% at the remainder. This conforms qualitatively to the concept of increased tidal mixing near shore and is of the right order of magnitude as indicated by computed eddy coefficients.

Probably there is a net seaward advection in the surface layer and a corresponding movement toward shore in the bottom layer. There is also an exchange by horizontal diffusion. Ketchum and Keen (1955) simplified the problem in their treatment of coastal salinity balance by calculating the flushing rate in terms of diffusion alone. Their method will be continued here. Computed values for the coefficient of horizontal eddy diffusivity ranged from 0.58 to 4.96×10^6 cm²/sec. In round numbers, a 2% horizontal interchange per day between successive stations will be equivalent to an eddy coefficient of 2.9×10^6 cm²/sec, which is near the mean of computed values.

The general equation for this type of distribution can be written

$$\frac{\delta N}{\delta t} = R + \frac{\delta}{\delta x} \cdot \frac{A_x}{\varrho} \cdot \frac{\delta N}{\delta x} + \frac{\delta}{\delta z} \cdot \frac{A_z}{\varrho} \cdot \frac{\delta N}{\delta z}, \tag{1}$$

in which the local time rate of change of a nutrient N is determined by a biological rate of change R and exchanges by eddy diffusion along the x (horizontal) and z (vertical) axes of the profile. Simple methods of analyzing this type of problem have been described by Steele (1958). Under the conditions stipulated above, the equation for the surface layer at Station 2, for example, can be written

$$\frac{\delta N}{\delta t} = R + .02\,N_1 + .02\,N_3 + .05\,N_2' - .09\,N_2, \qquad (2)$$

where N_1, N_2, N_3 are nutrient concentrations in the surface layer at Stations 1, 2 or 3, and N_2' is the concentration in the lower layer at Station 2.

Further assumptions are that the whole system exists in a steady state and that the rate of production of the phytoplankton is controlled by the nutrient concentration. During the spring and summer, when the coastal water is in a quasi-steady state, the quantity of nutrients present in the surface layer at any one time is commonly sufficient for three or four days growth. There is, however, a certain amount of regeneration in the surface layer by zooplankton and bacterial activity, so that the net daily utilization is likely to be of the order of 10°/o of the concentration of a limiting nutrient. In the present example $R = -.10N_2$, and under these conditions equation (2) reduces to

$$.02\,N_1 + .02\,N_3 + .05\,N_2' - .19\,N_2 = 0. \qquad (3)$$

If the biological system is in perfect balance, the net production of the surface layer will be removed to the lower layer, and its nutrient content will be regenerated in the lower part of the water column or on the bottom as rapidly as it is utilized in the surface layer. Hence an equation can be written for the lower layer at Station 2 that is analogous to equation (3) but includes the effect of regeneration. It is

$$.02\,N_1' + .02\,N_3' + .15\,N_2 - .09\,N_2' = 0, \qquad (4)$$

where N_1', N_2', N_3' are nutrient concentrations in the lower layer at Stations 1, 2, and 3.

Similar equations may be written for Stations 3 to 5. Station 1 requires the boundary condition that there is no further horizontal exchange in the landward direction. At Station 6 the bottom layer is arbitrarily assigned a nutrient value typical of deep water at the edge of the shelf. A sufficient boundary condition for the surface layer at this station is to assume that the nutrient concentration is constant in a seaward direction. A simultaneous solution then may be obtained with the series of inter-related equations for all of the designated points on the profile, using an appropriate method such as that given by Southwell (1946).

CALCULATED RESULTS

Results are shown in the uppermost of the families of curves in Fig. 1 *A* and and 1 *B,* which are labeled 100 to signify a true biological balance in which regeneration is 100°/₀ of utilization. The deepwater nutrient concentration at the edge of the shelf is designated arbitrarily as 100°/₀, and the remainder of the curves illustrates relative values in other parts of the profile. The gradient

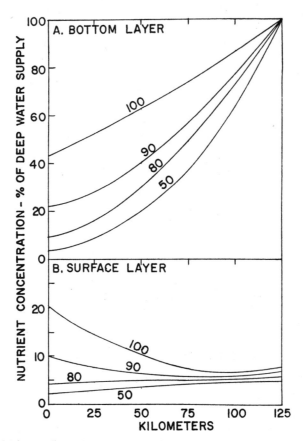

Figure 1. Nutrient gradients along a coastal water profile, arbitrarily rated on a percentage scale relative to the deepwater concentration at the edge of the continental shelf. Calculations are based on equations (3) and (4). The families of curves show gradients that result when regeneration in the bottom layer is variously rated from 50 to 100 °/₀ of nutrient utilization in the surface layer. Abscissas represent distance from shore in kilometers.

in the bottom water insures movement of nutrients, by diffusion, from the edge of the shelf toward shore. Despite the decrease in nutrients in the lower layer near shore, the concentration in the surface layer is maintained at a fairly high level because of increased vertical exchange. Biological productivity is therefore maintained at a higher level. This kind of distribution is often seen in coastal waters, and suggests that the distribution is due wholly or in part to enrichment by drainage, but the model shows that an offshore nutrient source plus increased vertical mixing near shore will provide a satisfactory explanation.

It is doubtful whether conditions in nature ever entirely conform to the perfect biological balance postulated above. During the quasi-steady state of the spring and summer season, the total concentration of nutrient elements in the water column, inorganic and combined, is less than that observed at the time of the midwinter maximum, suggesting that some portion of the nutrients which are sedimented on the bottom in combined, particulate form are un-available for further utilization until after the end of the spring-summer grow-ing season, or that there is a lag between utilization and regeneration which keeps a fraction of the nutrient stock out of circulation at all times.

This situation is readily presented in a model by postulating that regenera-tion in the bottom layer is some stipulated fraction of utilization in the surface layer, and by altering the coefficient of N_2 in equation (4) accordingly. The curves in Fig. 1 show nutrient concentrations that result when regeneration in bottom waters is varied between 50 and 100% of surface utilization. It is apparent now that biological cycling of nutrients during their transport into the coastal zone will affect the form of the curves and the overall level of productivity. If any considerable part of the nutrients is removed from circula-tion, there will be that much less available for enrichment of the inshore waters.

Nitrogen concentrations commonly are low and relatively uniform in the New England coastal zone in summer, corresponding to a regeneration rate of about 80% of utilization as pictured in Fig. 1 B. Phosphate tends to be higher near shore. This implies a higher regeneration rate for phosphate, which is in accord with the earlier hypothesis that phosphorus is cycled more rapidly than nitrogen.

A final model will investigate N:P relations in somewhat more detail. First it will be assumed that the nitrogen regeneration rate is 80% of utilization and that the source of supply at the edge of the shelf has a concentration of 15 μg-at N/l. The concentration at all stations then can be determined immediately from the relative curves in Fig. 1. It is further postulated (a) that the deepwater phosphate concentration is 1 μg-at P/l, (b) that it is utilized by phytoplankton in a ratio of 1 atom of P to 15 atoms of N, and (c) that its regeneration rate is 100% of utilization. Intuitively it is recognized that the phosphate concen-tration will be relatively high compared with nitrogen. Hence postulate (b) has the effect of setting nitrogen as the limiting factor, and the coefficient of phosphate utilization will be $0.10 N/15$, where N is the concentration of nitrogen in the surface water at any given station. Main features of the results, shown in Fig. 2, are a shoreward increase in surface phosphate and an accom-panying decrease in the N:P ratio to a minimum value of 2.2:1. Also the ver-tical gradient in phosphate is relatively slight, so that the surface concentration at Station 1 is 87°/₀ of the bottom concentration, as compared with 61°/₀ in the case of nitrogen. All of these characteristics are commonly observed in coastal waters in summer.

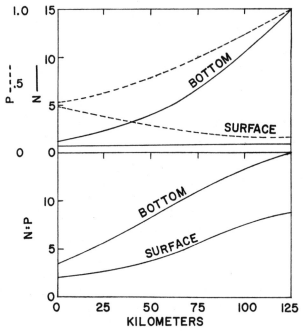

Figure 2. Relationship between soluble nitrogen and phosphorus in a coastal water profile in which nitrogen, with a regeneration rate set at 80°/₀ of utilization, is a limiting factor, and phosphate regeneration is postulated to equal utilization. Abscissas represent distance from shore in kilometers.

DISCUSSION

In the particular model that has been postulated, there is little difference in productivity between inshore and offshore waters. A slightly higher rate of nitrogen regeneration would be sufficient to establish a clear-cut gradient. Observations by Ryther and Yentsch (1958), however, show no indication of a pronounced gradient in either nitrate or productivity during the summer season, and the situation that has been postulated seems realistic as far as present knowledge goes.

The slightly higher annual productivity of the inshore waters observed by these authors was mainly the result of higher production levels during the period from December to April. A steady state model is inapplicable to this period, but the summer model has a bearing on autumn and winter nutrient conditions that should be discussed briefly.

The model illustrated in Fig. 2 requires, at the station nearest the coast, a daily removal of 0.14 μg-at N/l from the surface layer in excess of the amount regenerated. This would total 25 μg-at/l during a six-month growing season. Post season return of some or all of this nitrogen to the water column in soluble form would lead to a high concentration in the inshore waters in winter. This is indeed a typical situation in the New England coastal area, where the nitrate concentration reaches a high level, and the N:P ratio increases in winter.

Some of the nitrate undoubtedly is lost from the coastal zone by horizontal eddy diffusion in winter when the concentration exceeds that offshore. This is a slow process, however, as indicated by the work of Ketchum and Keen (1955) on diffusion of freshwater drainage. Thus a large quantity of nutrients remains inshore at the time of the diatom flowering in late winter or spring, leading to a higher level of production than that found in offshore waters.

The model allows for no post-season phosphate regeneration and in this respect is an oversimplification. Observations show an increase in phosphate in summer, as might be expected from the model, and a further increase in autumn which indicates that phosphate regeneration is not 100% as postulated. But the autumn increase in phosphate progresses more slowly than the increase in nitrate so that the N:P ratio gradually rises.

The aim here has been a limited one, namely, to achieve a realistic although admittedly simplified model of nutrient conditions in New England coastal waters. Similar models could be developed to fit other situations, the main variables being the width of the continental shelf, the magnitude of diffusion processes, and the relative importance of freshwater drainage. These variations would be expected to have considerable effect on details of areal distribution and seasonal change, but they probably would not alter the general features that have been described. Most coastal waters are richer than the open sea, and most of them show evidence that nitrogen is the most important limiting factor. The model provides a simple explanation of these features.

The basic hypothesis that nitrogen is cycled more slowly than phosphorus finds some support in experimental work on rates of bacterial decomposition, although the subject has not been documented as thoroughly as might be desired. The main evidence is of an indirect nature; no other way can be found to explain the observed distribution of these elements, and the model bolsters the hypothesis by showing that a quantitative formulation of the hypothesis leads to realistic results.

REFERENCES

HARRIS, EUGENE
  1959. The nitrogen cycle in Long Island Sound. Bull. Bingham oceanogr. Coll., *17* (1): 31–65.
HARVEY, H.W.
  1945. Recent advances in the chemistry and biology of sea water. Cambridge Univ. Press, 164 pp.

KETCHUM, B. H. and D. J. KEEN

1955. The accumulation of river water over the continental shelf between Cape Cod and Chesapeake Bay. Deep-Sea Res., *3* (suppl.): 346–357.

RILEY, G. A.

1951. Oxygen, phosphate, and nitrate in the Atlantic Ocean. Bull. Bingham oceanogr. Coll., *13* (1): 1–126.

1959. Oceanography of Long Island Sound 1954–1955. Bull. Bingham oceanogr. Coll., *17* (1): 9–30.

RILEY, G. A. and S. M. CONOVER

1956. Oceanography of Long Island Sound, 1952–1954. III. Chemical oceanography. Bull. Bingham oceanogr. Coll., *15*: 47–61.

RYTHER, J. H. and C. S. YENTSCH

1958. Primary production of continental shelf waters off New York. Limnol. & Oceanogr., *3*: 327–335.

SOUTHWELL, R. V.

1946. Relaxation methods in theoretical physics. Oxford Univ. Press, 248 pp.

STEELE, J. H.

1958. Plant production in the northern North Sea. Mar. Res. [Scotland] *1958* (7): 36 pp.

25

Reprinted from *Investigación Pesquera*, **35**(1), 309–330 (1971)

A simulation model of the nitrogen flow in the peruvian upwelling system*

by

JOHN J. WALSH ** and RICHARD C. DUGDALE **

INTRODUCTION

The present simulation model of nitrogen flow through the Peruvian upwelling system is based on data from an area off Punta San Juan, Perú (WALSH, KELLEY, DUGDALE, and FROST, 1971). The upwelling area appeared to be in quasi-steady state over at least a three-week period in March and April 1969. In response to the northerly wind stress, water upwells within a 10-20 km band off the coast carrying a seed population of phytoplankton and high nutrients. As the upwelled water drifts offshore in a persistent plume, the phytoplankton biomass increases and nutrients are depleted. Energy is passed up the food chain in the Peru system to support the world's largest fishery (RYTHER, 1969).

The model assumes that there is a continuous, steady gradient of biological properties from the rich inshore upwelling areas to the impoverished offshore regions. The simulation involves numerical solution of a series of coupled differential equations describing the behavior with time of nutrients and phytoplankton down the plume. The standing crop results predicted by the simulation are then compared with the observed distribution of nutrients and phytoplankton down the plume.

We would like to acknowledge the help of Mr. Perkins Bass in conversion of the simulation program for the IBM 1130. Drs. JAMES O'BRIEN and ALYN DUXBURY made helpful suggestions on treatment of the physical variables. NSF Grant GB-8648 provided financial support.

* Contribution 567, Department of Oceanography, University of Washington, Seattle, Washington.
** Department of Oceanography, University of Washington, Seattle, Washington 98105.

METHODS

The model was developed for use on the IBM 1130 computer aboard the R/V *Thomas G. Thompson* for comparison of the simulation results with the actual incoming data at sea. Restricted by the 8 k core storage of the IBM 1130, only five spatial blocks are included in this two layer model. The upper layer consists of distinct blocks 11 km wide, 11 km

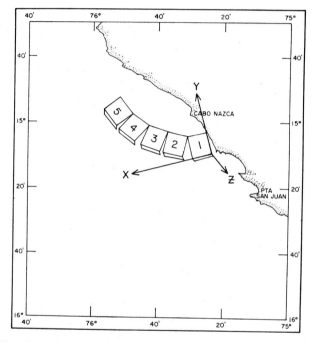

Fig. 1. — Location of the five spatial blocks down the plume.

long, and 10 m deep, which extend across the width of the plume and are distributed sequentially down the axis (figure 1). The lower layer extends below the plume from the 10 m depth to the bottom of the water column. At varying rates downstream, water upwells from the lower layer through the bottom face of each block into the upper layer (z-direction) and then flows downstream through the front and back faces (x-direction) of these blocks. Longshore water transport (y-direction) is considered to be negligible. For purposes of a simple coordinate system in the model, the plume is treated as a line perpendicular to the coast (x-direction) in the simulation calculations.

There are three biological compartments, nutrients, phytoplankton, and herbivores, within each spatial block. Figure 2 outlines in black box notations the inputs and outputs of each compartment within a block and the links to the upstream or downstream spatial blocks. N_1, P_1, and H_1 are the nutrient, phytoplankton, and herbivore compartments or

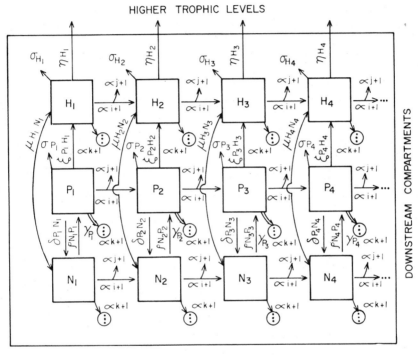

Fig. 2. — Compartment model of an upwelling ecosystem.

standing stocks in the first spatial block ; N_2, P_2, and H_2 are the standing stocks in the second downstream block ; and N_i, P_i, and H_i are the standing stocks in the ith downstream block. Fluxes between the compartments are nutrient uptake (ρ), grazing of the phytoplankton (ζ), and excretion of the phytoplankton (δ) and herbivores (μ) ; losses to the outside world are respiration (σ) and herbivore predation (ν) ; and fluxes between blocks are sinking of the phytoplankton (γ) and downstream, lateral, and vertical advection and diffusion ($\alpha_{i\ j\ k}$). The circles with three dots in the center indicate transport between the surface spatial block and lower levels of the water column.

The differential equations which describe the balance of fluxes controlling the standing crops of nutrients and phytoplankton at any point in the plume are the same for each of the five blocks, but individual values of the fluxes in each area depend on the position of the spatial block. Lack of data on herbivore biomass prevents inclusion of a budget equation for the herbivores in the present model, but herbivore interaction is simulated as an input term for the nutrients and a loss term for the phytoplankton.

In word form, the budget equation for nutrients is

(1) d nutrients$/dt$ = — advection + diffusion — nutrient uptake
                      + herbivore excretion

and for phytoplankton

(2) d phytoplankton$/dt$ = — advection + diffusion + nutrient up-
                      take — grazing — sinking

At each iteration of the simulation, the terms of equations (1) and (2) are calculated as a function of previous values of the variables, and then these terms are summed to give the standing crops of nutrients and phytoplankton at that time in the model. The non-linear form of the terms follows.

The advection and diffusion terms for fluxes between blocks of the model are taken from the general state equation for change of a quantity, c, at any point.

(3) $\partial c/\partial t + (u)(\partial c/\partial x) + (v)(\partial c/\partial y) + (w)(\partial c/\partial z) - \partial([K_x][\partial c/\partial x])/\partial x -$
      $- \partial([K_y][\partial c/\partial y])/\partial y - \partial([K_z][\partial c/\partial z])/\partial z - R = 0$

where at steady state $\partial c/\partial t = 0$, the local time change
and R = the biological terms.

The expression, $(w)(\partial c/\partial z)$, or $(w)(N_z)/dz$ in finite difference form for the nutrient flux, is the vertical advection term for the surface layer of the model, while $\partial([K_z][\partial c/\partial z])/\partial z$ is the vertical diffusion term. K_z is the vertical eddy coefficient and its contribution to the vertical flux is considered negligible compared to w, the upwelling velocity of the advection term. The vertical velocity, w, is a function of the wind stress and distance from shore and is calculated in the model from YOSHIDA's (1955) expression

(4) $w = - (k)(\tau_y)(e^{kx})/(\rho)(f)$

with

$$k = [f][(g)(z)(\Delta\rho)/\rho]^{-\frac{1}{2}}$$

where f = the Coriolis parameter $(2\omega \sin \theta)$; ω = angular velocity, θ =
  = latitude
  g = the gravitational field strength
  z = depth of the nutrient compartment
  $\Delta\rho$ = density difference between N_i and N_z
  ρ = density of the N_i compartment
  τ_y = the wind stress parallel to the coast, = $(\rho_{air})(C_D)(|U|)(U)$
    where ρ_{air} = density of air
      C_D = dimensionless drag coefficient, 0.0024 for winds
        > 15 knots and 0.0015 for winds < 15 knots
      U = the surface wind velocity
  x = distance down the plume.

The other variables of the vertical advection term are dz = the depth of the block (10 m) and either N_z = the boundary layer nutrient concentration in the z direction (21.0 mg-at NO_3/m^3) or P_z = the boundary layer phytoplankton concentration in the z direction (2.5 mg-at particulate nitrogen/m³).

Using an average wind velocity of 4.6 m/sec for the Punta San Juan area (SMITH et al., 1969) and equation (4), one obtains an upwelling velocity, w, of 6×10^{-2} cm/sec at the shoreward boundary of the plume, i.e., $x=0$; 1×10^{-2} cm/sec in the middle of the first block ($x=5.5$ km); and 3×10^{-3} cm/sec at the interface between the first and second block ($x=11$ km). O'BRIEN (1971) has pointed out that decay of w offshore may occur in 5-10 km instead of the 35 km postulated by YOSHIDA (1955). However, O'Brien has suggested that vertical upward transport may still occur in offshore regions as a result of mixing rather than upwelling. For purposes of input fluxes to blocks in the model, mixing and upwelling transport are considered to be the same.

The plume curves along the coast rather than moving directly offshore, and one must integrate the values of w over the assumed homogeneous 11 km length of each block. Therefore a linear decay of w was assumed as a first-order approximation of YOSHIDA's (1955) exponential decay down the plume with 7×10^{-3} cm/sec in the center of the bottom face of the first spatial block, 5.6×10^{-3} cm/sec in the second block, 4.2×10^{-3} cm/sec in the third, 2.8×10^{-3} cm/sec in the fourth, and 0.0 in the fifth block.

The sinking loss of the phytoplankton is of the same form as the upwelling term but of opposite sign. $(w_s)(P_i)/dz$ is the sinking term where w_s = the sinking velocity as determined by SMAYDA and BOLEYN (1965, 1966 a, b) in the laboratory sinking experiments with diatom cultures. Their measured values of w_s ranged from 0.29 m/day to 0.73 m/day (0.3×10^{-3} cm/sec to 0.7×10^{-3} cm/sec) depending on the species and age of the culture.

The expression $(u)(\partial c/\partial x) - \partial([K_x][\partial c/\partial x])/\partial x$ of equation (3) is the downstream advection and diffusion term of the model and in finite difference form for the nutrient flux becomes

$$(\tfrac{1}{2})(u_{i-1})(N_{i-1})/dx - (\tfrac{1}{2})(u_i)(N_i)/dx - (K_x)(N_{i-1} + N_{i+1} - 2N_i)/(dx)(dx)$$

where u = dowstream velocity ;

    dx = the length of the block ;

    K_x = the downstream eddy diffusivity. It is determined by the scale length of the block according to PEARSON'S (1956) $K_x = 0.01$ (length)$^{'/_4}$ cm^2/sec. It is$= 1 \times 10^6$ for the present model, and the diffusion term is negligible compared to the advection term in the downstream direction.

The velocity in the x direction, u, increases with distance down the plume as an additive output at the downstream face of each block resulting from the upwelling velocity input at the bottom face of a

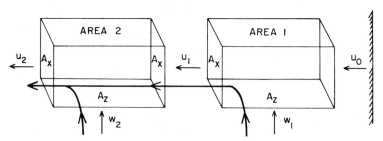

FIG. 3. — Water flow and continuity of mass in the first two spatial blocks.

block and the u input at the upstream face of a block (figure 3). The downstream u can be calculated from w of equation (4) and the upstream u through the mass continuity equation

(5) $$(u_i)(A_x) = (w_i)(A_z) + (u_{i-1})(A_x)$$

where $(u_i)(A_x)$ is the mass transport through the downstream face of the ith block, $(w_i)(A_z)$ is the mass transport upward through the bottom face of the ith block, and $(u_{i-1})(A_x)$ is the mass transport through the upstream face of the ith block. A_x and A_z are the areas of the x, z faces of the ith block. In figure 4, the downstream velocity from the second block away from the coast becomes

$$(u_2)(A_x) = (w_2)(A_z) + (u_1)(A_x)$$
$$u_2 = (w_2)(A_z)/(A_x) + (u_1)$$

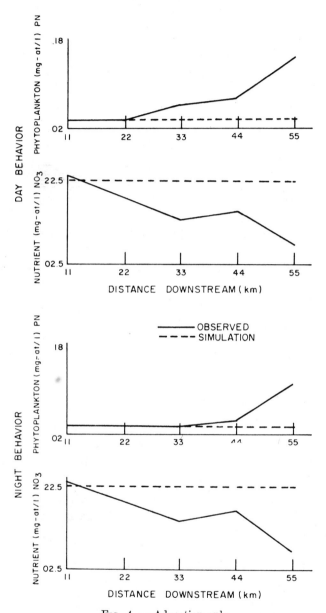

Fig. 4. — Advection only.

and the downstream velocity in the first block is

$$u_1 = (w_1)(A_z)/(A_x) + (u_0) = (w_1)(A_z)/(A_x)$$

because u_0 = zero at the shore boundary condition, i.e., no water flows out of the land.

Alternatively as a check to YOSHIDA's (1955) assumptions, the downstream u can be calculated from the EKMAN (1905) expression

(6) $$u_x = (Vo)(e^{-(\pi)(z/D)})(\cos [45° — (\pi)(z/D)])$$

where $Vo = \tau_y/[(\rho)(K_z)(f)]^{1/2}$ and τ_y, ρ, and f were defined previously. K_z, the vertical eddy diffusivity, can be estimated from the wind velocity by $K_z = 1.02\ U^3$ for winds < 6 m/sec and $K_z = 4.3\ U^2$ for winds > 6 m/sec. D is depth of frictional resistance at which the current is reversed from that cum sole the wind in the surface layer. $D = \pi/[(K_z)/(\rho)(1/2)(f)]^{1/2}$ and can be estimated from the wind by $D = 3.67\ ([U^3]^{1/2})/[\sin \theta]^{1/2}$ for winds < 6 m/sec and $D = 7.6\ U/[\sin \theta]^{1/2}$ for winds > 6 m/sec. Using equation (6) for the Ekman velocity, one obstains a u of 14 cm/sec at the surface and 9 cm/sec at a depth of 10 m. With the mass continuity equation (5) and the assumed upwelling velocities in each block, one gets $u = 7$ cm/sec at a depth of 10 m in the first block, 13 cm/sec in the second, 17 cm/sec in the third, 20 cm/sec in the fourth, and 20 cm/sec in the fifth. Calculated downstream velocities from the two methods agree fairly well with each other and with the observed drogue measurements of 12-24 cm/sec surface currents in the plume area.

The last expression $(v)(\partial c/\partial y) — \partial([K_y][\partial c/\partial y])/\partial y$ of equation (3) is the lateral advection and diffusion term of the model and in finite difference form for the nutrient flux becomes $(2)(v)(N_y — N_i)/dy + (2)(K_y)(N_y — N_i)/(dy)(dy)$, in which

v = velocity in the y direction, considered to be negligible in the model

dy = width of the block

N_y = boundary layer nutrient concentration in the y direction (3.0 mg-at NO_s/m^3)

P_y = boundary layer phytoplankton concentration in the y direction (2.5 mg-at PN/m^3)

K_y = the lateral eddy diffusivity and is determined by a variable 4/3 expression as a function of distance from the source of diffusing material (BROOKS, 1959), where

(7) $$K_y = (K_0)(L/dy)^{4/3}$$

with

$K_0 = K_x$ at $x = 0.0$, or 1×10^6 cm^2/sec

$L = (dy)(1 + (^2/_3)(J)(x/dy))^{3/2}$ and

$J = (12)(K_0)/(u)(dy)$

The variable eddy coefficients in the y direction which were calculated from BROOKS' (1959) model are 0.36 km^2/hr for the first spatial block, 0.59 km^2/hr for the second, 0.87 km^2/hr for the third, 1.20 km^2/hr for the fourth, and 1.59 km^2/hr for the fifth.

R of equation (3) represents the biological fluxes between compartments of the model. The poorly understood role of excretion in regeneration of nutrients (WHITLEDGE and PACKARD, 1970) and inadequate data on ammonia concentrations and uptake rates down the Peru plume require that an excretion term and its subsequent influence on the uptake term be implicit in the model without as yet an exact mathematical formulation.

The nutrient uptake term for nitrate utilization in the model, equation (8),

(8) $$\rho_{NO3} = (V_{max})(N_i)(P_i)/(K_T + N_i)$$

is the Michaelis-Menten expression for nutrient uptake in a nutrient-limited system (DUGDALE, 1967). The term, V_{max}, is the maximum uptake rate (hr^{-1}) at nutrient-saturated conditions. It is allowed to vary sinusoidally in the present model over 12 hours of daylight, with V_{max} assumed zero during the night. The full expression for V_{max} is $(1.43)(V_{max})(\sin 0.2168\ t)$ where t is the cumulative time in the model up to each iteration and 1.43 adjusts observed V_{max} mean values to be mid-day maxima at the peak of the sinusoid. At dusk, V_{max} is set equal to zero so that nutrient uptake will not be a negative term as the sinusoid becomes negative in the night period. The term, K_T, is the nutrient concentration (1 mgAt NO$_3$/m^3) at which the flux, ρ_{NO3} is half that at the maximum rate of uptake, when P_i is held constant.

In the model, the nutrient uptake term reflects both the distribution of nitrate down the plume and the estimated effect of the regenerated nutrient, ammonia, on nitrate uptake. V_{max} values of NO_3 uptake were entered in the model from field data collected by DUGDALE and MACISAAC (1971) down the Peru plume (table 1). With these values for each spatial block and equation (8), the nutrient uptake term of budget equation (1) was calculated as a loss from the nitrate concentration at each iteration.

The V_{max} of equation (8) was modified, however, for the uptake gain in budget equation (2) for the phytoplankton at each iteration. The implicit effect of excretion and regenerated nutrients is added in the

TABLE 1

Observed mean values of the variables in each of the five downstream spatial areas over 0-10 m depth.

Area	VARIABLES				
	NO_3 mgAt/m³	PNO_3 mgAt/m³/day	Chlorophyll a mg/m³	Phytoplankton particulate N mgAt/m³	V_{max} NO_3 hr⁻¹
1	22.37	1.018	2.30	2.76	0.032
2	16.00	0.710	2.01	3.28	0.019
3	12.20	1.355	3.32	6.03	0.020
4	14.42	2.126	5.01	7.71	0.017
5	5.99	1.639	13.13	14.39	0.011

model through a combined V_{max} which includes both the NO_3 and an estimate of the NH_3 contribution to total nitrogen uptake. A gradient of NH_3 V_{max} down the plume was estimated on the basis of possible grazing activities and NH_3 inhibition of NO_3 uptake.

Diel zooplankton data are inadequate for any firm interpretation of grazing patterns (WALSH, KELLEY, DUGDALE, and FROST, 1971), but there appears to be a diel variation in chlorophyll concentrations at the lower end of the plume and not at the upper. It is possible that higher grazing activity and resultant higher NH_3 excretion may occur in the downstream area as opposed to the upstream areas near the coast. Recent field experiments in the Mediterranean Sea (DUGDALE and MACISAAC, 1971) and laboratory studies of continuous cultures (CONWAY, personal communication) indicate that NH_3 is a preferential nitrogen source and in high concentrations inhibits NO_3 uptake. The downstream decline of NO_3 V_{max} in table 1 suggests that this hypothetical gradient of grazing, NH_3 concentration and uptake, and consequent inhibition of NO_3 uptake may actually occur down the plume. For purposes of this preliminary model, V_{max} values for NH_3 down the plume were assumed to be 0.001 in the first block, 0.005 in the second, 0.030 in the third, 0.040 in the fourth, and 0.050 in the fifth. The NH_3 values were then combined with NO_3 V_{max} values in each area and used to calculate the nutrient uptake for the phytoplankton at each iteration.

No data were available from the Peru area on grazing fluxes. If herbivore biomass figures were reliable, mathematical expression of a grazing flux could take the form, $G_{max}(H_i)(P_i - P^*)$ with $(P_i - P^*) \geq 0$, where G_{max} is the grazing rate, varying nocturnally as a cosine and zero during the day. G_{max} would be $(1.43)(G_{max})(\cos 0.2168\,t + 1.57)$ in analogy to the sinusoidal variation of V_{max} in equation (8), with $t =$ the cumulative time interval. $P^* =$ the threshold value below which herbivore grazing induces no change in the phytoplankton population, possibly

2.5 mg/m³ chlorophyll in the Peru plume (WALSH, KELLEY, DUGDALE, and FROST, 1971). The grazing loss in budget equation (2) for the phytoplankton was calculated at steady state, i.e., $dP_i/dt = 0$, by fitting the above grazing expression to the difference between observed and calculated night phytoplankton standing crop. Such a procedure is tenuous at best and independent measurements must be made in the future to provide data on the grazing flux.

The components of the model are then :

a)   constant wind along the coast
b)   upwelling velocity decreases down the plume
c)   downstream velocity increases down the plume
d)   eddy diffusivity in the y direction increases down the plume
e)   a sinusoidal NO_3 V_{max} decreases down the plume
f)   an adjusted NH_3 V_{max} increases down the plume
g)   a cosine grazing term increases down the plume.

Initial conditions of the model are 2.5 mgAt/m³ phytoplankton PN and 3.0 mgAt/m³ NO_3 in each block. Boundary conditions are 2.5 mgAt/m³ phytoplankton PN at the bottom, lateral, and downstream walls of the plume, 3.0 mg-at/m³ NO_3 at the lateral and downstream walls, and 21.0 mg-at/m³ NO_3 at the bottom wall. No fluxes occur through the upstream wall of the first block, i.e., the shore boundary. With the exception of the grazing simulation, the model was started at initial conditions and allowed to go to steady state during each simulation.

Differential equations (1) and (2) were solved with the Euler method of numerical integration utilizing a simulation program developed by BLEDSOE and OLSON (1968) and modified for the IBM 1130 computer. Simulations begin at 6 a.m. with the iterative time step equal to 1 hour, i.e., the losses and gains in equations (1) and (2) were summed every hour, and total simulated time was 10 days.

RESULTS

Simulations were run in a sequential fashion, adding one variable at a time, to see which fluxes were important in controlling the distributions of nutrient and phytoplankton concentrations down te plume. The output of each simulation was compared with the observed day gradients of nutrient and phytoplankton (table I) and with night gradients estimated from the underway map data presented by WALSH, KELLEY,

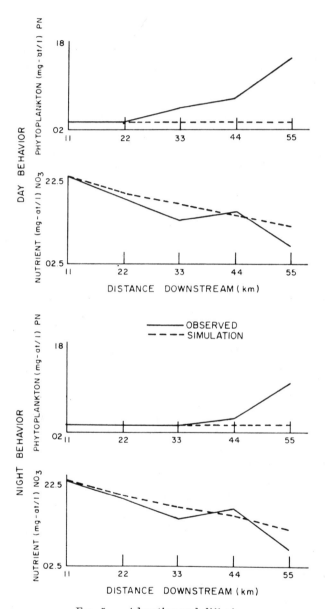

FIG. 5. — Advection and diffusion.

DUGDALE, and FROST (1971). The simulated day results in each block ($t = 228$ hours) are noon phytoplankton and nutrient values of the tenth day of the simulation, and night results ($t = 240$ hours) are midnight values of the variables during the same day. The importance of each flux was judged by how close the simulation results matched observed data.

Advection

Only the advective terms, w and u, were included in the first simulation. All other terms of equations (1) and (2) were set equal to zero. Figure 4 compares the simulation results (dotted line) and the observed data (solid line) down the plume. After 72 hours, the boundary concentration of 21.0 mg-at/m³ NO_3 in the upwelled water raised the initial concentration of 3.0 mg-at/m³ in each NO_3 compartment to the steady values of 21.0 mg-at/m³ NO_3. No losses were present in budget equation (1), and the upwelled input of nutrient simply filled up the NO_3 compartments during the transient state, i.e., $dN_i/dt \neq 0$.

Phytoplankton PN (Particulate Nitrogen) down the plume did not change during the simulation because both the initial and upwelled phytoplankton concentrations were the same, 2.5 mg-at/m³ PN, and no phytoplankton growth was allowed. As a result of the continuity equation, input = output for the phytoplankton mass transport, no transient state was observed, i.e., $dP_i/dt = 0$.

Advection and diffusion

When diffusion is added to the model (figure 5), simulated phytoplankton distribution is the same as in the advective case. There is still no nutrient uptake, and the gradient terms of equation (3) ($P_y - P_i$) are equal to zero, i.e., no diffusion loss occurs. With no inputs, no losses, and the mass transport balanced there is no time change in the phytoplankton compartments.

The simulated nitrate distribution down the plume matches fairly closely that of the observed data. The nutrient term, ($N_y - N_i$) $\neq 0$, allows for a diffusion loss to be balanced by the upwelling input, and figure 5 shows the resultant steady state values of the NO_3 compartments down the plume. This simulation suggests that the purely physical diffusion flux away from a high nutrient source along the coast may be a very strong factor in contributing to the depletion of NO_3 down the plume.

21

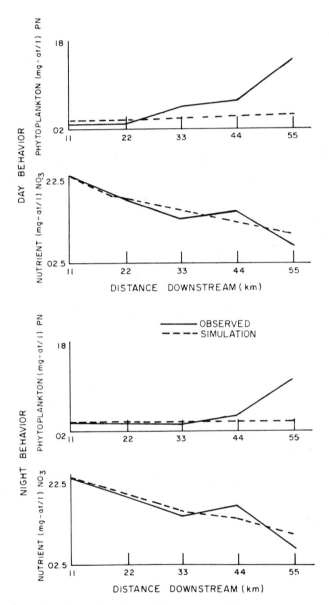

Fig. 6. — Advection, diffusion, and NO_3 uptake.

Advection, diffusion, and NO_3 uptake

If the phytoplankton are allowed to grow in the model at the observed NO_3 V_{max} uptake rates (figure 6), simulated NO_3 distribution converges on the observed data. Day and night differences in the steady state values of the NO_3 compartments are introduced by the sinusoidal input of equation (8). A balance of the upwelling input, diffusion loss, and uptake loss in budget equation (1) for NO_3 matches very well the observed surface NO_3 gradient down the plume and suggests that these are the major factors controlling NO_3 standing stocks in the Peru plume.

The simulated phytoplankton biomass of figure 6 does not come very close to the observed data, however. These results suggest that while the nitrate loss of equation (1) is well estimated, the uptake gain of the phytoplankton in equation (2) is not. The growth of the phytoplankton in the model must be supplemented from another nitrogen source.

Advection, diffusion, NO_3, and NH_3 uptake

With the hypothetical NH_3 uptake rates added to the model, a very close fit between the simulated and observed day phytoplankton distribution is obtained (figure 7). The simulated night phytoplankton distribution is higher than that postulated from the 10 underway night maps. Grazing has not been included in this simulation, however, and such a loss would lower the simulated night values. In contrast, if a constant 0.020 NH_3 V_{max} in each compartment is used down the plume for the supplemental nitrogen source (figure 8) instead of a gradient, the simulated steady state phytoplankton standing crop is higher than that observed.

The present fit of the gradient NH_3 model to observed data suggests that there may be a gradient in NH_3 concentration and utilization down the plume. The actual gradients may be higher than those used in the present model, however. If a day time grazing loss due either to zooplankton or fish is an important factor in the upwelling system, it could balance a higher gross nitrogen production than predicted by the model and still yield the same observed distribution of net phytoplankton PN standing crop.

Advection, diffusion, NO_3 + NH_3 uptake, and sinking

A sinking loss was then added to the model. SMAYDA and BOLEYN's (1965, 1966 a, b) smallest sinking rate, 0.29 m/day, produced lower

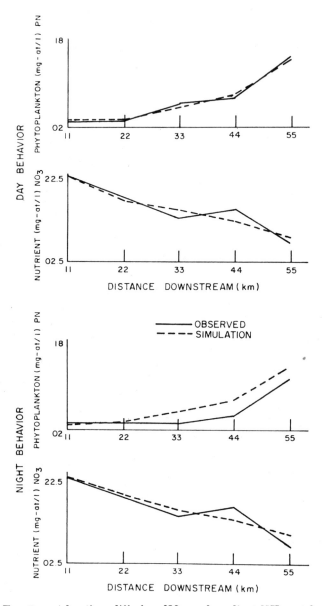

FIG. 7. — Advection, diffusion, NO_3, and gradient NH_3 uptake.

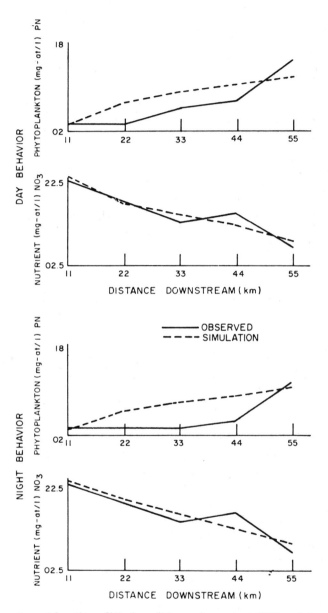

FIG. 8. — Advection, diffusion, NO₃, and constant NH₃ uptake.

simulated phytoplankton biomass than that observed (figure 9). The sinking term creates too large a loss flux in budget equation (2) and does not allow sufficient build-up of phytoplankton biomass in the model during the transient state. If the combined NO_3 and NH_3 uptake flux is a reasonable estimate of total nitrogen based production, inclusion of this low sinking rate in the model suggests that the actual sinking rates of phytoplankton in the Perú plume may be less than those found for a homogeneous laboratory water column by SMAYDA and BOLEYN (1965, 1966 a, b). Upwelling water and density stratification may impede phytoplankton sinking in the Peru plume.

Advection, diffusion, NO_3 + NH_3 uptake, grazing

A nocturnal grazing term was introduced at steady state conditions of the model as a replacement of the sinking loss (figure 10). If the herbivore term was introduced at the beginning of the transient state as were all other fluxes, phytoplankton growth would not occur in the model. One must have a phytoplankton build-up or time lag in the system before the grazing stress can be applied in equation (2). With this adjusted cosine term, there is no grazing in the day time, and the simulated phytoplankton distribution is the same as that of figure 7. Night phytoplankton distribution in the model now approximates that postulated from the underway data.

CONCLUSIONS

WALSH, KELLEY, DUGDALE, and FROST (1971) described a cyclic process of investigations of total systems which involves simulation models both as feedback control and as a predictive tool. Our present understanding of the Peru upwelling system is in the model building and validation stage. The assumptions and fit of the present non-linear model to observed and estimates data must be tested by return field studies to the Peru area.

This model is rudimentary and was constructed primarily as a tool for teaching ourselves how to build a useful spatial model of upwelling processes. However, the results encourage us to think that some of the processes are reasonably accurately postulated. The assumed upwelling velocities induce a steady state residence time of 44 hours for a phytoplankton cell to travel through the first block, 22 hours in the second, 18 in the third, 13 in the fourth, and 13 in the fifth. Growth rates in the model and residence times in the blocks lead to phytoplankton

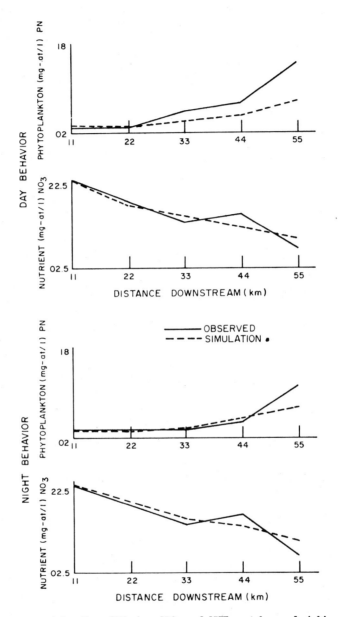

Fɪɢ. 9 — Advection, diffusion, NO_3 and NH_3 uptake, and sinking.

357

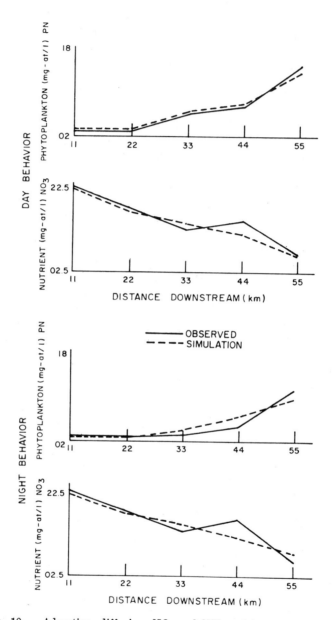

FIG. 10. — Advection, diffusion, NO_3 and NH_3 uptake, and grazing.

doubling times of 2 days at the upper end of the plume and ½ day at the lower end. These doubling rates are within the observed range of phyto-plankton growth in the Peru area (BARBER, personal communication).

In the last simulation (figure 10), however, an unknown input, regenerated nutrient as NH_3, is balancing an unknown output, grazing, and both are functions of each other. We must now validate the model with a future field study to see if at steady state a gradient of NH_3 concentration and utilization does exist down the plume and to see if the grazing flux of the model approximates the actual daily input to herbivore populations. Time series are also needed to document sugges-tions of nocturnal grazing and diel nutrient uptake.

When the present or second-generation model is validated, there will exist a working analog of the Peru system. Experiments can be run on the model as in figures 7 and 8 with two different sets of NH_3 V_{max} down the plume, or sensitivity analyses can be made on the components of the model, i.e., how sensitive is the phytoplankton biomass of the fifth compartment to a unit change of NO_3 concentration in the first compartment. Finite-difference approximation of continuous processes assumes partition of the process into finitely small discrete units. Five units 11 km long are a very crude approximation of a presumably continuous upwelling plume. We plan to expand the present model to as much as 1000 blocks to cover the same plume, and to consider species interactions, higher tropic levels, fluxes of other elements, and long-term variations of the system.

RESUMEN

UN MODELO DE SIMULACIÓN DEL FLUJO DE NITRÓGENO EN EL SISTEMA DE AFLORAMIENTO DEL PERÚ. — El modelo se basa en que existe un gradiente estacionario de las pro-piedades biológicas desde las áreas de afloramiento próximas a la costa hasta las regiones exteriores pobres. El sistema entero se divide en bloques, y dentro de cada bloque se consideran diversos compartimentos, constituidos por los nutrientes, el fitoplancton y los herbívoros. La simulación consiste en resolver numéricamente una serie de ecuaciones diferenciales expresando las relaciones entre los distintos com-partimentos de los diversos bloques. El modelo se va complicando por etapas con la inclusión de más factores, hasta conseguir un ajuste aceptable entre el cálculo y la observación.

REFERENCES

BLEDSOE, L. J., and OLSON, J. S. — 1968. Comsys 1: A stepwise compartmental simulation program. ORNL TM (in preparation).

BROOKS, N. H. — 1959. Diffusion of sewage effluent in an ocean current, pp. 246-267. *In* E. A. Pearson (ed.), *Waste disposal in the marine environment*. Pergamon Press, New York.

DUGDALE, R. C. — 1967. Nutrient limitation in the sea: dynamics, identification, and significance. *Limnol. Oceanogr.*, 12(4): 685-695.

DUGDALE, R. C., and MACISAAC, J. J. — 1971. A computational model for the uptake of nitrate in the Peru upwelling region. *Inv. Pesq.*, 35: 299-308.

EKMAN, V. W. — 1905. On the influence of the earth's rotation on ocean currents. *Ark. f. Mat. Astr. och Fysik. K. Sv. Vet. Ak.*, Stockholm, 1905-06, v. 2, n. 11, 1905.

O'BRIEN, J. J. — 1971. A two-dimensional physical model of the North Pacific. *Inv. Pesq.*, 35: 331-349.

PEARSON, E. A. — 1956. An investigation of the efficacy of submarine outfall disposal of sewage and sludge. *State Water Pollution Control Board, Publ. 14*, Sacramento, California.

RYTHER, J. H. — 1969. Photosynthesis and fish production in the sea. *Science*, 166: 72-76.

SMAYDA, T. J., and BOLEYN, B. J. — 1965. Experimental observations on the flotation of marine diatoms. I. *Thalassiosira* CF. *nana, Thalassiosira rotula*, and *Nitzschia seriata. Limnol. Oceanogr.*, 10(4): 499-509.

SMAYDA, T. J., and BOLEYN, B. J.—1966a. Experimental observations on the flotation of marine diatoms. II. *Skeletonema costatum* and *Rhizosolenia setigera. Limnol. Oceanogr.*, 11(1): 18-34.

SMAYDA, T. J., and BOLEYN, F. J.—1966b. Experimental observations on the flotation of marine diatoms. III. *Bacteriastrum hyalinum* and *Chaetoceras lauderi. Limnol. Oceanogr.*, 11(1): 35-43.

SMITH, R. L.; MOOERS, C. N. K., and ENFIELD, D. B. — 1969. Mesoscale studies of the physical oceanography in two coastal upwelling regions: Oregon and Peru. *Paper presented at the International Conference on the Fertility of the Sea, Sao Paulo, Brazil, December 1969.*

WALSH, J. J.; KELLEY, J. C.; DUGDALE, R. C., and FROST, B. W. — 1971. Gross features of the Peruvian upwelling system with special reference to diel variation. *Inv. Pesq.*, 35: 25-42.

WHITLEDGE, T. E., and PACKARD, T. T. — 1971. Nutrient excretion by anchovies and zooplankton in Pacific upwelling regions. *Inv. Pesq.*, 35: 243-250. zooplankton in Pacific upwelling regions. *Inv. Pesq.*

YOSHIDA, K. — 1955. Coastal upwelling off the California coast. *Rec. Oceanogr. Wkr., Japan*, 2(2): 1-13.

References

Barnes, H. 1959. *Apparatus and Methods of Oceanography,* Vol. I, Chemical. George Allen & Unwin Ltd., London.

Browne, C. A. 1942. Justus von Liebig—Man and Teacher. In F. R. Moulton (ed.), *Liebig and after Liebig.* American Association for the Advancement of Science, Washington, D.C., p. 1–9.

Clarke, F. W. 1924. *The Data of Geochemistry.* U.S. Geol. Survey Bull. 770.

Deevey, E. S. 1960 The hare and the haruspex: a cautionary tale. *Amer. Scientist 48:* 515– 430.

Denigès, G. 1921. Détermination quantitative des plus faibles quantités de phosphates dans les produites biologiques par la méthode céruléomolybdique. *Compt. Rend. Soc. Biol. Paris 84:* 875– 877.

Gardiner, A. C. 1937. Phosphate production by planktonic animals. *J. Cons. Int. Explor. Mer. 12:* 144– 146.

Harvey, H. W. 1926. Nitrate in the sea. *J. Marine Biol. Assoc. U.K. 14:* 71– 88.

Juday, C., and E. A. Birge. 1931. A second report on the phosphorus content of Wisconsin lake waters. *Trans. Wisconsin Acad. Sci. Arts Letters 26:* 353– 382.

Kuhn, T. S. 1962. *The Structure of Scientific Revolutions.* University of Chicago Press, Chicago.

Liebig, J. 1840. *Organic Chemistry and Its Applications to Agriculture and Physiology.* Taylor and Walton, London.

Lindemann, R. L. 1942. The trophic-dynamic aspect of ecology. *Ecology 23:* 399– 418.

Matthews, D. J. 1916. On the amount of phosphoric acid in the sea water off Plymouth Sound. *J. Marine Biol. Assoc. U.K. 11:* 122– 130.

Michaelis, L., and M. L. Menten. 1913. Die Kinetik der Invertinwirkung. *Biochem. Z. 49:* 333– 369.

Monod, J. 1942. *Recherches sur lar croissance des cultures bacteriennes.* Hermann & Cie, Paris.

Mortimer, C. H. 1941. The exchange of dissolved substances between mud and water in lakes: I and II. *J. Ecol. 29:* 280– 329.

Mortimer, C. H. 1942. The exchange of dissolved substances between mud and water in lakes: III and IV. *J. Ecol. 30:* 147– 201.

Odum, H. T. (ed.). 1970. *A Tropical Rain Forest. A Study of Irradiation and Ecology at El Verde, Puerto Rico.* U.S. Atomic Energy Comm. Tech. Info. Doc. 24270.

Odum, H. T., and E. P. Odum. 1955. Trophic structure and productivity of a windward coral reef community on Eniwetok Atoll. *Ecol. Monograph 25:* 291– 320.

Pomeroy, L. R. 1970. The strategy of mineral cycling. *Ann. Rev. Ecol. Systems 1:* 171– 190.

Pomeroy, L. R., H. M. Mathews, and H. S. Min. 1963. Excretion of phosphate and soluble organic phosphorus by zooplankton. *Limnol. Oceanog. 8:* 50– 56.

Redfield, A. C. 1958. Biological control of chemical factors in the environment. *Amer. Scientist 46:* 205– 221.

Redfield, A. C., B. H. Ketchum, and F. A. Richards. 1963. The influence of organisms on the composition of sea water. In M. N. Hill (ed.), *The Sea,* Vol. 2. John Wiley & Sons, Inc., New York, pp. 26– 77.

Riley, G. A. 1947. Factors controlling phytoplankton populations on George's Bank. *J. Marine Res. 6:* 54– 73.

Riley, G. A. 1951. Oxygen, phosphate, and nitrate in the Atlantic Ocean. *Bull. Bingham Oceanog. Collection 13:* 1–126.

Riley, G. A., H. Stommel, and D. F. Bumpus. 1949. Quantitative ecology of the plankton of the western North Atlantic. *Bull. Bingham Oceanog. Collection 12:* 1–169.

Rodin, L. E., and N. I. Bazilevich. 1956. *Production and Mineral Cycling in Terrestrial Vegetation,* translation by G. E. Fogg, 1967. Oliver & Boyd Ltd. Edinburgh.

Smith, F. E. 1970. Analysis of ecosystems. In D. E. Reichle (ed.), *Analysis of Temperate Forest Ecosystems.* Springer-Verlag New York, New York, pp. 7–18.

Steele, R. 1972. *Tracer Probes in Steady State Systems.* Charles C Thomas, Publisher, Springfield, Ill.

Stumm, W. 1972. The acceleration of the hydrogeochemical cycling of phosphorus. In D. Dyrssen and D. Jagner (eds.), *The Changing Chemistry of the Oceans.* Almqvist & Wiksell, Stockholm.

Sverdrup, H. U., M. W. Johnson, and R. H. Fleming. 1942. *The Oceans.* Prentice-Hall, Inc., Englewood Cliffs, N.J.

Tansley, A. G. 1935. The use and abuse of vegetational concepts and terms. *Ecology 16:* 284–307.

Waksman, S. A. 1916. Studies on soil protozoa. *Soil Sci. 1:* 135–152.

Watt, K. E. F. 1968. *Ecology and Resource Management.* McGraw-Hill Book Company, New York.

Yonge, C. M. 1930. Studies on the physiology of corals: I. Feeding mechanisms and food. *Great Barrier Reef Expedition 1:* 13–57.

Yonge, C. M. 1930. Studies on the physiology of corals: II. Digestive enzymes. *Great Barrier Reef Expedition 1:* 59–81.

Yonge, C. M. 1940. The biology of reef-building corals. *Great Barrier Reef Expedition 1:* 353–391.

Author Citation Index

Subject Index

371